Climate Change, Vulnerabilities and Adaptation

Abha Lakshmi Singh • Saleha Jamal
Wani Suhail Ahmad
Editors

Climate Change, Vulnerabilities and Adaptation

Understanding and Addressing Threats with Insights for Policy and Practice

Springer

Editors
Abha Lakshmi Singh
Aligarh Muslim University
Aligarh, India

Saleha Jamal
Aligarh Muslim University
Aligarh, India

Wani Suhail Ahmad
University of Ladakh
Kargil, India

ISBN 978-3-031-49641-7 ISBN 978-3-031-49642-4 (eBook)
https://doi.org/10.1007/978-3-031-49642-4

This Springer imprint is published by the registered company Springer Nature Switzerland AG
The registered company address is: Gewerbestrasse 11, 6330 Cham, Switzerland

Paper in this product is recyclable.

Foreword

According to Sixth Assessment Report on Climate Change 2022: Impacts, Adaptation and Vulnerability of the Intergovernmental Panel on Climate Change (IPCC), nearly half of the world's population (8 billion) have witnessed severe water scarcity due to climate change, and during the last five decade about 44% of all disaster are flood related. In contemporary society, there exists a prevailing perception that various calamities, such as landslides, floods, storms, and droughts, are increasingly regarded as manifestations of climate change induced by the synergy between technological progress and human agency, rather than being solely attributable to natural occurrences. The global climatic change and changes in land use/land cover, and water pollution are the major drivers of the loss and deterioration of freshwater ecosystems. Climate change and global warming may lead to large-scale water and air health risks to people across the world, especially in the third-world countries. According to estimations, when global warming reaches a range of 1.5–2 °C, localized and gradual adaptation will be required. However, it is anticipated that negative consequences will extend extensively and intensify, leading to reduced food grain production and slow economic growth, heightened inequality and poverty, diminished biodiversity and ecosystem services, and increased instances of illness and death among populations in various African and Asian nations. The estimate shows that limiting global warming to 1.5 °C needs global emissions to be slashed by 45% by 2030, compared to 2010 levels. This would benefit us with reducing water-related risks across regions and sectors.

It is worth mentioning that climate change is projected to significantly increase population exposure to heat waves, heat-related morbidity, and mortality as well as drought. It is seen that several parts of Asia may experience some of the worst drought in the last two decades by the end of June 2021. All such problems in some way or the other are closely connected with urbanization and urban growth. According to world urbanization prospects, a significant proportion of the Earth's inhabitants (nearly 4.2 billion) reside in urban regions. The phenomenon of urbanization engenders susceptibility and exposure, which synergistically interact with climatic change perils, thereby fostering urban risk and a multitude of consequences.

At global scale, the swift proliferation of urban vulnerability and exposure has been witnessed in modestly sized towns and cities, where adaptive capabilities are restricted, particularly within unstructured and non-official settlements situated in economically less affluent and middle-income nations possessing smaller to medium-sized urban centers. The repercussions of climate change and associated perils exhibit heterogeneity in both spatial and temporal dimensions. This variability arises from disparities in the extent of exposure and susceptibility exhibited by diverse systems, encompassing ecosystems, economic circumstances, and societal cohorts. Consequently, evaluating the vulnerability of a system assumes paramount importance in discerning suitable adaptation strategies for addressing climate change risks. Furthermore, understanding how to effectively manage existing climate change risks assumes great significance.

In this context, the book *Climate Change, Vulnerabilities and Adaptation* is much needed in the present-day context. The book comprises a collection of 20 research articles (Book chapters), authored by prominent scientists, scholars, and specialists representing various academic fields. The chapters are dealing with wide range of interesting topics like climate change and cyclones its associate consequences, climate change and forest fire in Himalaya, climate change and weather extremes and livelihood, impact of land use/land cover on microclimate with reference to thermal variability, urban heat island on local climatic zones using remote sensing and statistical analysis. The other sets of papers are on extreme climatic events on sustainable urban development, modeling Himalayan glaciers using GIS and Machine Learning (ML). Risk and resilience to the floods in Indian cities, unveiling precipitation and temperature patterns in Kashmir valley, variability of rainfall in the evidence of climate change scenario, and climate change and human resources, are some of the other sets of papers of the book.

I observe that the primary aim of this book volume is to tackle and comprehend climate change, which stands as the most widely discussed and intricate challenge of the twenty-first century. The book also seeks to cope with problems arising from it and what are the regions and communities which are exposed, sensitive, and vulnerable to it? and how to adapt in the prevailing situation with climate change in the present situation? The book represents a significant endeavor aimed at generating current and comprehensive knowledge regarding the state and progression of climate change across different levels. It strives to bridge existing research deficiencies by incorporating diverse advanced remote sensing datasets, Geographic Information System (GIS), Artificial Intelligence (AI), and Machine Learning (ML) techniques. In this regard, the current initiative is a great scientific effort and significant in the era of changing climate. I am pleased that the editors possess the visionary insight and enthusiasm to compile this book, encompassing a subject of utmost significance. Their diligent research endeavors have contributed to enhancing our understanding of vulnerabilities associated with climate change and the corresponding measures of adaptation.

I am extremely delighted and feel proud to introduce the said book volume *Climate Change, Vulnerabilities and Adaptation: Exposure, Sensitivity and Adaptation*. In fact, it's my proud privilege to write the foreword of the book which is the brainchild of Late Prof. Dr. Abha Lakshmi Singh, my mentor, my torch bearer, my guide of academic life with whose blessing and training, I am writing, what I am writing today. What all I could achieve today it's solely because of her training in my academic as well as in personal life.

I congratulate the editors for their foresighted effort in bringing out this book at a very right time on such a pertinent and burning topic of climate change. My best wishes for their future academic endeavors and wish that this book will be very useful to the students, researchers, and scholars across the age and disciplines.

Department of Geography
Jamia Millia, Islamia
New Delhi, India

Atiqur Rahman

Contents

Part I
Climate Change Adaptation in Urban Areas: Linkage Between Research, Perspectives and Practice

Chapter 1
Risk and Resilience in Indian Cities: Floods, Heat Islands and the Work of Professor R.B. Singh

Guy M. Robinson

Introduction

Nature often poses a threat to the citizenry of Indian cities, as seen graphically in the destruction wrought by the floods in Chennai in 2015 and the earthquake in Gujarat in 2001, which destroyed fifty multi-storey buildings in Ahmedabad. Several Indian cities face regular risks from the physical environment, through flood hazards, with the growth of megacities, such as Bengaluru, Hyderabad and Mumbai, exposing ever more residents to flood risk. Climate change may increase the frequency of extreme flood events by further intensifying the hydrological cycle. This could also contribute substantially to global increases in urban flood risk, thereby affecting safety, health and livelihoods (Gupta, 2020).

Vulnerability to flooding tends to be the greatest amongst the poorest in society, living in high-density areas of cities, where frequent flood impact can lead to increased poverty in a vicious cycle affecting the ability of communities to demonstrate resilience and limiting possibilities for recovery from flooding episodes. Exposure to flood risk has grown due to increased settlement on flood plains and urban in-fill development that increases population densities while sealing surfaces and disrupting natural drainage channels (Ramachandra & Mujumdar, 2009). Often, the poorest in society have little option but to live in the most flood-prone areas where housing is the cheapest. There may be few regulations or planning controls to limit the construction of low-quality dwellings. This can also be seen in cities in wealthy and developed countries, as in the case of Hurricane Katrina in New Orleans in 2005, in which the areas of the city suffering the greatest damage were

G. M. Robinson (✉)
Department of Geography, Environment and Population, University of Adelaide, Adelaide, SA, Australia

Department of Land Economy, University of Cambridge, Cambridge, UK
e-mail: guy.robinson@adelaide.edu.au

A. L. Singh et al. (eds.), *Climate Change, Vulnerabilities and Adaptation*,
https://doi.org/10.1007/978-3-031-49642-4_1

disproportionately occupied by African Americans who were then also the slowest to return to the city during the reconstruction phase (Fussell et al., 2010).

This notion of the environment as hazard is part of a complex interaction between city dwellers and nature, and hence there are multiple, and often conflicting, relationships between people and nature within and on the fringes of cities. These interactions can pose difficult choices for city planners and represent a key challenge in the evolution towards more sustainable communities (Robinson, 2018). Geographers have long studied these physical risks affecting urban dwellers in India, with one of the leading researchers being the late Professor Ram Babu (R.B.) Singh at the University of Delhi (for details of his life and career, see Singh, 2021).

This chapter draws upon the work of R.B. Singh, a major figure in Indian geography in the last three decades, to examine two particular hazards he studied, closely associated with climate change, namely increased frequency of extreme events producing severe floods in various Indian cities and urban heat islands (UHIs). The chapter briefly refers to the contributions of R.B. Singh to understanding the impact of climate change on people and livelihoods in India, and then focuses on the two specific risks affecting urban residents. In both cases, special attention is directed at attempts to adapt and mitigate risk. With respect to increased incidence of floods, such attempts are often having to address poor planning decisions that have put citizenry at risk by building on flood plains, as in the case of Chennai's floods in 2015, or failing to maintain urban drainage systems, as for recent flooding in Mumbai. With respect to urban heat islands, the chapter draws upon the work by Singh and other Indian researchers, which has revealed long-term trends and possibilities for both adaptation and mitigation. As part of this chapter a bibliography of R.B. Singh's principal published work is presented as an Appendix.

R.B. Singh and Studies of Hazards, Livelihoods and Climate Change in India

R.B. Singh was the first Indian geographer to hold both the positions of Secretary-General to the International Geographical Union (IGU) and Scientific Committee Member of the International Council for Science (ICSU). He was Professor of Geography in the School of Economics in the University of Delhi, having been educated at Banaras Hindu University, and was elected as the Vice-President of the IGU for successive terms from 2012. He was probably the most well-known Indian geographer in the last two decades, partly because he was quite prolific, with over 230 papers in academic journals and more than 35 edited and 15 single- and co-authored books (Roy, 2021) (see Appendix below). His work covered a wide range of topics, but primarily focused on disaster management, remote sensing and geographical information systems (GIS), environmental studies, climate change, and urban and regional development in India. He also had an outstanding reputation as a mentor and guide to PhD students and early career researchers.

This chapter focuses on just two areas of Singh's work: floods and urban heat islands, but these are ones in which he made a distinctive contribution. In both cases one strand of research focused on New Delhi where he worked and lived for many years. His writing shows a growing concern for the consequences of the rapid growth of the city, as it approaches a population of 30 million by the end of the decade. Urban development has had huge impacts on the environment and Singh lamented the growing inability of the city's infrastructure to deal with traffic congestion, air pollution, sewerage, water supply and maintaining agriculture in the face of urban sprawl. He noted the impact of urban growth on human health and well-being and presented a two-pronged approach in his research: use of the latest techniques to measure and analyse environmental processes to chart the unfolding problems, and following the analysis with prescriptions for policies to promote both mitigation and adaptation. Techniques included the use of satellite remote sensing and Geographical Information Systems (GIS) to map key characteristics of heat islands and deteriorating environmental conditions in the city. The prescriptions often included recommendations for major infrastructural improvements, which have rarely seemed to be forthcoming. However, there have also been some positive advances in the form of more green space in the city, which can help improve well-being and reduce some of the worst effects of urban heating.

In his work on urban flooding, Singh linked long-term analysis of climate change affecting glacier melt in the Himalayas and variations in the monsoonal rainfall pattern, which have contributed to increased flood propensity in some of India's major cities. He also acknowledged the problems associated with creation of more impervious hard surfaces in urban areas and changes to groundwater that have affected patterns of run-off and recharge. However, it has not only been a growth of concrete and tarmac that has altered run-off and assimilation of rainfall, there has also been wholesale removal of natural storage systems, such as naturally occurring lakes, and disruption of river channels by engineering works with little consideration regarding the long-term effects of changing natural water distribution systems. In highlighting this, Singh has suggested that there needs to be a single authority charged with overseeing hydro-engineering systems, with attention given to protecting small lakes and instigating improved systems of waste management, sewerage and provision of drinking water. Preserving and enabling replenishment of groundwater is another vital concern as New Delhi and many other major Indian cities are rapidly depleting available groundwater reserves, creating an uncertain future in terms of availability of future drinking water supplies.

While a central theme in Singh's work was the need for planning of urban development that worked in harmony with the natural environment and available resources, he also acknowledged the growing impact of human-created climate change. Some of his early work focused on the headwaters of the major Indian rivers, recognising changes occurring to weather patterns in the Himalayas, with major consequences for glacier melt and downstream river flows. He examined the impacts of changing water regimes upon agricultural production, not only in the mountain regions themselves but in the heartlands of Indian agriculture, where he warned of the need for more efficient use of irrigation water as part of a more broadly-based approach to natural resource management. He acknowledged that climate change was affecting hydrological regimes, with impacts felt by residents of the major cities in the form of floods and drought, as well as having major consequences for the nation's farmers. He also noted that the changing weather patterns associated with climate change were exacerbating urban heat island effects in some cities, and hence, the need for more imaginative urban planning that emphasised provision of green space and adopted urban methods such as use of green roofs and blue-green space rather than just high-rise and relentless urban sprawl.

The rest of this chapter explores two of the key areas in which Singh made a major contribution: floods (including related issues of overuse of groundwater and pollution of watercourses) and urban heat islands in Indian cities. It examines both the risks posed by these hazards and the attempts made to generate greater resilience through mitigation and adaptive actions.

Floods

Climatic extremes are increasingly resulting in major disasters associated with substantial damage to infrastructure and economy, loss of life and destruction of the natural environment. River systems have spread damage from floods over wide areas in India, as in the case of the Uttarakhand floods (2012, 2013), the Mumbai floods (2005), and the Kashmir floods (2014). Similarly, the heat wave in 2017 affected a large area across northern India (Mal et al., 2018, p. 3) resulting in increased death rates. These growing extremes are accentuating hazards associated with the seasonal changes to atmospheric flow patterns between the Indian sub-continent and the sea, which are related to the monsoons that frequently produce droughts, floods and mass movements over substantial areas of India. Indeed, India has been one of the world's most disaster-prone countries, even before climate change began to exacerbate weather-related hazards. In particular, coastal cities may be at increasing risk from flooding through climate change-induced sea-level rise, with the IPCC predicting more severe inundations (Oppenheimer et al., 2019). For example, low-lying areas in Mumbai are seriously threatened by floods even when there is only minor rainfall. Grover and Singh (2020) note that there were 111 places in the city (26 in Mumbai city district, 73 in the eastern suburbs and 12 in the western suburbs) that were identified in 1993 as flood-prone areas.

Some of the worst examples of catastrophic floods in the country have occurred in Chennai, which has experienced recurrent flooding affecting tens of millions of people across the Adyar basin in the monsoon months, with floods in 2015 especially severe. Kumaran et al. (2020) highlight that poor flood management systems have played a key part in these floods, with little by way of adaptive measures. The severe floods have been compounded by the growth of the city that has reduced the capacity of the existing waterways at the same time as more impervious surfaces have contributed to increased run-off. Drainage of wetlands has reduced the natural

Table 1.1 Proposed measures to address flood problems in Chennai

Better understanding of weather through improvements in forecasting, partly through creation of a more finely grained system of reporting and recording.
Placement of water-gauging stations allied to more detailed flood modelling.
Construction of storages rather than building chaotically in areas where storage should occur.
Public awareness needs to be increased through better education about climate change and associated risks and vulnerabilities, including community-based information initiatives.
Clearing off encroachments in and near waterbodies, which may involve some resettlement of people.
Desilting and conservation of water bodies.
Rapid assessment of flood inundation, e.g., through community mapping.
Flood mitigation programmes.
Watershed management to reduce siltation.
Increasing green cover.
Scientific study of flooding patterns for a period of at least 30 years, including use of modelling techniques to help guide strategic planning.
Riverfront water development that involves key stakeholders in managing flood control.

Source: Kumaran et al. (2020)

capacity of the area around the city to soak up water, contributing to lack of proper water storage and management. Kumaran et al. (2020) propose a series of measures that could contribute to addressing Chennai's flood problems (see Table 1.1). These are typical of the types of measures that need to be applied to address the problem of urban floods in India's major cities.

Most flooding in India occurs between June and September, reflecting the presence of the south-west monsoon, but in southern parts the northeast monsoon can bring floods from October to December. The combination of extreme rainfall events and the growth of built-up areas has exacerbated the severity of floods, as illustrated by Gupta (2020, p. 2): 'Rapid urbanization has resulted in increased paved areas, decreased water bodies, reduced groundwater recharge and reduced capacity of urban drainage channels.' The huge influx of rural migrants to many Indian cities has placed great strains on the sewerage systems, with increased amounts of untreated wastewater often discharged into the stormwater drains and rivers. Flooding then sends this contaminated mixture pouring through the streets and houses, as was observed in the Chennai floods of 2005 and 2015. This flooding has also been exacerbated by urban drainage systems in which investment into storm drainage has been inadequate and largely intended for events whose intensity is regularly being exceeded, as also highlighted by Gupta (2020, ibid.): 'A compilation of the high-intensity rainfall (greater than 50 mm day^{-1}) and subsequent flooding incidents during 2018–2019 in the capital cities of the 28 Indian states and the 9 Union Territories indicates the severity of the rainfall-related flooding in these cities.'

Singh and Singh (2011) noted the importance of impervious land cover types in promoting flooding in the planned city of Noida (population approaching three-quarters of a million), immediately south-east of Delhi in Uttar Pradesh, with 'direct runoff from urban areas ... responsible for a flood event compared with runoff from other land use areas' (p. 147). Repeated heavy floods in the city reflect its unsuitable terrain in a low-lying former river basin readily subject to waterlogging. Using remote sensing and modelling techniques, the authors showed how micro-scale planning can be formulated to alleviate flooding. Yet, this has largely been something that has been eschewed by city planners, with an emphasis on reacting after serious flooding has occurred rather than using modelling to identify flood-prone areas and taking measures to focus on future flood-proofing. Simple measures such as re-zoning to prevent future sprawl into undesirable areas (e.g., the most vulnerable parts of flood plains) could prevent destruction of property and loss of life.

Control of floods offers opportunities for socio-economic growth and development by application of new approaches such as building more raised platforms, drainage channels, culverts, sluices and embankments (Singh et al., 2014b). However, Singh et al. (2016) recommend that regional cooperation between states and neighbouring countries is required to improve water resource management through mitigation of floods to ensure greater stability of food production and power generation.

While climate change has been associated with greater irregularity of monsoon rains, it has also contributed to flooding from glacier melt in the Himalayas coupled with increasing population pressure, overgrazing, deforestation, road and dam construction, and over-development of agriculture and horticulture on steep slopes. Global warming has produced ice melting leading to overflows from glacial lakes while more severe storm events have also contributed to destructive floods.

One of the most flood-prone rivers in India is the Brahmaputra in Assam, where 3.2 million ha are subjected to periodic flooding, including towns such as Dhubri, Goalpara and Guwahati. Projections suggest rainfall will increase across Assam in the twenty-first century. Added to more glacier melt in the Himalayas, this could exacerbate existing flood problems. However, major dam-building schemes upstream in China may reduce water flows while further anthropogenic intervention in the form of deforestation may increase water flows. Pradhan et al. (2021) suggest several ways in which the flood hazard might be alleviated, echoing ideas developed earlier by Singh et al. (2014b). One key aspect is to obtain transboundary co-operation as exists at present between India and Bangladesh. However, such agreement is currently harder to obtain with China given the ongoing dispute over territory. A community-based flood early warning system (CBFEWS) is urgently needed and has been conceptualised by local agencies. Some initial implementation is in place but this needs upscaling across the river basin.

To lessen the impact of floods on communities adjacent to the Brahmaputra there is a need for more flood-resilient infrastructure. One example would be the greater use of embankments around raised constructions (such as local houses), which may be strengthened through use of native plant species to avoid landslides triggered by soil erosion. The introduction of simple ecological sanitation (eco-san) toilets can help prevent pollution of the river, while also producing fertilizer for agriculture. Improved wetland conservation and management is another important consideration to ensure that the wetlands act as water reservoirs which retain water during dry periods and help sustain stream flow. 'The hydrological connectivity among wetlands and with the rivers are of crucial importance for ecohydrological functioning of both rivers and wetlands' (Pradhan et al. 2021, p. 16). However, that connectivity is threatened by severe ecological degradation associated with ongoing urbanisation, pollution, drainage for agriculture and over-exploitation of fish and macrophytes. With regard to the latter, there is a need to create freshwater fish safe zones (FFSZ).

The scale of investment needed to reduce the flood hazard along the Brahmaputra is substantial, especially when set against additional problems associated with climate change. Hence, there is a need for special initiatives to bring about change. One such example is the US$120 million invested by the Asian Development Bank since 2010 into the Assam Integrated Flood and Riverbank Erosion Risk Management Investment Program. This Program has been implemented by the Flood and River Erosion Management Agency of Assam (FREMAA), aiming 'to improve flood and riverbank erosion management by adopting a community-based approach built around institutional reform and the strengthening of the knowledge base of the water sector' (p. 17). Another major project developed by the Government of Assam is addressing modernisation of flood and erosion management, investing US$200 million since 2016 and involving the building of infrastructure and preparing an integrated erosion, flood and sediment management strategy.

Climate change is also likely to mean increased sediment loads through the river basin, perhaps a 40% increase in annual sediment load by 2075–2100 compared with 1986–1991 (Fischer et al., 2017). Increased siltation could have major impacts on agriculture, aquatic ecology, hydroelectric power generation and water storage in reservoirs (used for both irrigation and drinking). Climate change is also likely to increase water scarcity in the dry season from November to May, with potential to disrupt agriculture, food security and industry (Gain & Wada, 2014).

Overuse of Groundwater Supplies and Pollution of Watercourses

One aspect of Singh's work was on the United Nations' Sustainable Development Goal 6, 'ensure availability and sustainable management of water and sanitation for all'. He observed that India fell short of desirable standards with regard to many aspects of water management. Key issues are excessive demand on available supplies of water, partly associated with significant drought in some regions, contamination of water supplies and increased occurrence of floods. The World Health Organization estimates that close to 100 million Indians do not have access to safe drinking water. In addition, around 21% of communicable diseases in the country are water-borne (Singh, 2017). However, there are two major problems affecting water supply, both of which are related to climate change. The first is drought, which in recent years has affected around one-third of all administrative districts in India, extending across ten states and affecting 330 million people. This is having a profound effect on availability of water in some cities, e.g., in Hyderabad the four major reservoirs had dried up in 2017, while in Pune, water tankers have been used to maintain essential supplies. Elsewhere industry has had to shut down because of lack of water (p. 324).

Ironically, at a time when many Indian cities are being subjected to more severe flood episodes, some, including Bhatinda, Chennai, Jaipur, Lucknow, Mumbai and Nagpur are experiencing severe water shortages. Severe depletion of groundwater resources has been reported for 21 major cities affecting over 100 million people. This reflects the rapid growth in demand for water as the cities have grown so rapidly in recent decades, with 40% of the country's population projected to be living in urban areas by 2030. Indeed, several cities have already begun to ration water in the summer months, including Bhatinda, Jaipur, Lucknow, Mumbai and Nagpur (Ghosh, 2021). Between 40% and 50% of India's urban water supplies are derived from groundwater, generally considered a safer option than reliance on surface water, with one estimate reporting that 80% of surface water is polluted by domestic sewage because of an absence of sanitation facilities and lack of policies addressing wastewater and sanitation (Sengupta, 2019).

The annual extraction rate of groundwater in India is now the highest in the world (Saha, 2019, p. 24), with urban use accompanying extraction for irrigation and other rural use. In Punjab, Rajasthan, Haryana and Delhi, extraction from groundwater exceeds recharge (see also Khetwani & Singh, 2018). In addition, groundwater is becoming polluted, with rising levels of arsenic, chloride, fluoride, iron, nitrate and uranium. Urban sprawl has often removed water bodies, reducing both infiltration and groundwater recharge, and also contributing to urban flooding. This has been reported for Chennai, where 150 water bodies had been reduced to 27 by 2010 (Gupta & Nair, 2011).

A similar trend in the decline of water bodies was reported for Delhi by Singh et al. (2013). Their research showed that in 1970, the total number of water bodies was 807 with an area of 14.41 km^2, but that there had been a decline of 21% to 640 by 2008, with an area of 8.51 km^2 (a 41% decline in area). Moreover, there was deterioration of water quality in the remaining water bodies through eutrophication, inflow of wastewater from nearby residential areas and garbage dumping around them. These water bodies (locally known as *baolies, johads* and *hauz*) were part of traditional forms of water management, with artificial ponds and reservoirs created to harvest water. These were managed over the centuries by local communities to ensure equitable and fair distribution of water, but as population numbers grew, the protectors of the water became the polluters and destroyers. The loss of agricultural land to urban sprawl has been another major reason behind the loss of rural water bodies, with no clear laws to protect them and their catchments. Hence, 'the introduction of piped water supply from far off sources and extraction of groundwater through tube-wells has changed the human interaction and dependence on traditional water bodies' (ibid, p. 365).

Singh et al. (2013) argue that the myriad of rural water bodies played a key part in recharging aquifers, maintaining groundwater levels, controlling groundwater salinity and supporting various flora and fauna. Their decline is not only having

Table 1.2 Measures to improve the water quality of India's major rivers

Build sewage treatment plants at the mouth of major drains.
Construct low-cost toilets and crematoriums to counter non-point pollution.
Maintain minimum flow required in the river.
Build reservoirs to store water.
Apply optimum application of fertilizers.
Relocate factories still operating on the banks of rivers.
Use treated effluents, both domestic and industrial, for irrigation.
Create special areas of need where funeral ashes and garlands can be processed with respect and without affecting the sanctity of the river.
Use traditional clay rather than baked clay for idols to be immersed in the river water during festivals.
Discourage construction of any type along the river bed.

Source: Singh (2017, p. 327)

noticeable impacts on biodiversity but also affecting groundwater. The traditional water bodies acted as recharge zones, but their gradual decline coupled with increased extraction of groundwater has meant that the groundwater level of Delhi is declining by an alarming 2 m every year. Seven out of nine groundwater districts for the city had been declared as 'over-exploited' by 2012. The situation was worsened by construction related to Delhi hosting the Commonwealth Games in October 2010. This directly reduced the groundwater level. Similarly, major road construction has removed water bodies in the south of the city. Yet, city planning has largely ignored the importance of these traditional water bodies and there has been little protection afforded to them. One issue has been the fragmentation of responsibility for overseeing their protection. Hence, Singh et al. (2013) proposed that all water bodies should come under the jurisdiction of a single authority.

Excessive pollution of water sources has become an increasing problem in many Indian cities. Singh (2017) attributes this in part to the by-product of insufficient and delayed investment in urban water treatment facilities. There is a lack of water treatment plants for reuse and recycling of waste water, encroachment of settlement on water bodies and no proper pipe distribution for households. For example, in 2010 45% of the city remained unconnected to the sewer network, and one-quarter of the city was unconnected to the water supply system (Hughes, 2013, p. 7). Singh also notes that there is no proper pricing mechanism for industrial and commercial use of water. Moreover, little use has been made of water-saving technologies commonly in place in many countries around the world. He highlights the squalid condition of the river Yamuna which runs through Delhi, into which 2000 million litres of raw sewage are pumped every day. He offers a set of measures that need to be implemented to address these water quality issues (Table 1.2). However, while some of these are quite simple and relatively easy to introduce, others affecting location of industry and construction of sewage treatment plants involve substantial costs to government.

Given the scale of the urban water quality problem, it is evident that major investment is needed in all of India's major cities. At present, the two key water quality initiatives in place nationally are the National Rural Drinking Water Programme (NRDWP) and the Accelerated Urban Water Supply Programme (AUWSP), which target provision of universal and equitable access to safe and affordable drinking water for all by 2030.

The NRDWP, which was launched in 2009, aims to provide safe and adequate water for cooking, drinking and other domestic needs to every rural person on a sustainable basis, and specifically ensuring that 35% of rural households have water connections and about two buckets of water per person per day. However, despite spending INR 89,956 (US$1199.44) × 10 million between 2012 and 2017 the Government Auditor in 2018 reported that it had 'failed to meet its targets', referring to 'poor execution' and 'weak contract management' (Paliath, 2018). The AUWSP dates to 1993, but applies primarily to small towns rather than large cities, for whom municipal authorities are responsible for water supplies.

Singh and Chandna (2011) analysed the pollution levels in the Yamuna River pre- and post-monsoon periods, revealing the very high levels of pollution as the river flows through metropolitan Delhi. The post-monsoon period dilutes the pollution levels a little, but there is evidence that reduced run-off from the Himalayas because of climate change may reduce this dilution effect, whereas short-term flooding will pose more challenges for water quality in terms of increasing concentrations of pollutants. Hence, we can conclude that without significant and fundamental government interventions the problem of floods, water shortages and major pollution of water bodies in Indian cities will remain unresolved.

Urban Heat Islands

Veena et al.'s (2020) survey of urban heat island (UHI) studies in India recorded 37 academic papers for different parts of major cities since 2000, with about 30% dealing with Delhi. Bengaluru was the next most studied city. They concluded that numerical modelling studies of UHIs had been quite limited in India compared with many countries, so a greater focus on this topic is needed. For India's megacities, Qureshi (2018) quotes UHIs of between 4 and 8 °C higher than the temperatures in the surrounding urban hinterland.

UHIs are concentrations of higher surface temperatures in cities, in comparison with surrounding rural areas. UHIs are associated with heat-absorbing surfaces in the built elements of the urban landscape and disturbed airflow patterns (often termed the 'canyon effect'). These generate radiative, thermal, moisture and aerodynamic modifications in the surrounding environment. UHIs are associated with radiative trapping and screening through the canyon effects of building geometry and anthropogenic heat sources. The latter include automobile engines and air conditioning units. In addition, urban surface roughness can lower albedo and increase thermal mixing, which, in turn, can impact local-level natural solar and hydrological balances (see Yow, 2007). The decreased turbulence that occurs in the urban canyons between high-rise buildings results in reduced heat loss. The bigger and more densely settled the city, especially with concentrations of high-rise buildings, the greater is the likelihood of a substantial UHI effect (Dos Santos et al., 2017). However, there may also be a cooling effect exerted by oceanic influence for coastal cities. Kotharkar et al. (2018) argue that UHIs have played a role in increasing urban floods by enhancing convective rainfall.

Various factors contribute to the development of UHIs, notably human-created changes in the urban environment; greater use of energy sources by industry, transport and for residential heating/cooling; changes in land use and land cover; impervious surfaces; urban geometry, especially with high-rise buildings; local weather and locational factors (Grover & Singh, 2020, p. 156). UHIs are not just a minor meteorological phenomenon, with poor air quality allied to rising temperatures contributing to cardiovascular and respiratory illnesses and diseases.

Roy et al.'s (2011) analysis of Delhi's urban heat island noted how the highest urban temperatures, especially in the late afternoon, were concentrated in the northern and western sections of the city, and coincided with industrial and high-density residential land uses. Here, there were extremely high daytime temperatures of prolonged duration, especially in the summer months, reflecting predominantly clear-sky conditions. The areas of high temperatures were generally more localised during the coolest months. However, the severity of the UHI was contributing to heat-related stresses in the local population, with the UHI up to 10 °C in some cases. The presence of green space exerted an ameliorating effect on the UHI, notably in the centre of New Delhi, which was created as a fully planned city in the later years of British colonial rule and incorporated 'green' parkland.

Green cover in Delhi has been substantially upgraded in the last three decades, with a social forestry programme commencing in the 1980s and a joint forestry programme in 1990. Some of this cover has been developed as part of the Bhagidari System initiated in 2003. This is a Delhi Government initiative promoting broadly based citizen participation in local governance, and especially to improve the quality, efficiency and delivery of public services (Chakrabarty, 2012; Ghertner, 2011; Hughes, 2013). The green cover area of Delhi was increased from 26 km^2 in 1996–1997 to more than 300 km^2 in 2009. Bhagidari means 'partnership', and the aim of the System is to empower and provide locales for residential welfare associations (RWAs) across the city to establish priorities for their own neighbourhoods and to communicate these to the Delhi government. However, the RWAs are largely property owners rather than a broader cross-section of the community, though there are unofficial RWAs representing informal settlements that can be incorporated into the planning process, especially regarding planning for climate change adaptation (Hughes, 2013, p. 7). Yet, 'Delhi's tree planting program has successfully partnered with RWAs to revitalise parks and green spaces throughout the city' (p. 8), with the Parks and Gardens Society embarking on a substantial programme of tree planning. 'The high level of greenness in the capital city is accountable for maintaining relatively low temperatures and the absence of clear UHI even during the hottest month of the year in May' (Grover & Singh, 2020, p. 169).

Grover and Singh (2015) compared the UHIs of Mumbai and Delhi, the two largest metropolitan cities of India, using Landsat 5 Thematic Mapper imagery (see also Singh & Grover, 2014, 2015). For Delhi, they showed that the Delhi Ridge forests and the River Yamuna, which cut across the city, are important in moderating surface temperatures and limiting the UHI. In contrast, coastal Mumbai has much less tree cover and is more densely populated, with a strong UHI in its city centre. Indeed, according to Bherwani et al. (2021), the urban sprawl of the city in the last two decades has increased the UHI temperatures by between 3.4 and 8.8 °C. This research illustrates the importance of city planning that incorporates norms which emphasise green cover in order to ameliorate the UHI.

Additional research by Singh et al. (2014a) on Delhi for all four seasons further illustrated the significant impact of complex land use and land cover patterns on temperature, again emphasising the cooling effects of vegetation cover and water bodies. Unusually, they found that Delhi did not follow the standard pattern for UHIs, in which the highest temperatures occur in the densely built-up city centre. In Delhi's case, it was bare soil arable land on the city's peripheries that had the highest temperatures, especially in summer and autumn. Indeed, they did not record a high UHI in Delhi at any season. In part, this was because 'the healthy forests and tree cover and especially the ridge forests act as thermal moderators and cooling agents in the climate of Delhi' (p. 1824). However, the lack of a cluster of tall skyscrapers is also a key factor.

Roy and Singh (2015) focused on the role of local level atmospheric humidity levels across the Delhi Metropolitan Region, reinforcing the significance of green cover in restricting the development of the UHI. In contrast, different factors were found to influence concentrations of pollutants in the atmosphere. The lowest pollution levels for SO_2, NO_2 and O_3 occur in the summer months, but higher values are observed in the second half of the year. The strength of the diurnal cycle is greatest in the first half of the year, because clear skies are more prevalent (O'Shea et al., 2016). Peak occurrence of NO_2 is around midnight, whereas for O_3 this is the time of lowest concentrations, with a peak in the early afternoon. The lowest concentration levels have been observed in the mixed-use residential areas, e.g., NSIT Dwarka and Shadipur, with the highest levels of pollutants consistently found at two of the stations, Civil Lines and IGI Airport.

Guha et al.'s (2017) analysis of changes in the UHI of Raipur City (Chhattisgarh) over time showed a steady increase in its magnitude between 1995 and 2016 (2.6 °C in 1995, 2.85 °C in 2006, 3.42 °C in 2009 and 3.63 °C in 2016). Key hot spots were recorded in association with high-density transport parking lots, industrial areas and metal roofs in the north-west of the city. A similar increase in intensity of the UHI was recorded for Ludhiana City in the Punjab by Singh and Kalota (2019): up to 3 °C between 2009 and 2016, reflecting the growth of the urban area and the concentration of high-density built-up areas. With reference to the UHI of Hyderabad, Salahuddin (2021) notes that it is the more pronounced nocturnal heat island growth that has been most hazardous to health.

Kumar et al. (2017) refer to an urban cool island (UCI) effect, which they claim affects around 60% of India's major cities in the daytime. Their observations point to

UCIs being attributable to limited moisture and vegetation in non-urban areas relative to the cities. By contrast, urban areas in extensively irrigated landscapes may experience positive UHI cool effects. UHI warming usually intensifies at night, and occurs across circa90% of the country's urban areas. The extent and difference of temperature contrasts between rural and urban areas are primarily determined by agriculture and moisture availability from irrigation. However, they contend that atmospheric aerosols may also play an important role.

Singh's work on UHIs, especially in conjunction with S.S. Roy and A. Grover, has been part of a growing body of work by Indian scholars that has broadened the understanding of this phenomenon both meteorologically and in terms of its impacts on the urban population. It has shown that climate change may be exacerbating some of the high temperatures associated with UHIs, adding to the urgent need for implementation of policies designed to reduce UHIs and other negative outcomes of rapid urban growth, such as atmospheric pollution. The latter is closely linked to the UHI as are potential solutions, with Menon and Sharma (2021) noting that mitigating air pollution may lessen differences between urban and rural in terms of incoming radiation, thus mitigating the UHI effect. In accordance with some of Singh's arguments, they urge the adoption of multifunctional, cost-effective nature-based solutions, which can contribute to reducing costs of energy and health while supporting biodiversity conservation (Singh and Prokop, 2016). There are signs that such solutions are beginning to be incorporated into urban planning in India. For example, there are 'greening' strategies in the National Mission for Green India through the National Action Plan on Climate Change (NAPCC) and via city clean air action plans and smart city plans, though they need better integration into the national urban planning agenda.

Conclusion

This chapter concludes with an Appendix that lists R.B. Singh's principal publications in terms of those most widely cited, plus some recent books published within twelve months of his death on July 22, 2021 (aged 66). This list illustrates the wide canvas of Singh's work and key threads interwoven through his work. Among his most cited works there are various papers that deal with floods and UHIs, as referenced and quoted above. It is a corpus of writing that gets to the heart of key problems facing Indian cities today, especially the risk and resilience in face of the growing impact of climate change. Singh consistently highlighted that urban development had insufficiently considered the need to accommodate natural systems, thereby increasing risk and reducing ability to recover from severe events such as major floods. In addition, other problems such as massive pollution of watercourses, declining air quality and the growing concentrations of the poorest people in areas most susceptible to risk are presenting a growing list of issues for governments to address.

Much of Singh's writing used careful data collection and methods, often involving the latest technologies, to indicate the nature and extent of particular problems, but turning analysis into remediation measures adopted in policy has proved less successful. In part, this reflects the particular dilemma that India faces, with a need to grow its economy to raise people out of poverty while at the same time recognising a major requirement to improve the standards of environmental management.

This was illustrated at the COP26 conference in late 2021 where both India and China drew criticism for refusing to phase out coal production within an acceptable timeframe. Instead, they agreed to 'phase down' production over time and pointed to a lack of adequate climate financing from developed countries as a more serious issue. Yet India and China are among the highest polluters on the planet, and controlling worldwide climate change rests heavily on both countries reducing use of fossil fuels. In India's case, there is a pledge to meet a net-zero emissions target by 2070, but some fear this will be too late and catastrophic damage will occur in the intervening 50 years. Ironically, India is one of the most heavily impacted already by catastrophic floods, droughts and temperature rises. So, the conundrum highlighted in Singh's writing remains: how to reduce poverty through economic growth while taking meaningful pro-environmental measures that themselves can contribute to poverty alleviation and address the poor urban planning of previous decades?

Appendix: Professor R B Singh – Principal Publications

This list includes all of Singh's publication which have achieved at least ten citations according to Google Scholar plus recent books published.

Mishra, R.K., Singh, R.B. and Dubey, A. (eds.), 2021. *Sustainable climate action and water management.* Springer, Cham, Switz.

Grover, A. and Singh, R.B., 2020. *Urban health and wellbeing: Indian case studies.* Springer, Singapore.

Singh, R.B., Srinagesh, B. and Anand, S. (eds.), 2020. *Urban health risk and resilience in Asian cities.* Springer Nature, Singapore.

Mal, S., Singh, R.B., and Huggel, C. (eds.), 2018. *Climate change, extreme events and disaster risk reduction: Towards sustainable development goals.* Springer, Cham, Switz.

Singh, R.B. and Majoral, R. (eds.), 2018. *Marginal regions: Processes, technological developments and societal reorganizations.* Routledge, New York and Abingdon, Oxon. First published by Westview Press, 1996.

Schickhoff, U., Singh, R.B. and Mal, S. 2016. Climate change and dynamics of glaciers and vegetation in the Himalaya: an overview. In: R.B. Singh, U. Schickhoff and S. Mal (eds.), *Climate change, glacier response, and vegetation dynamics in the Himalaya: Contributions toward future Earth initiatives.* Springer, Cham, Switz., pp. 1–26.

Singh, R.B. (ed.), 2016. *Progress in Indian geography. A country report, 2012–2016*. The 33rd International Geographical Congress, Beijing, China (August 21–25, 2016). Indian National Science Academy, New Delhi.

Singh, R.B. and Prokop, P. (eds.), 2016. *Environmental geography of South Asia: Contributions towards a future earth initiative*. Springer, Tokyo.

Singh, R.B. and Grover, A. 2015. Spatial correlations of changing land use, surface temperature (UHI) and NDVI in Delhi using Landsat satellite images. In: R.B. Singh (ed.), *Urban development challenges, risks and resilience in Asian mega cities*. Springer Japan, Tokyo, pp. 83–97.

Singh, R.B. and Kumar, A. 2015. Climate variability and water resource scarcity in drylands of Rajasthan, India. *Geoenvironmental Disasters* 2(1), 1–10.

Grover, A. and Singh, R.B. 2015. Analysis of urban heat island (UHI) in relation to normalized difference vegetation index (NDVI): A comparative study of Delhi and Mumbai. *Environments* 2(2), 125–138.

Singh, R.B. (ed.) 2015. *Urban development challenges, risks and resilience in Asian mega cities*. Springer, Tokyo.

Singh, R.B. and Grover, A. 2015. Spatial correlations of changing land use, surface temperature (UHI) and NDVI in Delhi using Landsat satellite images. In: R.B. Singh (ed.), *Urban development challenges, risks and resilience in Asian mega cities*. Springer, Tokyo, pp. 83–97.

Ma, M., Haapanen, T., Singh, R.B. and Hietala, R. 2014. Integrating ecological restoration into CDM forestry projects. *Environmental Science & Policy* 38, 143–153.

Singh, R.B. and Grover, A. 2014. Remote sensing of urban microclimate with special reference to urban heat island using Landsat thermal data. *Geographia Polonica* 87 (4), 555–568.

Singh, R.B., Grover, A. and Zhan, J. 2014. Inter-seasonal variations of surface temperature in the urbanized environment of Delhi using Landsat thermal data. *Energies* 7(3), 1811–1828.

Singh, R.B. and Hietala, R. (eds.), 2014. *Livelihood security in Northwestern Himalaya*. Springer Verlag, Japan, Tokyo.

Singh, R.B. and Singh, S. 2014. Human-induced biome and livelihood security. In: R.B. Singh and R. Hietala (eds.). *Livelihood security in Northwestern Himalaya*. Springer Verlag, Japan, Tokyo, pp. 53–66.

Singh, R.B. and Kumar, P. 2014. Geographic and socio-economic realities of Himachal Pradesh, North-western Himalaya. In: R.B. Singh and R. Hietala (eds.), *Livelihood security in Northwestern Himalaya*, Springer Verlag, Japan, Tokyo, pp. 11–26.

Singh, R.B. and Mal, S., 2014. Trends and variability of monsoon and other rainfall seasons in Western Himalaya, India. *Atmospheric Science Letters*, 15(3), pp. 218–226.

Singh, R.B. and Shi, C. 2014. Advances in observation and estimation of land use impacts on climate changes: improved data, upgraded models, and case studies. *Advances in Meteorology*, 748169.

Sahu, N., Singh, R.B., Kumar, P., Silva, R.V.D. and Behera, S.K. 2013. La Niña impacts on Austral Summer extremely high-streamflow events of the Paranaíba River in Brazil. *Advances in Meteorology*, 461693.

Singh, R.B. and Anand, S. 2013. Geodiversity, geographical heritage and geoparks in India. *International Journal of Geoheritage* 1(1), 10–26.

Anand, A., Chandan, P. and Singh, R.B. 2012. Homestays at Korzok: Supplementing rural livelihoods and supporting green tourism in the Indian Himalayas. *Mountain Research and Development* 32(2), 126–136.

Ma, B., Singh, R.B. and Hietala, R. 2012. Human driving forces for ecosystem services in the Himalayan region. *Environmental Economics* 3(1), 53–57.

Singh, R.B. 2012. Climate change and food security. In: N. Tuteja, S. S. Gill and R. Tuteja (eds.), *Improving crop productivity in sustainable agriculture.* Wiley-Blackwell, Weinheim, pp. 1–22

Singh, R.B. and Kumar, D. 2012. Remote sensing and GIS for land use/cover mapping and integrated land management: case from the middle Ganga plain. *Frontiers of Earth Science* 6(2), 167–176.

Roy, S.S., Singh, R.B. and Kumar, M. 2011. An analysis of local spatial temperature patterns in the Delhi Metropolitan Area. *Physical Geography* 32(2), 114–138.

Singh, R.B. and Singh, S. 2011. Rapid urbanization and induced flood risk in Noida, India. *Asian Geographer* 28(2), 147–169.

Sati, V.P. and Singh, R.B. 2010. Prospects of sustainable livestock farming in the Uttarakhand Himalaya, India. *Journal of Livestock Science*, 1(1), 9–16.

Singh, R.B., Mal, S. and Kala, C.P. 2009. Community responses to mountain tourism: A case in Bhyundar Valley, Indian Himalaya. *Journal of Mountain Science* 6(4), 394–404.

Abhineety, G. and Singh, R.B. 2006. Sustainable forestry in mega-cities of India for mitigating carbon sequestration: case study of Delhi. *Advances in Earth Science* 21(2), 144–150.

Singh, R.B. (ed.), 2006. *Sustainable urban development.* Concept Publishing Company, New Delhi.

Singh, R.B. and Mishra, D.K. 2004. Green tourism in mountain regions-reducing vulnerability and promoting people and place centric development in the Himalayas. *Journal of Mountain Science* 1(1), 57–64.

Gardner, J., Sinclair, J., Berkes, F. and Singh, R.B. 2002. Accelerated tourism development and its impacts in Kullu-Manali, HP, India. *Tourism Recreation Research* 27(3), 9–20.

Sen Roy, S. and Singh, R.B. (eds.), 2002. *Climate variability, extreme events and agricultural productivity in Mountain Regions.* Oxford & IBH Pub. Co., New Delhi.

Singh, R.B. and Sen Roy, S. 2002. Climate variability and hydrological extremes in a Himalayan catchment. *ERB and Northern European FRIEND Project 5 Conference, Demänovská dolina, Slovakia.*

Singh, R.B. (ed.) 2001. *Urban sustainability in the context of global change: Towards promoting healthy and green cities.* Science Publishers, Enfield, NH.

Singh, R.B. 2001. Impact of land-use change on groundwater in the Punjab-Haryana plains, India. *Impact of Human Activity on Groundwater Dynamics* (Proceedings

of a symposium held during the Sixth IAHS Scientific Assembly at Maastricht, The Netherlands, July 2001). IAHS Publ. no. 269, pp. 117–122.
Singh, R.B., Fox, J. and Himiyama, Y. (eds.). 2001. *Land use and cover change*. Science Publishers, Enfield, NH.
Singh, R.B. 2000. Environmental consequences of agricultural development: a case study from the Green Revolution state of Haryana, India. *Agriculture, Ecosystems & Environment* 82 (1–3), 97–103.
Singh, R.B. 2000. Monitoring of dust pollution by higher groups of plants around dust polluted habitats in Sonbhadra, Uttar Pradesh. *Indian Journal of Environmental Ecoplanning* 3, 163–166.
Singh, R.B. 1999. Urban impacts on groundwater quality in the Delhi region. In: *Impacts of Urban Growth on Surface Water and Groundwater Quality* (Proceedings of IUGG99 Symposium HS5, Birmingham, July 1999). IAHS Publ. no. 259.
Duffield, C., Gardner, J.S., Berkes, F. and Singh, R.B. 1998. Local knowledge in the assessment of resource sustainability: case studies in Himachal Pradesh, India, and British Columbia, Canada. *Mountain Research and Development*, 18(1), 35–49.
Haigh, M.J., Singh, R.B. and Krecek, J. 1998. Headwater control: matters arising. In: M.J. Haigh, G.S. Rajwar, J. Krecek and M. Kilmartin (eds.), *Headwaters: water resources and soil conservation*. CRC Press, Boca Raton, Fla.
Singh, R.B. (ed.), 1998. *Sustainable development of mountain environment in India and Canada: CIDA-SICI Project Experience*. Oxford & IBH Publishers, New Delhi.
Singh, R.B. 1998. Land use/cover changes, extreme events and ecohydrological responses in the Himalayan region. *Hydrological Processes* 12(13-14), 2043–2055.
Singh, R.B. (ed.), 1996. *Disasters, environment and development*. Taylor & Francis, London.
Singh, R.B. (ed.) 1995. *Global environment change: Remote sensing and GIS perspectives*. A.A. Balkema, Rotterdam.
RB Singh 1995. Land use change, diversification of agriculture and agro-forestry in Northwest India. In: R.B. Singh (ed.), *Global environment change: Remote sensing and GIS perspectives*. A.A. Balkema, Rotterdam, pp. 313–321.
Singh, R.B. and Haigh, M.J. (eds.), 1995. *Sustainable reconstruction of highland and headwater regions: Proceedings of the third international symposium, New Delhi 6–8 October 1995*. A.A. Balkema, Rotterdam.
Singh, R.B. and Pandey, B.W. 1995. Common resources and sustainable livelihoods of mountain environments: A micro-level experience of Upper Kulu Valley. In: R.B. Singh and M.J. Haigh (eds.), *Sustainable reconstruction of highland and headwater regions: Proceedings of the third international symposium, New Delhi 6–8 October 1995*. A.A. Balkema, Rotterdam.
Singh, R.B. 1986. *Geography of rural development: The Indian micro-level experience*. Stosius Incorporated/Advent Books Division.

References

Bherwani, H., Kumar, S., Kumar, N., Singh, A., & Kumar, R. (2021). Geospatial analysis to understand the linkage between urban sprawl and temperature of a region: Micro- and meso-scale study of Mumbai City. In R. K. Mishra, R. B. Singh, & A. Dubey (Eds.), *Sustainable climate action and water management* (pp. 181–189). Springer.

Chakrabarty, B. (2012). Citizen–government partnership: Bhagidari in Delhi. *International Journal of Urban Sciences, 16*(1), 1–22.

Dos Santos, A. R., de Oliveira, F. S., da Silva, A. G., Gleriani, J. M., Gonçalves, W., Moreira, G. L., Silva, F. G., Branco, E. R. F., Moura, M. M., da Silva, R. G., & Juvanhol, R. S. (2017). Spatial and temporal distribution of urban heat islands. *Science of the Total Environment, 605*, 946–956.

Fischer, S., Pietron, J., Bring, A., Thorslund, J., & Jarsj, J. (2017). Present to future sediment transport of the Brahmaputra River: Reducing uncertainty in predictions and management. *Regional Environmental Change, 17*, 515–526.

Fussell, E., Sastry, N., & Van Landingham, M. (2010). Race, socio-economic status, and return migration to New Orleans after Hurricane Katrina. *Population and Environment, 31*(1–3), 20–42.

Gain, A., & Wada, Y. (2014). Assessment of future water scarcity at different spatial and temporal scales of the Brahmaputra River Basin. *Water Resources Management, 28*, 999–1012.

Ghertner, D. A. (2011). Gentrifying the state, gentrifying participation: Elite governance programs in Delhi. *International Journal of Urban and Regional Research, 35*(3), 504–532.

Ghosh, P. (2021). Water stress and water crisis in large cities of India. In R. K. Mishra, R. B. Singh, & A. Dubey (Eds.), *Sustainable climate action and water management* (pp. 131–138). Springer.

Grover, A., & Singh, R. B. (2020). *Urban health and wellbeing: Indian case studies*. Springer.

Grover, A., & Singh, R. B. (2015). Analysis of urban heat Island (UHI) in relation to normalized difference vegetation index (NDVI): A comparative study of Delhi and Mumbai. *Environments, 2*(2), 125–138.

Guha, S., Govil, H., & Mukherjee, S. (2017). Dynamic analysis and ecological evaluation of urban heat islands in Raipur city, India. *Journal of Applied Remote Sensing, 11*(3), 036020.

Gupta, A. K., & Nair, S. (2011). Urban floods in Bangalore and Chennai: Risk management challenges and lessons for sustainable urban ecology. *Current Science, 100*(11), 1638–1645.

Gupta, K. (2020). Challenges in developing urban flood resilience in India. *Philosophical Transactions of the Royal Society A, 378*(2168), 20190211.

Hughes, S. (2013). Justice in urban climate change adaptation: Criteria and application to Delhi. *Ecology and Society, 18*(4), 48.

Khetwani, S., & Singh, R. B. (2018). Groundwater dynamics in Marathwada region: A spatio-temporal analysis for sustainable groundwater resource management. *International Journal of Conservation Science, 9*(3), 537–548.

Kotharkar, R., Ramesh, A., & Bagade, A. (2018). Urban Heat Island studies in South Asia: A critical review. *Urban Climate, 24*, 1011–1026.

Kumar, R., Mishra, V., Buzan, J., Kumar, R., Shindell, D., & Huber, M. (2017). Dominant control of agriculture and irrigation on urban heat Island in India. *Scientific Reports, 7*(1), 1–10.

Kumaran, T. V., Murali, O. M., & Senthamarai, S. R. (2020). Chennai floods, 2005, 2015: Vulnerability, risk and climate change. In R. B. Singh, B. Srinagesh, & S. Anand (Eds.), *Urban health risk and resilience in Asian cities* (pp. 73–100). Springer Nature.

Mal, S., Singh, R. B., Huggel, C., & Grover, A. (2018). Introducing linkages between climate change, extreme events, and disaster risk reduction. In S. Mal, R. B. Singh, & C. Huggel (Eds.), *Climate change, extreme events and disaster risk reduction: Towards sustainable development goals* (pp. 1–16). Springer.

Menon, J. S., & Sharma, R. (2021). Nature-based solutions for co-mitigation of air pollution and urban heat in Indian cities. *Frontiers in Sustainable Cities, 93*.

Oppenheimer, M., et al. (2019). Sea level rise and implications for low-lying islands, coasts and communities. In H.-O. Pörtner et al. (Eds.), *IPCC special report on the ocean and cryosphere in a changing climate*.

O'Shea, P. M., Roy, S. S., & Singh, R. B. (2016). Diurnal variations in the spatial patterns of air pollution across Delhi. *Theoretical and Applied Climatology, 124*(3–4), 609–620.

Paliath, S. (2018). National Rural Drinking Water Programme 'failed' to achieve targets: Government Auditor. Here's why. *IndiaSpend*, 26 November.

Pradhan, N. S., Das, P. J., Gupta, N., & Shrestha, A. B. (2021). Sustainable management options for healthy rivers in South Asia: The case of Brahmaputra. *Sustainability, 13*(3), 1087.

Qureshi, S. (2018). Perspectives on urban climate change and policy measures in India. *National Geographical Journal of India, 64*(1–2), 166–173.

Ramachandra, T. V., & Mujumdar, P. P. (2009). Urban floods: Case study of Bangalore. *Disaster Development, 3*(2), 1–98.

Robinson, G. M. (2018). Nature as threat and opportunity in the peri-urban fringe. In T. K. Marsden (Ed.), *The Sage handbook of nature* (pp. 1403–1432). Sage.

Roy, S. S. (2021). Professor R. B. Singh. In B. W. Pandey & S. Anand (Eds.), *Water science and sustainability* (pp. 1–3). Springer.

Roy, S. S., & Singh, R. B. (2015). Role of local level relative humidity on the development of urban Heat Island across the Delhi Metropolitan Region. In R. B. Singh (Ed.), *Urban development challenges, risks and resilience in Asian mega cities* (pp. 98–118). Springer.

Roy, S. S., Singh, R. B., & Kumar, M. (2011). An analysis of local spatial temperature patterns in the Delhi Metropolitan Area. *Physical Geography, 32*(2), 114–138.

Saha, D. (2019). Tainted dream: Discourse in India ignores deterioration of groundwater quality, both by natural and anthropogenic causes. In *State of India's environment* (6th ed.). Centre for Science and Environment.

Salahuddin, G. (2021). Urban heat Island growth and health hazard in the megacity of Hyderabad. In B. W. Pandey & S. Anand (Eds.), *Water science and sustainability* (pp. 43–51). Springer.

Sengupta, S. (2019). Towards day zero: India is fast running out of water due to pollution, overextraction, and changing climate. In *State of India's environment* (6th ed.). Centre for Science and Environment.

Singh, R. B., & Kalota, D. (2019). Urban sprawl and its impact on generation of Urban Heat Island: A case study of Ludhiana City. *Journal of the Indian Society for Remote Sensing, 47*, 1567–1576.

Singh, R. B. (2017). Ensuring clean water accessibility and mitigating water induced disasters-contributing towards sustainable development goals. *Productivity, 57*(4), 323–331.

Singh, R. B., & Chandna, V. (2011). Spatial analysis of Yamuna River water quality in pre-and post-monsoon periods. In: *Water quality: Current trends and expected climate change impacts. Proceedings of symposium H04 held during IUGG2011 in Melbourne, Australia, July 2011.* IAHS-AISH Publication, 348, pp. 8–13.

Singh, R. B., Gahlot, S., & Singh, A., (2013). Ecohydrological perspectives of declining water sources and quality in traditional water bodies in Delhi. In *Proceedings of H04, IAHS-IAPSO-IASPEI Assembly, Gothenburg, Sweden, July 2013*, 4, pp. 361–368.

Singh, R. B., & Grover, A. (2015). Spatial correlations of changing land use, surface temperature (UHI) and NDVI in Delhi using Landsat satellite images. In R. B. Singh (Ed.), *Urban development challenges, risks and resilience in Asian mega cities* (pp. 83–98). Springer.

Singh, R. B., & Grover, A. (2014). Remote sensing of urban microclimate with special reference to urban heat Island using Landsat thermal data. *Geographia Polonica, 87*(4), 555–568.

Singh, R. B., Grover, A., & Zhan, J. (2014a). Inter-seasonal variations of surface temperature in the urbanized environment of Delhi using Landsat thermal data. *Energies, 7*(3), 1811–1828.

Singh, R. B., Kumar, A., & Haque, S. (2016). Environment and resource, 2012–16. In R. B. Singh (Ed.), *Progress in Indian geography. A country report, 2012–2016. The 33rd International Geographical Congress, Beijing, China (August 21–25, 2016)* (pp. 59–76). Indian National Science Academy.

Singh, R. B., Pandey, B. W., & Prasad, A. S. (2014b). Living with floods and sustainable livelihood development in Lower Brahmaputra River Basin, Assam. *Golden Research Thought, 3*(10), 1–10.

Singh, R. B., & Prokop, P. (Eds.). (2016). *Environmental geography of South Asia: Contributions towards a future earth initiative*. Springer.

Singh, R. B., & Singh, S. (2011). Rapid urbanization and induced flood risk in Noida, India. *Asian Geographer, 28*(2), 147–169.

Singh, R. P. B. (2021). Professor R.B. Singh (1955–2021), an icon of Indian geography: A passage on the path of lineage, legacy and liminality. *Space and Culture, India, 9*(2), 6–49.

Veena, K., Parammasivam, K. M., & Venkatesh, T. N. (2020). Urban Heat Island studies: Current status in India and a comparison with the international studies. *Journal of Earth System Science, 129*(1), 1–15.

Yow, D. M. (2007). Urban heat islands: Observations, impacts, and adaptation. *Geography Compass, 1*(6), 1227–1251.

Chapter 2
Impacts of Extreme Climatic Events on Sustainable Urban Development in Coastal Regions: Selected Case Studies of India

Anwesha Haldar, Surajit Kar, Swarnendu Paul, and L. N. Satpati

Introduction

The terms 'climate change' and 'global warming' are now used almost interchangeably (Nda et al., 2018), although climate change refers to the change in weather phenomena by natural and anthropogenic causes (Swain et al., 2020) and global warming relates to human-induced earth warming (Vogel et al., 2019). The surface temperature has risen by 1.09 °C between 2011 and 2020 compared to the last few decades (Nagai et al., 2020). The number of cold days and nights has decreased and warm days and nights have increased (Toros et al., 2019). The global weather pattern has been oscillating over the last century in a systematic mode, but the contemporary rate of climate change is distinctly more rapid due to the ever-increasing of greenhouse gases (mostly carbon dioxide and methane) from the urban and industrial centres into the atmosphere.

Anwesha Haldar, Surajit Kar, Swarnendu Paul and L. N. Satpati contributed equally with all other contributors.

A. Haldar
Department of Geography, East Calcutta Girls' College, West Bengal State University, Kolkata, India

S. Kar · S. Paul
Department of Geography, University of Calcutta, Kolkata, India

L. N. Satpati (✉)
UGC – Human Resource Development Centre, University of Calcutta, Kolkata, India

A. L. Singh et al. (eds.), *Climate Change, Vulnerabilities and Adaptation*,
https://doi.org/10.1007/978-3-031-49642-4_2

Indian Coastline and Climatic Hazards

As the earth–atmosphere complex feedback mechanism is accelerating over spatio-temporal scales (local to regional), the various dimensions of the present climate system are rapidly changing, especially with the occurrences of extreme climatic events and their increasing frequencies as well as intensities across the South Asian region, especially surrounding the Indian subcontinent (Mandal et al., 2022). The geography of the Indian subcontinent has a unique distribution of land and sea in which the peninsular landmass protrudes towards the Indian Ocean, making it seem as if the two arms of the Indian Ocean, namely the Bay of Bengal and the Arabian Sea are projecting towards the north (Hijioka et al., 2014). Although much of the region is lying north of the Tropic of Cancer to signify its extra-tropical regime, the presence of the gigantic Himalayas, running for almost 2500 km with an average altitude of 4–6 km, has separated India from the Asian mainland, thereby giving this subcontinent a more monsoonal climatic character. The continental and maritime landmasses of the region are both affected by extreme climatic conditions in the form of heat and cold waves, floods, droughts, tropical cyclones, high-intensity rainfalls, etc. However, the ex situ influence of the El Niño Southern Oscillation (ENSO) and the Indian Ocean Dipole (IOD) play an important role not only on the climate of the region but also on the large climate and socio-economic effects worldwide (Ashok and Saji, 2007; Meyers et al., 2007). The coastal cities along the eastern coast of India are in general thickly populated because of the wide extent of the coastal plains, availability of fertile soil and water for irrigation, concentration of settlements and subsequent urban development. On the other hand, the west coast cities usually suffer from the impacts of heat and high-intensity rainfall at different occasions. However, during the last about one decade the trend of cyclonic activity, intensifying to attain greater magnitudes to become severe and very severe cyclones are more common in these coastal regions and the Arabian coastline is increasingly being affected in the recent times due to global warming–induced positive change in sea surface temperature (SST), which is quite evidenced from the recent satellite data.

The Indian coastline, inclusive of the major indentations and island shores, is roughly about 9000 km, while the mainland coast without the estuaries is 6631.53 km. As per the official records of the National Institute of Ocean Technology under the Ministry of Earth Sciences and the National Centre for Coastal Research in 2019, the west coast of India experienced only 27% erosion and 25% accretion while in the east coast 38% of the land area has eroded and 24% accreted over the period of 1990–2016. The lengths of the coastline facing more than 5 m/year erosion or accretion are classified as high zones while 3–5 m/yr. as moderate, 0.5-3 m/yr. as low zone and less than 0.5 m/year change as stable zones. The data shows that the erosion of coastlines is higher than its accretion levels in West Bengal, Pondicherry, Kerala and Tamil Nadu and is negligible in Maharashtra (Table 2.1 and Figs. 2.1, 2.2 and 2.3).

Table 2.1 Shoreline change map along the Indian coast (1990–2016)

Coastline character	Erosion (%)	Accretion (%)	Stable (%)	Coastline length (km)
West Bengal	63	24	13	534.35
Odisha	28	51	21	549.5
Andhra Pradesh	27	42	31	1027.58
Puducherry	57	8	35	41.66
Tamil Nadu	41	23	36	991.47
Kerala	45	21	34	592.96
Karnataka	22	30	48	313.02
Goa	12	20	68	139.64
Maharashtra	24	12	64	739.57
Gujarat (including Daman and Diu)	31	26	43	1701.78

Source: Kankara et al. (2018)

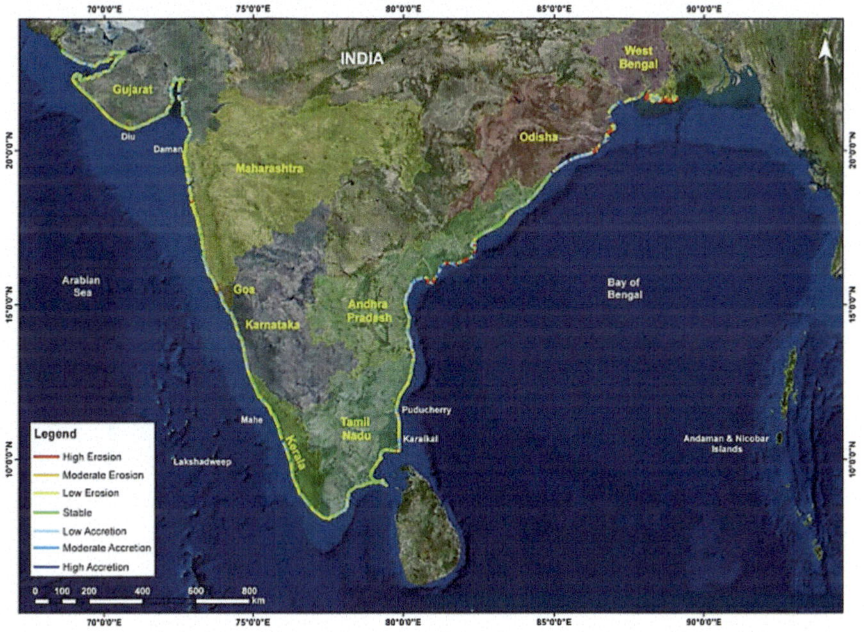

Fig. 2.1 Shoreline change map along the Indian coast (1990–2016). (*Source*: Kankara et al., 2018)

In the coastal cities, the main climatic effects are cyclonic storms of different types, intensities and duration associated with thunderstorms and high-intensity rainfall leading to floods and water-logging. Cyclone-induced hazards affect the cities and towns both directly by gusting winds leading to various damages to the infrastructure and service utilities, like transport, electricity, water supply and

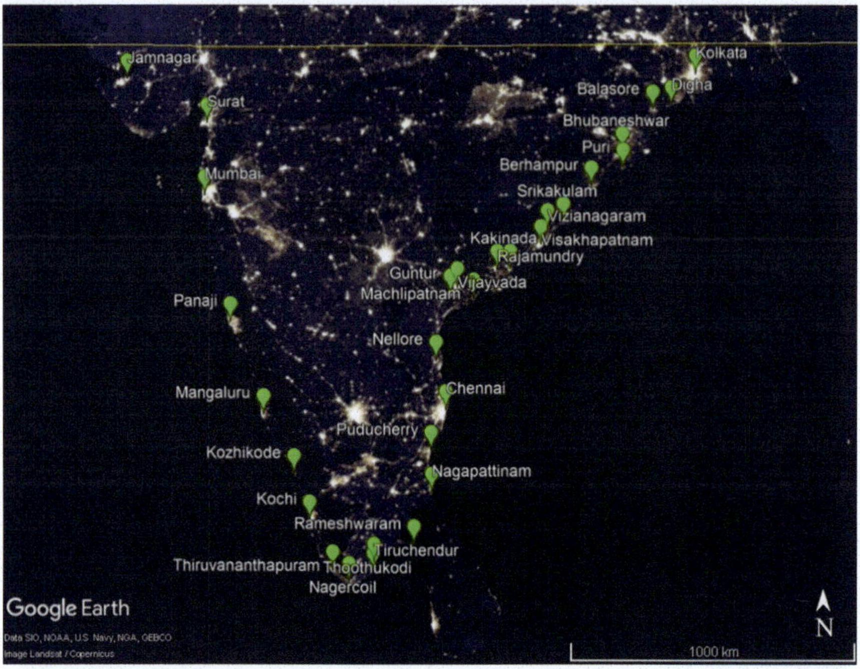

Fig. 2.2 Coastal urban centres of India selected for study. (*Source*: Computed by the authors based on night-time satellite image procured from https://worldview.earthdata.nasa.gov, dated 5 December 2016)

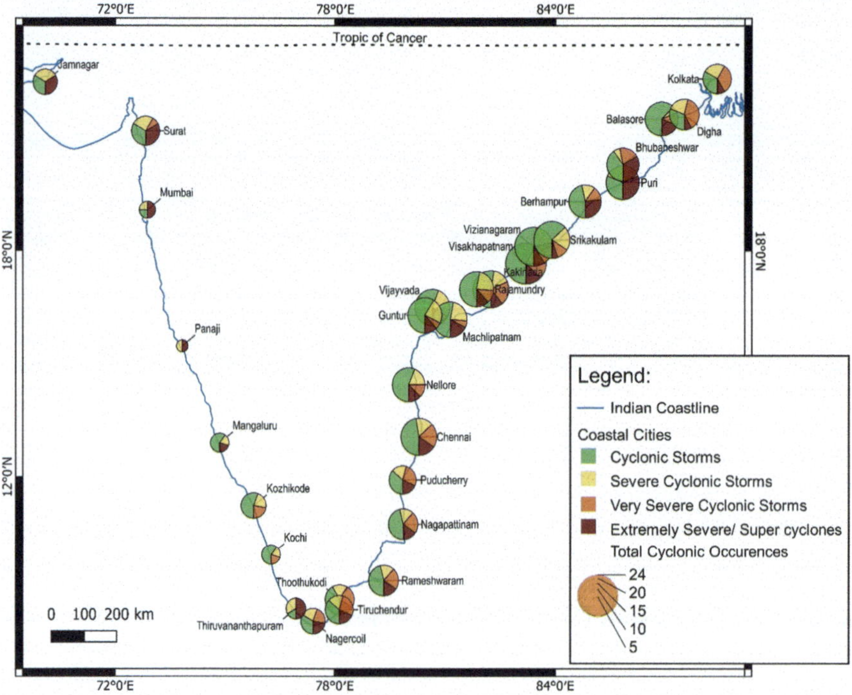

Fig. 2.3 Status of cyclonic activity in the selected coastal urban centres. (*Source*: Compiled by the authors based on different government and non-government sources)

breaking down of weak structures, e.g., buildings and towers, and indirectly by storm surges, floods, water-logging, with associated man-days loss and health hazards. Other climatic hazards affecting the cities include the thermal impacts of urban heat pockets/islands, severe heat stress (hyperthermia) and the associated health impacts. High temperatures originating from in-situ heat budget or ex-situ heat advections cause severe heat stress to the exposed populations working outdoors for long hours during the summer days. Heat island effects are often found to be caused by the size, morphology and land use and land cover types of the towns and cities. Usually the larger the cities are, the more the chances of heat pockets/ islands on the surface, canopy and boundary layer types. To cope with the heat stress, more artificial weather conditioning is required, necessitating more consumption of electricity in the indoor environment. But this eventually leads to an ambient increase of heat generated from air-conditioning machines at the locations of concentration of consumers. This is a cause of discomfort for the population exposed to such artificial ambient heating. In cold seasons, the average temperatures are found to be increasing and with the increase in temperature more moisture in the atmosphere cause blocking of re-radiation through positive feedback. This raises the number of heat pockets even in the smaller cities.

In the coastal areas, there is a two-way local wind system, called land and sea breezes, for which maritime cities are expected to have a moderate climate, but due to accumulation of heat and moisture in the cities, relatively low pressures develop and high temperature continues till late night where ingression of moisture laden winds over the cities increases the temperature and moisture in the city region. This facilitates frequent occurrences of late-evening thunderstorms in the summer months. The pollution load from the urban industrial areas in the form of suspended particulate matter and hydrocarbons also help the development of thunder clouds. These thunderclouds, usually of cumulonimbus type, are characterized by high-intensity rainfall and lightening, causing reported deaths within the city region or its adjoining areas. Heat waves are cumulative weather hazards mostly affecting the poorer section of the people. Under this condition the demand for water is more, leading to occasional water crisis and pressure on municipal water supply services.

The coastal cities and towns of India are now connected through rail and roadways running along the coastlines to facilitate the development and growth of urban centres in between the erstwhile leap-frog development of large urban centres in the region, e.g., the capital cities of the coastal states. Therefore, in many occasions, the urban centres of different sizes and extensions have formed an almost contiguous linear development along the two coastlines of India, both east and west, leading to the formation of urban corridors which have definite impacts on transformation of local climates, and in turn, several impacts of cyclonic hazards on them. The urban corridors thus formed look like a garland on the motherland of India (Fig. 2.4).

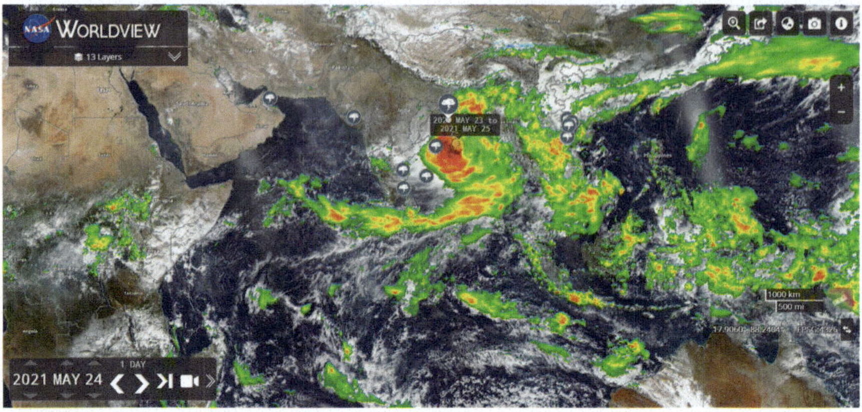

Fig. 2.4 Yaas cyclonic activity over the North Indian Ocean region. (*Source*: https://worldview.earthdata.nasa.gov/)

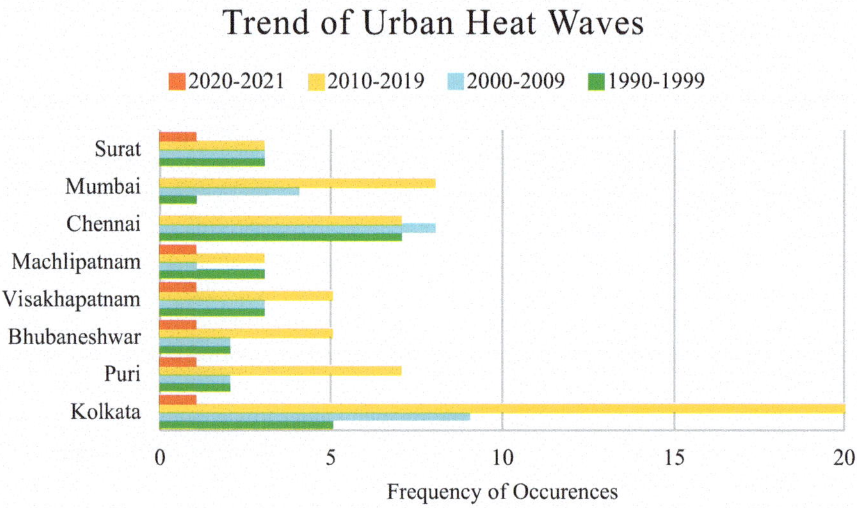

Fig. 2.5 Heatwave frequency in the coastal urban centres. (*Source*: Computed by the authors based on the IMD data)

Cyclones ranging from deep depressions to super cyclones have vast regional impacts due to their extent, damaging potential and associated devastating weather features compared to the other extreme climatic effects occurring in the coastal cities. For example, urban heat islands and heat stress are confined to urban locations only, whereas cyclones jeopardize the regional geographical set-up which requires many years of restoration after the cyclone (Figs. 2.5 and 2.6). It is also found that by the time some recoveries are made against the previous cyclone, a series of subsequent cyclonic events batter the coastal locations including the urban areas, where a high concentration of population accelerates the miseries even though early forecasting

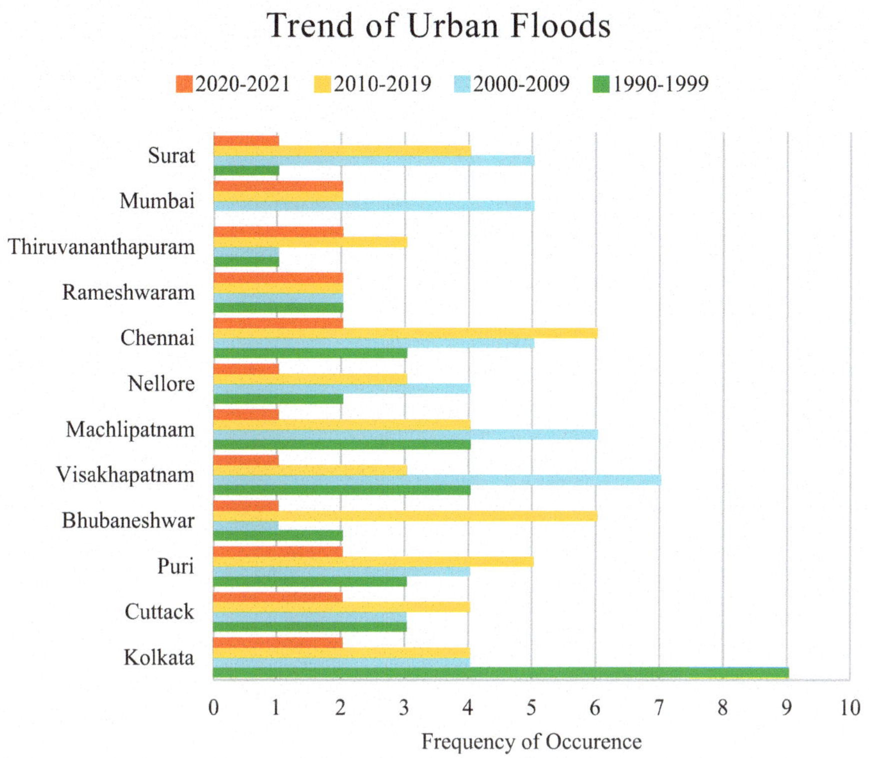

Fig. 2.6 Flood frequency in the coastal urban centres. (*Source*: Computed by the authors based on the IMD data)

with a considerable lead time and some other precautionary measures and reliefs are given to the affected people. Loss of lives has definitely been averted in recent times, but damage to the infrastructure is still very high depending on the nature of cyclones and the entities exposed to them.

It is clear (Table 2.2 and Fig. 2.3) that the eastern coast of India is more prone to tropical cyclones than the western coast. The Bay of Bengal creates the perfect condition for the development of tropical cyclones. So, the cities located along the eastern coast of India are more vulnerable to the cyclones. Cities which have emerged along the shoreline are at greater risk. Odisha and Andhra Pradesh are mainly affected by tropical cyclones. Puri, Berhampore, Srikakulam, Vizianagaram, Visakhapattanam and Kakinada are the main affected cities. Even Kolkata, the capital city of West Bengal, which is located nearly 150 km away from shoreline, is not safe from tropical cyclones of the Bay of Bengal. In the recent past, more cyclones hit Kolkata and/or near regions i.e., the frequency has increased. Cyclones also frequently hit Chennai, Nagapattinam, Thoothukodi and Puducherry. Although cyclones hit less frequently in Thiruvananthapuram, the southernmost capital city of the eastern coast, it is notable that the frequency of arrival of cyclones is much less in

Table 2.2 Major geographical identities of the selected urban centres

Town/city	Area and approx. distance from the seashore/shoreline in km	Population (2011)	Share of slum population	Principal land use and land cover
Kolkata (W.B.)	1886.67 km^2 (KMA)/ 206.08 km^2 (KMC) Located approximately 150 km away from seashore	4,496,694 (KMC)/ 14,617,882 (KMA)	32.54659% in KMC area (2011)	Built-up area
Digha (W.B.)	3153 km^2 Located approximately 2.5 km away from sea coast	6916 (Digha Census Town)	–	Agriculture
Balasore (Odisha)	17.48 km^2 Straight line distance between Bhubaneswar and Balasore is 175 km	144,373	–	Agricultural land use
Cuttack (Odisha)	192.5 km^2 Located approximately 41 km away from seashore	606,007	26.84% (2011)	Waste land/ residential land use (CMC area)
Puri (Odisha)	16.84 km^2 Located along the shoreline of the Bay of Bengal	201,026	35.13% (2011)	Built-up area
Bhubaneswar (Odisha)	186 km^2 Locate approximately 53 km away from shoreline	837,321	18.5% (2011)	Built-up area
Berhampore (Odisha)	86.82 km^2 Located approximately 9 km away from shoreline	355,823	25.75% (2011)	
Srikakulam (Andhra Pradesh)	20.89km^2 Located approximately 9 km away from shoreline	228,025	23.88% (2011)	Agriculture
Viziyanagaram (Andhra Pradesh)	29.27 km^2 Located approximately 30 km away from shoreline	228,025	42.94%	Agriculture
Visakhapattanam (Andhra Pradesh)	681.96 km^2 Located along the shoreline	1,728,128	44.61% (2011)	Built-up area
Kakinada (Andhra Pradesh)	30.51 km^2 Located along the shoreline	384,128	41% (2011)	Built-up area

(continued)

Table 2.2 (continued)

Town/city	Area and approx. distance from the seashore/shoreline in km	Population (2011)	Share of slum population	Principal land use and land cover
Rajahmundry (Andhra Pradesh)	44.50 km^2	376,333	38.02% (2011)	Agriculture
	Approximately 70 km away from shoreline			
Vijayawada (Andhra Pradesh)	61.88 km^2	1,021,806	Approximately 43.62% (2011)	Built-up area
	Approximately 86 km away from seashore			
Guntur (Andhra Pradesh)	168.41 km^2	743,354	39.77% (2011)	Built-up area
	Distance between Guntur and seashore is approximately 60 km			
Nellore (Andhra Pradesh)	150.48 km^2	404,775	37.32% (2011)	Agricultural land
	Distance between Nellore and seashore is approximately 22 km			
Machlipatnam (Andhra Pradesh)	26.67 km^2	170,000	53.73% (2011)	Agriculture
	Located along shoreline			
Chennai (Tamil Nadu)	426 km^2	7,088,000	26% (2011)	Built-up area
	Located along shoreline			
Puducherry	483 km^2	1,394,467	17.43% (2011)	Built-up area
Rameswaram (Tamil Nadu)	55 km^2	44,856	33% (approximately	
	An island town in Gulf of Munnar			
Nagapattinam (Tamil Nadu)	17.92 km^2	102,905	34.85% (2011)	Built-up area
	Located along shoreline			
Thoothukodi (Tamil Nadu)	90.633 km^2	237,830	–	Agricultural land and agro-based industries
	Located approximately 5 km away from shoreline			
Tiruchendur (Tamil Nadu)	10.05 km^2	32,171	–	Agriculture
	Located along shoreline			
Nagercoil (Tamil Nadu)	61.36 km^2	289,916	4.76% (2011)	Built-up land
	Located approximately 11 km away from shoreline			

(continued)

Table 2.2 (continued)

Town/city	Area and approx. distance from the seashore/shoreline in km	Population (2011)	Share of slum population	Principal land use and land cover
Thiruvananthapuram (Kerala)	214 km^2 Located along shoreline	957,730	0.42% (2011)	Built-up land with agricultural land
Kozhikode (Kerala)	118 km^2 Straight line distance between Thiruvananthapuram and Kozhikode is 329.5 km	609,224	9.15% (2011)	Agricultural land use
Kochi (Kerala)	94.88 km^2 Straight line distance between Thiruvananthapuram and Kochi is 172 km	677,381	0.82% (2011)	Agricultural land use
Mangalore (Karnataka)	170 km^2 Straight line distance between Bangalore (capital) and Mangalore is 333 km	484,785	1.55% (2011)	Built-up land
Panaji (Goa)	8.27 km^2 Located along shoreline	40,017	–	Agricultural land and built-up area.
Mumbai (Maharashtra)	603 km^2 Located along shoreline	12,478,447	Nearly 62% (2011)	Built-up land
Surat (Gujarat)	474.185 km^2 Located approximately 17 km away from shoreline	617,600	10.46% (2011)	Agricultural land
Jamnagar (Gujarat)	125 km^2	479,920	11.90% (2011)	

Source: Compiled by the authors based on different government and non-government sources

the cities, located in the southern part of it. Formation of tropical cyclone is much less in the Arabian Sea than the Bay of Bengal; as a result, the cities of the western coast of India, like Panaji, Mumbai, Surat and Jamnagar receives less tropical cyclone and related hazards, but in the recent years devastating high magnitude cyclones have become quite common in this region.

The configuration of the coastline, along with the genesis, movement and landfall of the depressions and cyclones of various categories affect the coastal regions of

India and its adjoining countries adversely. However, the nature of depressions and cyclones of the Bay of Bengal and the Arabian Sea are significantly different in terms of their number, seasons of occurrence, magnitude and movement leading to landfall; and therefore, the impacts are also divergent at various locations of the country. Usually it is found that the impacts of cyclonic hazards are more along the east coast, and in the states of Andhra Pradesh, Odisha and West Bengal, in comparison to the west coast, especially Maharashtra and Gujarat. In the region in the southern tip of the country, comprising Tamil Nadu and Kerala, the impacts of the cyclones originating from and travelling through the Bay of Bengal are more than those of the Arabian Sea. From a 33-year analysis of cyclonic events and their life-time energy, it is found that the power dissipation index (PDI) is comparatively less in the North Indian Ocean rather than in the Pacific or the Atlantic Oceans, leading to high-energy cyclonic formations around the Indian coastline since 1980s (Patra et al. 2019). Also, the severe cyclones have become widely spaced in the Bay of Bengal over the last three decades with the monthly trends initiating from end of April between the Tamil Nadu coasts and the Andaman and Nicobar islands, which peaks in the last week of May having northern and north-eastern paths. The June to August cyclonic events are few and mainly follow a west and north-west alignment unlike the high-frequency depressions causing recurrent floods in the coastal regions. While the frequencies of severe cyclonic storms slightly rise again in September and end of November, depressions are very few from November to May (Haldar, 2016). It was also observed that even though the frequency and intensity of cyclones increased in the North Indian Oceans, the total deaths and damages are prominently decreasing with better forecasting and warning systems undertaken by the India Meteorological Department (IMD) along with the state governments and the disaster management authorities. Cyclone brings some of the massive disasters in coastal cities like flooding; water-logging situation; damages to agricultural industries, small-scale industries, fishing and public transport systems; food shortages; diseases; loss of lives and wreaths. Cities having a greater percentage of slum population face more difficulties. Cities like Kolkata, Visakhapatnam, Vizianagaram, Machlipatnam and Mumbai which have more share of slum population experiences more difficulties in the context of cyclone and extreme weather events. So, proper planning and development are needed.

Rapid urbanization in the coastal cities across the eastern and western coast of India results in Urban Heat Island (UHI) effects. Urban areas modify the atmospheric boundary layer (ABL) processes as the natural land surfaces are being increasingly replaced by artificial surfaces that have very different thermal properties resulting in urban air to be 2–10 °C warmer than those in surrounding periphery areas (Patra et al., 2017). The urban areas are experiencing enhanced low-level convergence due to mechanical turbulence from rough surfaces, increased surface sensible heat flux and elevated aerosol loading, and hence, urbanization and changes in settlement characters can have a significant feedback on the spatio-temporal patterns of precipitation. This phenomenon is also associated with extreme cases of high-intensity rainfall and heat waves. It has been observed that the number of rainy days and the mean monthly precipitation over Kolkata has significantly increased in the last

50 years (Patra et al., 2017). In India, heat waves typically occur from March to June, and in some cases, even extend till July. On an average, five to six heat wave events occur every year over the northern parts of the country. Single events can last weeks, occur consecutively and can impact large population. Some of the important heat waves in the major coastal cities in India since 1990 are shown in Figs. 2.5 and 2.6.

Climatic Hazards and Coastal Cities

The dynamics of climate change and its impacts on human society are aligned with each other, which differs from one geographical set-up to another. For example, the dry land regions are suffering from excessive heat waves and land degradation, deltaic and coastal region are primarily tormented with catastrophic floods and cyclonic storms, and the inland portion of any country faces extreme summer and winter temperatures. Apart from the different nature of climatic imprints on different geographical sets up, regional resource and their utilization are backed by provincial climatic characteristics. Due to the shifting nature of the monsoon, some of the regions in India have been facing acute water scarcity since the last few decades. Frequent cyclonic storms in parts of the eastern coastal region have become an anxious concern for the planners. In this context, cities particularly in the coastal zones are not only threatened by water-logging and seasonal flood risks, but the uncontrolled urban growth acts like a catalyst that provides the breeding ground of climate change–related adversities in the coastal zone. Effective urban planning and controlled governance prevent unwanted situations where both livelihood and urban ecosystem gets supported. The impact of climate change–related issues is associated with a few elements like calamities, resilience, adaptation and government regulations. The level of vulnerability also relies on certain aspects like how often catastrophic events have occurred and societal perspectives about the storm, along with familiarization of government regulation at community level. Urban expansion and increasing population bring difficulty for policy implementation due to the presence of a diverse range of the socio-culture and economic classes. Most of the million-plus cities are facing weak urban infrastructure which is not equally distributed, and thereby the marginalized communities suffer a lot from climatic hazards.

Among the traditional urban studies, disaster-related research are not being incorporated widely as these are usually considered natural events and are the task of policymakers. Few of the contemporary studies on disaster vulnerability shows how the frequency of disaster and their impacts are primarily triggered by unrestricted urban growth (Dhiman et al., 2019; Nigussie & Altunkaynak, 2019; Kateja & Jain, 2021; Pusdekar & Dudul, 2021; Kadaverugu et al., 2021). The relationship between urban systems and climate change impacts needs to assist through decisive academic assessment to identify allied factors to urban vulnerability and climate change.

Based on this symbiotic relationship between the urban system and climatic hazards, we are primarily focusing on three major sub-themes, namely nature of climate change, how coastal cities are vulnerable to climate change impacts and identification of some of the adaptive measures against climate change–related impacts in the contemporary Indian scenario.

Nature of Climate Change and City Region

Here, the emission of greenhouse gases has been regarded as the main driver of anthropocentric climate change (Vousdoukas et al., 2022) or contemporary global warming. After the Industrial Revolution, the economy shifted from primary to secondary and tertiary sectors (Gleason, 2018), which resulted in increasing the proportion of greenhouse gases in the atmosphere (Liu et al., 2012). In 2019, the total proportion of CO_2 and methane had increased by 48% and 160%, respectively, since 1750, which was comparatively higher over the last 800,000 years (WMO, 2017). The increasing demand for food leads to conversion of forest land into agricultural lands through deforestation, which has been considered as one of the prime causes of global warming–induced climate change (Gomes et al., 2019).

This also includes excessive use of fossil fuels, deforestation that accelerates land degradation, water scarcity and frequent extreme weather events (Ritchie, 2020). Land-use change is not only triggering the emission of greenhouse gases, but the change in cropping pattern also affects local temperatures (Singh, 2020). For example, in Tamil Nadu, degraded forest lands are primarily handed over to private enterprises where oilseed gets promoted (Baka, 2017). This changing political economy is not only restricted to local livelihood but causes regional transformation in weather patterns (Deepika et al., 2020) and depletion of natural resources like groundwater (Kishore et al., 2020). The inputs of climate change may originate from a single pathway, but the outcome of it is spread out in an extensive geographical set-up. In addition, most of the cities (and nations) that face the highest risks from the negative effects of climate change are those with almost negligible contributions to atmospheric greenhouse gases (Zheng et al., 2019), as the intensity of the impact depends upon the physical set-up and socio-economic background of the region. This observation also holds well in case of some micro-regions of high-emitting countries, like India and China.

Among the different geographical set-ups, cities are likely at greater risk for the consequences of climate change. This risk factor is not only due to the high amount of greenhouse gas emission from the city region but also due to their associated socio-cultural components, nature of governance and overcrowded population pressure creating an unfavourable situation to tackle the effect of climate change. Coastal regions are dynamic due to their unique geophysical ambience and growing concentration of human settlements (Douben, 2006). Moreover, the coastal zones are open to receiving the first blow of climate change impacts. Throughout the globe an average of 46 million people per year experience storm-surge flooding. This number may increase due to the exceeding rate of coast-ward migration of population all over the world (Balica et al., 2009). The developing nations, like India, are facing the biggest challenge due to climate change-related impacts over poorly developed infrastructure in many areas.

Climate Change and Vulnerability in the Coastal Cities

The impact of climate change on human livelihood depends on man–nature coupling within an environmental set-up. The massive influx of rural population in surrounding urban centres, demand for urban amenities, presence of environmentally degraded areas, and faulty infrastructure make the urban population extremely defenceless against extreme weather events. Till 2010, about 30% of people in India were residing in urban centres (Census of India, 2006). Owing to rapid urbanization and prevailing pulling factors from surrounding regions, Indian cities are experiencing rapid population growth. On estimation, about 70 Indian cities will accommodate more than 1 million population by 2025 (CSO, 2006). Three mega-urban regions, namely Mumbai–Pune (50 million), the national capital region of Delhi (more than 30 million), and Kolkata (20 million) will be among the largest urban concentrations in the world (Conservation Action Trust, 2006). In this context, the Intergovernmental Panel on Climate Change (IPCC) has noted in its overview how the rising amount of precipitation may cause a tremendous vulnerable situation for the cities in the global south (EM-DAT, 2010). Flood-related deaths have increased threefold from 1960 to 1990 (Dasgupta et al., 2013). The climate change–related issues or weather events and changing patterns in cities or urbanization create a contiguous system where one causes the other. The vulnerability of cities is much dependent on their geographical set-up, where coastal cities are regarded as very much threatened by climate change-related adversities. In many such cities, flooding and cyclonic storms are common phenomena that turned into a disaster due to unplanned growth and unsatisfactory infrastructural facilities, even though coastal cities like Kolkata and Mumbai experience overcrowded slum populations which experience negligible backup to prevent environmental threats (Roy & Sharma, 2015).

The climate change–related changing weather pattern has its widespread impacts. In India, the nature of climate change is primarily connected with temperature and precipitation, including the following factors: (1) increase of mean annual and monthly temperature (by 2–4 °C); (2) variability of monsoon rainfall, where some regions experience high rain and some regions experience low rain; (3) the number of rainy days decreases and about 50% of monsoon rainfall occurs within 15 days, thereby increasing the number of high-intensity rainfall days; (4) the regional temperature rise is the factor behind the fragile ecosystem in central India (dry land), where 5–10% of rainfall is declining. This changing weather pattern has a great impact on the coastal cities in India in multiple ways. Considering the climatic vulnerability perspectives, it is observed that cyclonic and coastal flooding largely affects the urban centres of Kolkata, Mumbai and Chennai and frequently causes mass destructions to the cities of Visakhapatnam, Surat, Bharuch, Bhavnagar and

Jamnagar, even though the Bay of Bengal sector is far more prone to such hazards than the Arabian Sea (Mishra et al., 2021; Bhardwaj et al., 2019). A sea level rise of between 30 and 80 cm has been projected over the century along India's coast affecting the regions around Gujarat, Mumbai and parts of the Konkan coast and South Kerala, as well as the deltas of the Ganga, Krishna, Godavari, Cauvery and Mahanadi. The low-elevation coastal zones on the western coastal region are more vulnerable than the eastern coast (Stern & Stern, 2007; Saintilan et al., 2020). In the recent years, the coastal metropolitan cities like Mumbai, Chennai and Kolkata have seen the ravages caused by urban flooding over long durations (Prasad & Singh, 2005; Malik et al., 2020). Another associated hazard of these urban centres is the problem of scarcity of safe drinking water and consequently water-borne health problems (Revi, 2008; Maity et al., 2018). Such risks greatly affect the economy of the region (Heger & Neumayer, 2019).

A major group of disaster studies has considered geographical location to be the main factor for climatic risks (Bai et al., 2018; Hobbie & Grimm, 2020). The coastal zones are considered as being the most vulnerable to cyclones, floods and tsunamis (Sinay & Carter, 2020). Apart from the location-specific factor on climatic hazard, another group of scholars has considered the social and economic status that causes different scales of vulnerability and risk factors mainly in a coastal city (Friedrich et al., 2020; Chan et al., 2018). Song et al. (2019) argues on location-specific factor and considered the level of resilience and different adaptive measure that are responsible for tackling the risk factor. Hari et al. (2021) have shown how marginalized people always become highly vulnerable during the storm or even its aftermath effect (Mehta et al., 2019). In many disaster-related literature, urban study and urban governance are considered the prime factor that tackles the risk factor by improving urban infrastructure, which on other hand, can be related to resilience and adaptation of the region (Dhiman et al., 2019; Singh et al., 2021). Therefore, there is not a single predominant element responsible for disaster risk in a coastal city; rather the combination of factors provides the foundation that makes coastal cities in India at risk. Based on the available literature, we are primarily focusing on five major factors which are associated with climate change-related disaster and risks. Most of the low deltaic and coastal areas are attaining high risk for climatic hazard (Grases et al., 2020), where the relatively wealthy community can bear the cost of hazard and revive quickly, while the poor communities suffer much due to already existing challenges in their livelihood (Sinha et al., 2021). In this context, Idris and Dharmasiri (2015) have advocated socio-economic integrity as a survival strategy for reducing the risks. Type and magnitude of disaster matters, e.g., large-scale disasters like tsunami and cyclones are riskier than local floods (Chittibabu et al., 2004; Gupta et al., 2019). Resilience and adaptation depend on the level of awareness about the events and community-level solidarity to match with the relevant government policies (Tajuddin, 2018; Dhiman et al., 2019; Singh et al., 2021). Therefore, to minimize the risks, integration between local, state and central governments, and adequate budgetary allocation for the geographical fragile areas along with equitable distribution of facilities are also very important (Singh et al., 2021; Chu, 2018).

The unprecedented urban growth in the large cities in South Asia accelerates the intensity of storm-induced calamities, where the urban population increases by 140,000/day (Fuchs, 2010). More importantly, this growth has become quite extraordinary in the port cities. The port side location provides high-risk factors through cyclones and floods. Even low-intensity storms become deadly for the community which is near the coast. For example, in India, Kolkata and Mumbai are ranked first and second where about 14.0 and 11.4 million of the population, respectively, are estimated to be vulnerable to coastal flooding in the upcoming years (Fuchs et al., 2011).

Resilience Against Vulnerability in Coastal Cities/Regions

The concept of vulnerability is connected with the 1970s environmental movement (Kroll-Smith et al., 1997) which denotes the ecological perspective of disaster and how much exposure the community is facing against a particular risk (Trickett, 1995). However, a group of scholars have considered the relative nature of vulnerability that depends on human perception (Heijmans & Victoria, 2001), as people act differently within the same set of environment due to their socio-cultural and economic individuality (Kollmuss & Agyeman, 2002). This dissimilar behaviour depends on wealth, social integration, cultural contrast and knowledge about that single environment phenomenon. Vulnerability or exposure to the disaster is conceptualized as the natural state of being (Ewald, 2002). The terms 'vulnerability' and 'risk' are often interchangeably used where 'risk' denotes anticipated losses from a particular hazard to a certain element(s) at a particular point of time in the future (Cardona, 2005). In climate change vulnerability research, two particular sets of vulnerabilities are yet to be properly addressed, one is the bio-physical vulnerability and another one is social vulnerability (Nyong et al., 2008). The biophysical vulnerability denotes the ultimate collision of a hazard effect, whereas social vulnerability denotes the response of human society to the shock (Vincent, 2004; Nyong et al., 2008). Ghosh and Ghosal (2021) show how different socio-economic backgrounds in the urban sector make unparallel risk factors during the storm. In most of the cases, marginalized or migrated people from the surrounded rural sector (and/or urban poor) are concentrated in economically and environmentally weaker zones of the city (Mukherjee et al., 2021). Affluence allows individuals and households to diminish risks factors; for instance, by having safer housing, choosing safer jobs or locations to live in, having assets that can be called on in emergencies and protecting their wealth by insuring assets that are at risk (Sinha et al., 2021). Generally in the urban sector, optimum sites which are less vulnerable to risk factors are primarily distributed to the wealthy communities, because they can afford the rent. Whereas low-income groups are far away to make their choices and as a results sites that are vulnerable to risk are distributed to low-income groups (Picciariello et al., 2021). For example, in the Sundarban region, cyclonic storm is a yearly event wherein an average of two or sometimes three cyclonic storms are common each year (Bhui et al., 2022). This is why for decades the primary activity in the Sundarban region gets distorted (Ghosh et al., 2018). This results from the working-age males migrating to surrounding urban centres (Kolkata and Howrah), and leaving women to take care of their families. This is how within the same set of environmental condition women and children becomes much vulnerable (Jalais, 2010). Considering the religious and cultural contexts, indigenous communities and Muslim-dominated portion of damaged sectors are exaggerated the most (Bardsley & Wiseman, 2012). For example, in the case of the Sundarban region, isolated islands are affected most by cyclonic storms particularly inhabited by indigenous people (Mehta et al., 2019). In big cities like Kolkata and Mumbai slums are being considered as the worst site for habitation due to degraded environmental settings primarily occupied by the marginal Muslim community (Das et al., 2021). Apart from the different human societies, there are the following elements and populations which are at risk due to climate change vulnerability: (1) slums and squatter settlements along with migrants from a rural sector (Ramesh & Iqbal, 2022); (2) urban ecosystem through the damage of wetland and canals (Chaudhuri et al., 2022); (3) high rise buildings which are explicitly vulnerable to gusty wind flow during cyclones (Revi, 2008); (4) daily workers and commuters (Rumbach, 2014) and (5) risk on urban basic amenities like drinking water, delivery system and associated services.

Three major factors that drive the intensity of vulnerability to society are exposure, susceptibility and resilience. Where exposure and susceptibility are the prevailing risky condition within the system, resilience is the capacity to adapt or recover from the risk (Balica & Wright, 2009). Here, exposure is the progression that estimates the intensity of any storm event (Balica et al., 2009). Fekete (2009) noted that exposure is the measurement of the vulnerable element within a system, while Penning-Rowsell et al. (2005) defined exposure as the likelihood of human and environment. On the other hand, susceptibility is the degree to which the system gets affected (Smit & Wandel, 2006). It is unavoidable that the susceptibility or sensitivity varies from place to place, therefore it is relative. The concept of susceptibility is still under process and yet to get mature. Hence, the perspective of susceptibility is debatable and creates misconception between the social scientists and physical scientists. Resilience acts like an intermediate for reducing risk factors as it is the method to take on some measure to tackle the effect and bounce back to the previous state (Tanner et al., 2009).

Resilience is the adaptive capacity by people and government to prevent negative consequences of climate change or any natural calamities. Capacity refers to the total strengths, attributes and available resources that society, community, or organization possesses to cope and decrease the risk from disaster and increase resilience (UNISDR, 2017). A culture of resilience could develop if we can enhance the disaster risk reduction capacity of individuals as well as organizations. On the application or the procedure to make an effective resilience or adaptive tool, different scholars put the diverse range of perspectives. For instance, Sterr et al. (2000) represent the process of coastal city resilience through maintaining the consecutive steps, namely protect, retreat and accommodate. Maintaining this synchronous procedure needs to incorporate proper governance. In this view, Revi (2008) points out how the adaptive measures of climate change need to assist through various scales (national, region, city and local). Patnaik and Narayanan (2009) pointed out a group-specific capacity-building approach, specifically among the poor community in cities, as they suffer the most. Focused group capacity building is regarded as the best possible way due to the over-concentration of marginal communities in the Indian cities. Muneerudeen (2017) considers rapid land-use change as the prime cause to enhance regional vulnerability in coastal cities through flooding. This needs to be supported by local adaptation strategies with the consideration of socio-ecological resilience (Tajuddin, 2018). Roy (2019) considered coastal infrastructure development which can cope with the dynamic nature of weather phenomenon without compromising the socio-economic activity of the commons. The perception of resilience can be divided into multiple-element within a system or geographical set-up. In this point, Di Mauro (2006) has considered analysing the pattern of resilience into the fourfold segment, i.e., political, administrative, environmental and social organizational.

Suggested Adaptive Measure to Tackle Flood and Cyclonic Risks in the Indian Coastal Cities/Regions

Among the different types of climatic hazards, coastal zones are primarily affected by cyclones and floods (Sundaram et al., 2021; Malakar et al., 2021). Scientists have suggested various adaptive measures against cyclonic storms and related flood hazards for different specific regions of the coastal states. Samaddar et al. (2015) and Chatterjee (2010) have prescribed Yonmenkaigi System Method (YSM), which is the capacity building method for community implementation of participatory approach called ALM (advanced locality management) in case of Mumbai; while Agnihotri and Patel (2008) are in favour of flood water detention pond, for the diversion of flood water to other rivers in Surat. In case of Cochin, Thiruvananthapuram, and Kottayam, infrastructural development, river basin management dam restoration and formation of flood emergency action team (FAAT) are some of the viable options (Sudheer et al., 2019). For Udupi and Mangalore, infrastructural development (Roy, 2019) and interlinking of rivers for water sharing to mitigate floods (Chakraborty, 2021) are advocated. Chong et al. (2018) have suggested eco-tagging and text-mining techniques to boost the resilience among communities, whereas Samuel et al. (2018) argued in favour of locality-specific cultural green social work for Chennai. The situation of Visakhapatnam and Machilipatnam are favourable for strengthening rural-urban linkage for the application of the National Cyclone Mitigation Program (Revi, 2008); Integrated Water Resource Management (IWRM) (Kalyani & Jayakumar, 2021) and reservoir and embankment management (Jain & Singh, 2022). In the northern part of the east coast, cyclonic hazards are more frequent and prominent and cause massive devastations; therefore, in case of Puri, Balasore and Cuttack, formation of social capital through the vision of bonding-bridging-linking (Behera, 2021) and restoration of mangroves in coastal areas (Das, 2022) are better options. In a similar context, development of sewer network, drainage infrastructure and financial resources (Dasgupta et al., 2013) and canal and wetland restoration (Saha et al., 2021) are suggested for Kolkata; besides mangrove restoration (Erwin, 2009), polder land management (Sung et al., 2018) and embankment management (Paszkowski et al., 2021) are advised in the Sundarban deltaic region where the cyclones first strike before its way to the Kolkata megacity region.

Conclusion

Although the share of urban population to total population of India is still low compared to the global north, the urban centres are increasing rapidly in terms of their total population and urban infrastructure development on both horizontal and vertical scales. The cities and towns of the coastal region of the country are no exception. During the last three decades, such urban centres have been witnessing climate change related to extreme weather events particularly in the form of heat stress and cyclone-induced hazards. Different adaptive measures as suggested earlier in this chapter need to be implemented on priority basis to tackle the problems in the near future to reduce the vulnerabilities, especially among the risk-prone sections of the society.

References

Agnihotri, P. G., & Patel, J. N. (2008). Study of food at Surat City and its remedial measures the 3rd IASME. *WSEAS International Conference on Water Resources, Hydraulics and Hydrology, 3*, 23–25.

Ashok, K., & Saji, N. H. (2007). On the impacts of ENSO and Indian Ocean dipole events on sub-regional Indian summer monsoon rainfall. *Natural Hazards, 42*, 273–285. https://doi.org/10.1007/s11069-006-9091-0

Bai, X., Dawson, R. J., Ürge-Vorsatz, D., Delgado, G. C., SalisuBarau, A., Dhakal, S., et al. (2018). Six research priorities for cities and climate change. *Nature, 555*(7694), 23–25.

Baka, J. E. (2017). Political-industrial ecologies of energy. In *Handbook on the geographies of energy* (pp. 477–489). Edward Elgar Publishing.

Balica, S. F., & Wright, N. G. (2009). A network of knowledge on applying an indicator-based methodology for minimizing flood vulnerability. *Hydrological Processes, 23*(20), 2983–2986.

Balica, S. F., Douben, N., & Wright, N. G. (2009). Flood vulnerability indices at varying spatial scales. *Water Science and Technology Journal, 60*(10), 2571–2580. ISSN 0273-1223.

Bardsley, D. K., & Wiseman, N. D. (2012). Climate change vulnerability and social development for remote indigenous communities of South Australia. *Global Environmental Change, 22*(3), 713–723.

Behera, J. K. (2021). A study on the Role of Social Capital in Disaster Risk Reduction and Resilience Building evidence from the coastal communities, Odisha, India. *European Journal of Molecular & Clinical Medicine, 8*(03).

Bhardwaj, P., Pattanaik, D. R., & Singh, O. (2019). Tropical cyclone activity over Bay of Bengal in relation to El Niño-Southern Oscillation. *International Journal of Climatology, 39*(14), 5452–5469.

Bhui, K., Hazra, S., Bhadra, T., & Venugopal, V. (2022). Spatio-temporal variability of tidal velocities in the rivers of the Indian Sundarban Delta: A hydrodynamic modelling approach. *Journal of The Institution of Engineers (India): Series, C*, 1–15.

Cardona, O. (2005). *Indicators of disaster risk and risk management: Summary report*. Inter-American Development Bank. Available at lib.riskreductionafrica. org accessed on 12.02.2022.

Census of India. (2006). *Population projections for India and States 2001–2026*. Office of the Registrar General and Census Commissioner.

Chakraborty, S. K. (2021). Ecobiopolitics, policies, and conservation strategies of rivers. In *Riverine ecology* (Vol. 2, pp. 609–654). Springer.

Chan, F. K. S., Chuah, C. J., Ziegler, A. D., Dąbrowski, M., & Varis, O. (2018). Towards resilient flood risk management for Asian coastal cities: Lessons learned from Hong Kong and Singapore. *Journal of Cleaner Production, 187*, 576–589.

Chatterjee, M. (2010). Slum dwellers response to flooding events in the megacities of India. *Mitigation and Adaptation Strategies for Global Change, 15*, 337–353.

Chaudhuri, A. S., Gaur, N., Rana, P., & Verma, P. (2022). Ecohydrological perspective for environmental degradation of lakes and wetlands in Delhi. In *Geospatial technology for landscape and environmental management* (pp. 143–163). Springer.

Chittibabu, P., Dube, S. K., Macnabb, J. B., Murty, T. S., Rao, A. D., Mohanty, U. C., & Sinha, P. C. (2004). Mitigation of flooding and cyclone hazard in Orissa, India. *Natural Hazards, 31*(2), 455–485.

Chong, W. K., Naganathan, H., Liu, H., Ariaratnam, S., & Kim, J. (2018). Understanding infrastructure resiliency in Chennai, India using Twitter's Geotags and texts: A preliminary study. *Engineering, 4*(2), 218–223.

Chu, E. K. (2018). Transnational support for urban climate adaptation: Emerging forms of agency and dependency. *Global Environmental Politics, 18*(3), 25–46.

Conservation Action Trust. (2006). Mumbai Marooned: An enquiry into Mumbai floods 2005, Mumbai. https://cat.org.in/portfolio/concerned-citizens-commission-an-enquiry-into-mumbais-floods-2005/accesses on 10.10.23.

CSO. (2006). *National accounts statistics*. Central Statistical Organization.

Das, S. (2022). Valuing the role of mangroves in storm damage reduction in coastal areas of Odisha. In *Climate change and community resilience* (pp. 257–273). Springer.

Das, M., Das, A., Giri, B., Sarkar, R., & Saha, S. (2021). Habitat vulnerability in slum areas of India – What we learnt from COVID-19? International Journal of Disaster Risk Reduction, 65, 102553.

Dasgupta, S., Gosain, A. K., Rao, S., Roy, S., & Sarraf, M. (2013). A megacity in a changing climate: the case of Kolkata. *Climatic Change, 116*(3), 747–766.

Deepika, R., Swaminathan, C., Sathiyamoorthy, N. K., & Kannan, P. (2020). Rainfall and crop diversity analysis for response farming in Annur block of Coimbatore, Tamil Nadu. *International Journal of Ecology and Environmental Science, 2*(4), 102–106.

Dhiman, R., VishnuRadhan, R., Eldho, T. I., & Inamdar, A. (2019). Flood risk and adaptation in Indian coastal cities: recent scenarios. *Applied Water Science, 9*, 1–16.

Di Mauro, C. (2006). Regional vulnerability map for supporting policy definitions and implementations. In ARMONIA conference "*Multi-hazards: Challenges for risk assessment, mapping and management*", Barcelona.

Douben, N. (2006). Characteristics of river floods and flooding: A global overview, 1985–2003. *Irrigation Drainage, 55*, S9–S21. (Published online in Wiley InterScience).

EM-DAT. (2010). *The international disaster database*. Prepared by the Centre for Research on the Epidemiology of Disasters – CRED http://www.emdat.be/database. Accessed on 14.02.2022.

Erwin, K. L. (2009). Wetlands and global climate change: the role of wetland restoration in a changing world. *Wetlands Ecology and Management, 17*(1), 71–84.

Ewald, F. (2002). The return of Descartes's malicious demon: An outline of a philosophy of precaution. In *Embracing risk: The changing culture of insurance and responsibility* (pp. 273–301).

Fekete, A. (2009). Validation of a social vulnerability index in context to river-floods in Germany. *Natural Hazards and Earth System Sciences, 9*, 393–403. http://www.nat-hazards-earthsystsci.net/9/393/2009/

Friedrich, J., Stahl, J., Fitchett, J. M., & Hoogendoorn, G. (2020). To beach or not to beach? Socio-economic factors influencing beach tourists' perceptions of climate and weather in South Africa. *Transactions of the Royal Society of South Africa, 75*(2), 194–202.

Fuchs, R. J. (2010). *Cities at risk: Asia's coastal cities in an age of climate change.*

Fuchs, R., Conran, M., & Louis, E. (2011). Climate change and Asia's coastal urban cities: Can they meet the challenge? *Environment and Urbanization ASIA, 2*(1), 13–28.

Ghosh, M., & Ghosal, S. (2021). Climate change vulnerability of rural households in flood-prone areas of Himalayan foothills, West Bengal, India. *Environment, Development and Sustainability, 23*(2), 2570–2595.

Ghosh, U., Bose, S., & Bramhachari, R. (2018). *Living on the edge: Climate change and uncertainty in the Indian Sundarbans* (STEPS Working Paper 101). IDS Sussex.

Gleason, N. W. (2018). Singapore's higher education systems in the era of the fourth industrial revolution: Preparing lifelong learners. In *Higher education in the era of the fourth industrial revolution* (pp. 145–169). Palgrave Macmillan.

Gomes, V. H., Vieira, I. C., Salomão, R. P., & Steege, H. (2019). Amazonian tree species threatened by deforestation and climate change. *Nature Climate Change, 9*(7), 547–553.

Grases, A., Gracia, V., García-León, M., Lin-Ye, J., & Sierra, J. P. (2020). Coastal flooding and erosion under a changing climate: implications at a low-lying coast (Ebro Delta). *Water, 12*(2), 346.

Gupta, S., Jain, I., Johari, P., & Lal, M. (2019). Impact of climate change on tropical cyclones frequency and intensity on Indian coasts. In *Proceedings of international conference on remote sensing for disaster management* (pp. 359–365). Springer.

Haldar, A. (2016). Trend analysis of cyclones, over Kolkata and South 24 Parganas District, West Bengal. In *Climate and society – A contemporary perspective*. University of Calcutta. ISBN: 978-81-923448-0-5.

Hari, V., Dharmasthala, S., Koppa, A., Karmakar, S., & Kumar, R. (2021). Climate hazards are threatening vulnerable migrants in Indian megacities. *Nature Climate Change, 11*(8), 636–638.

Heger, M. P., & Neumayer, E. (2019). The impact of the Indian Ocean tsunami on Aceh's long-term economic growth. *Journal of Development Economics, 141*, 102365.

Heijmans, A., & Victoria, L. (2001). *Citizenry-based and development-oriented disaster response* (pp. 1–171). Centre for Disaster Preparedness and Citizens' Disaster Response Centre. Available at: www.proventionconsortium.org. accessed on 11.02.2022

Hijioka, Y., Lin, E., Pereira, J. J., Corlett, R. T., Cui, X., Insarov, G. E., Lasco, R. D., Lindgren, E., & Surjan, A. (2014). Asia. In V. R. Barros, C. B. Field, D. J. Dokken, M. D. Mastrandrea, K. J. Mach, T. E. Bilir, M. Chatterjee, K. L. Ebi, Y. O. Estrada, R. C. Genova, B. Girma, E. S. Kissel, A. N. Levy, S. MacCracken, P. R. Mastrandrea, & L. L. White (Eds.), *Climate change 2014: Impacts, adaptation, and vulnerability. Part B: Regional aspects. Contribution of Working Group II to the fifth assessment report of the Intergovernmental Panel on Climate Change* (pp. 1327–1370). Cambridge University Press.

Hobbie, S. E., & Grimm, N. B. (2020). Nature-based approaches to managing climate change impacts in cities. *Philosophical Transactions of the Royal Society B, 375*(1794), 20190124.

Idris, S., & Dharmasiri, L. M. (2015). Flood risk inevitability and flood risk management in urban areas: A review. *Journal of Geography and Regional Planning, 8*(8), 205–209.

Jain, S. K., & Singh, V. P. (2022). Strategies for flood risk reduction in India. *ISH Journal of Hydraulic Engineering*, 1–10.

Jalais, A. (2010). *Forest of tigers: People, politics and environment in the Sundarbans*. Routledge.

Kadaverugu, A., Kadaverugu, R., Chintala, N. R., & Gorthi, K. V. (2021). Flood vulnerability assessment of urban micro-watersheds using multi-criteria decision making and InVEST model: A case of Hyderabad City, India. *Modeling Earth Systems and Environment*, 1–13.

Kalyani, K., & Jayakumar, K. V. (2021). Integrated water resources management of Thatipudi Command Area, Vizianagaram, Andhra Pradesh. In *Water resources management and reservoir operation* (pp. 1–12). Springer.

Kankara, R. S., Ramana Murthy, M. V., & Rajeevan, M. (2018). *National assessment of shoreline changes along Indian Coast – A status report for 1990–2016*. National Centre for Coastal Research.

Kateja, A., & Jain, R. (Eds.). (2021). *Urban growth and environmental issues in India*. Springer.

Kishore, P., Singh, D. R., Chand, P., & Prakash, P. (2020). What determines groundwater depletion in India? A meso level panel analysis. *Journal of Soil and Water Conservation, 19*(4), 388–397.

Kollmuss, A., & Agyeman, J. (2002). Mind the gap: Why do people act environmentally and what are the barriers to pro-environmental behavior? *Environmental Education Research, 8*(3), 239–260.

Kroll-Smith, S., Couch, S. R., & Marshall, B. K. (1997). Sociology, extreme environments, and social change. *Current Sociology, 45*(3), 1–18.

Liu, Z., Geng, Y., Lindner, S., & Guan, D. (2012). Uncovering China's greenhouse gas emission from regional and sectoral perspectives. *Energy, 45*(1), 1059–1068.

Maity, P. K., Das, S., & Das, R. (2018). Remedial measures for saline water ingression in coastal aquifers of South West Bengal in India. *MOJ Ecology & Environmental Sciences, 3*(1), 00061.

Malakar, K., Mishra, T., Hari, V., & Karmakar, S. (2021). Risk mapping of Indian coastal districts using IPCC-AR5 framework and multi-attribute decision-making approach. *Journal of Environmental Management, 294*, 112948.

Malik, S., Pal, S. C., Sattar, A., Singh, S. K., Das, B., Chakrabortty, R., & Mohammad, P. (2020). Trend of extreme rainfall events using suitable Global Circulation Model to combat the water logging condition in Kolkata Metropolitan Area. *Urban Climate, 32*, 100599.

Mandal, S., Patra, P., Haldar, A., & Satpati, L. N. (2022). Extreme climatic events: A review of trends, vulnerabilities and adaptations in the South Asia Region. In Kar, Mukhopadhyay, & Sarkar (Eds.), *South Asia and climate change: Unravelling the conundrum*. https://doi.org/10.4324/9781003045731-3

Mehta, L., Srivastava, S., Adam, H. N., Bose, S., Ghosh, U., & Kumar, V. V. (2019). Climate change and uncertainty from 'above' and 'below': Perspectives from India. *Regional Environmental Change, 19*(6), 1533–1547.

Meyers, G., McIntosh, P., Pigot, L., & Pook, M. (2007). The years of El Niño, La Niña, and interactions with the Tropical Indian Ocean. *Journal of Climate, 20*(13), 2872–2880. Retrieved Mar 13, 2022, from https://journals.ametsoc.org/view/journals/clim/20/13/jcli4152.1.x

Mishra, M., Acharyya, T., Santos, C. A. G., da Silva, R. M., Kar, D., Kamal, A. H. M., & Raulo, S. (2021). Geo-ecological impact assessment of severe cyclonic storm Amphan on Sundarban mangrove forest using geospatial technology. *Estuarine, Coastal and Shelf Science, 260*, 107486.

Mukherjee, S., Kar, S., & Pal, S. (2021). Environmental disaster management and risk reduction. In *Environmental management: Issues and concerns in developing countries* (pp. 221–252). Springer.

Muneerudeen, A. (2017). *Urban and landscape design strategies for food resilience in Chennai city*. Master's thesis.

Nagai, S., Saitoh, T. M., & Morimoto, H. (2020). Does global warming decrease the correlation between cherry blossom flowering date and latitude in Japan? *International Journal of Biometeorology, 64*(12), 2205–2210.

Nda, M., Adnan, M. S., Ahmad, K. A., Usman, N., Razi, M. A. M., & Daud, Z. (2018). A review on the causes, effects and mitigation of climate changes on the environmental aspects. *International Journal of Integrated Engineering, 10*(4).

Nigussie, T. A., & Altunkaynak, A. (2019). Modeling the effect of urbanization on flood risk in Ayamama Watershed, Istanbul, Turkey, using the MIKE 21 FM model. *Natural Hazards, 99*(2), 1031–1047.

Nyong, A., Dabi, D., Adepetu, A., Berthe, A., & Ibemegbulem, V. (2008). Vulnerability in the Sahelian zone of northern Nigeria: A household-level assessment. In N. Leary, C. Conde, J. Kulkarni, A. Nyong, & J. Pulbin (Eds.), *Climate change and vulnerability* (pp. 218–238). Earthscan.

Paszkowski, A., Goodbred, S., Borgomeo, E., Khan, M., & Hall, J. W. (2021). Geomorphic change in the Ganges–Brahmaputra–Meghna delta. *Nature Reviews Earth & Environment, 2*(11), 763–780.

Patnaik, U., & Narayanan, K. (2009). *Vulnerability and climate change: An analysis of the eastern coastal districts of India.*

Patra, P., Haldar, A., & Satpati, L. N. (2017). Precipitation trends in the city of Kolkata and its implication on urban flooding. *Geographical Review of India, 79.*

Patra, P., Haldar, A., & Satpati, L. N. (2019). Tropical cyclones and hazard potentiality: A case study of coastal regions around North Indian Oceans (NIO). In Chattoraj, Islam, & Chakraborty (Eds.), *Environmental change, human adaptation and sustainability*. Words Publishers. PRMS Mahavidyalya.

Penning-Rowsell, E., Floyd, P., Ramsbottom, D., & Surendran, S. (2005). Estimating injury and loss of life in floods: A deterministic framework. *Natural Hazards, 36*(1–2), 43–64.

Picciariello, A., Colenbrander, S., Bazaz, A., & Roy, R. (2021). *The costs of climate change in India.*

Prasad, A. K., & Singh, R. P. (2005). Extreme rainfall event of July 25–27, 2005 over Mumbai, West Coast, India. *Journal of the Indian Society of Remote Sensing, 33*(3), 365–370.

Pusdekar, P. N., & Dudul, S. V. (2021). *Overview of Recent developments in flood mitigation techniques with respect to Indian subcontinent.*

Ramesh, V., & Iqbal, S. S. (2022). Urban flood susceptibility zonation mapping using evidential belief function, frequency ratio and fuzzy gamma operator models in GIS: A case study of Greater Mumbai, Maharashtra, India. *Geocarto International, 37*(2), 581–606.

Revi, A. (2008). Climate change risk: An adaptation and mitigation agenda for Indian cities. *Environment and Urbanization, 20*(1), 207–229.

Ritchie, H. (2020). Sector by sector: Where do global greenhouse gas emissions come from? *Our World in Data*. Accessed on 14.02.2022.

Roy, A. (2019). *Making India's coastal infrastructure climate-resilient: Challenges and opportunities*. Observer Research Foundation.

Roy, A. K., & Sharma, S. (2015). Perceptions and adaptations of the coastal community to the challenges of climate change: A case of Jamnagar city region, Gujarat, India. *Environment and Urbanization Asia, 6*(1), 71–91.

Rumbach, A. (2014). Do new towns increase disaster risk? Evidence from Kolkata, India. *Habitat International, 43*, 117–124.

Saha, M., Sarkar, A., & Bandyopadhyay, B. (2021). Water quality assessment of East Kolkata Wetland with a special focus on bioremediation by nitrifying bacteria. *Water Science and Technology, 84*(10–11), 2718–2736.

Saintilan, N., Khan, N. S., Ashe, E., Kelleway, J. J., Rogers, K., Woodroffe, C. D., & Horton, B. P. (2020). Thresholds of mangrove survival under rapid sea level rise. *Science, 368*(6495), 1118–1121.

Samaddar, S., Choi, J., Misra, B. A., & Tatano, H. (2015). Insights on social learning and collaborative action plan development for disaster risk reduction: Practicing Yonmenkaigi system method (YSM) in food-prone Mumbai. *Natural Hazards, 75*(2), 1531–1554.

Samuel, M., Annadurai, P., & Sankarakrishnan, S. (2018). The 2015 Chennai foods: green social work, an emerging model for practice in India. In *The Routledge handbook of green social work* (pp. 281–292).

Sinay, L., & Carter, R. W. (2020). Climate change adaptation options for coastal communities and local governments. *Climate, 8*(1), 7.

Singh, S. (2020). Farmers' perception of climate change and adaptation decisions: A micro-level evidence from Bundelkhand Region, India. *Ecological Indicators, 116*, 106475.
Singh, C., Madhavan, M., Arvind, J., & Bazaz, A. (2021). Climate change adaptation in Indian cities: A review of existing actions and spaces for triple wins. *Urban Climate, 36*, 100783.
Sinha, M., Sendhil, R., Chandel, B. S., Malhotra, R., Singh, A., Jha, S. K., & Sankhala, G. (2021). Are multidimensional poor more vulnerable to climate change? Evidence from rural Bihar, India. *Social Indicators Research*, 1–27.
Smit, B., & Wandel, J. (2006). Adaptation, capacity and vulnerability. *Global Environmental Change, 16*, 282–292.
Song, J., Chang, Z., Li, W., Feng, Z., Wu, J., Cao, Q., & Liu, J. (2019). Resilience-vulnerability balance to urban flooding: A case study in a densely populated coastal city in China. *Cities, 95*, 102381.
Stern, N., & Stern, N. H. (2007). *The economics of climate change: The Stern review*. Cambridge University Press.
Sterr, H., Klein, R., & Reese, S. (2000). *Climate change and coastal zones: An overview of the state-of-the-art on regional and local vulnerability assessment.*
Sudheer, K. P., Bhallamudi, S. M., Narasimhan, B., Thomas, J., Bindhu, V. M., Vema, V., & Kurian, C. (2019). Role of dams on the floods of August 2018 in Periyar River Basin, Kerala. *Current Science (00113891), 116*(5).
Sundaram, S., Devaraj, S., & Yarrakula, K. (2021). Modeling, mapping and analysis of urban floods in India – A review on geospatial methodologies. *Environmental Science and Pollution Research*, 1–17.
Sung, K., Jeong, H., Sangwan, N., & Yu, D. J. (2018). Effects of flood control strategies on flood resilience under sociohydrological disturbances. *Water Resources Research, 54*(4), 2661–2680.
Swain, D. L., Singh, D., Touma, D., & Diffenbaugh, N. S. (2020). Attributing extreme events to climate change: A new frontier in a warming world. *One Earth, 2*(6), 522–527.
Tajuddin, N. (2018). *Leveraging socio-cultural networks: Local adaptation strategies to bring about food resilience in Chennai metropolitan area, India.* Thesis. TU Delft Architecture and the Built Environment.
Tanner, T., Mitchell, T., Polack, E., & Guenther, B. (2009). Urban governance for adaptation: Assessing climate change resilience in ten Asian cities. *IDS Working Papers, 2009*(315), 01–47.
Toros, H., Mokari, M., & Abbasnia, M. (2019). Regional variability of temperature extremes in the maritime climate of Turkey: A case study to develop agricultural adaptation strategies under climate change. *Modeling Earth Systems and Environment, 5*, 857–865.
Trickett, E. J. (1995). The community context of disaster and traumatic stress: An ecological perspective from community psychology. In *Extreme stress and communities: Impact and intervention* (pp. 11–25). Springer.
UNISDR. (2017). *United Nations International Strategy for Disaster Reduction (UNISDR): Terminology on Disaster Risk Reduction 2017, Geneva, Switzerland.*
Vincent, K. (2004). *Creating an index of social vulnerability to climate change for Africa* (Technical Report 56). Tyndall Center for Climate Change Research, University of East Anglia.
Vogel, M. M., Zscheischler, J., Wartenburger, R., Dee, D., & Seneviratne, S. I. (2019). Concurrent 2018 hot extremes across Northern Hemisphere due to human-induced climate change. *Earth's Future, 7*(7), 692–703.
Vousdoukas, M. I., Clarke, J., Ranasinghe, R., Reimann, L., Khalaf, N., Duong, T. M., et al. (2022). *African heritage sites threatened as sea-level rise accelerates* (pp. 1–7). Nature Climate Change.
World Meteorological Organization. (2017). *WMO statement on the state of the global climate in 2016*. World Meteorological Organization (WMO).
Zheng, X., Streimikiene, D., Balezentis, T., Mardani, A., Cavallaro, F., & Liao, H. (2019). A review of greenhouse gas emission profiles, dynamics, and climate change mitigation efforts across the key climate change players. *Journal of Cleaner Production, 234*, 1113–1133.

Chapter 3
Climate Change and Cyclones: Its Associated Consequences in Tamil Nadu Coastal Cities of Chennai and Thoothukudi

Annaidasan Krishnan and Jaganathan Ramasamy

Introduction

The urban area covers less than 2% of the land surface in the world, but houses just over 50% of the global population (UN, 2018), and it is expected to increase to more than 50% by 2030 (UN, 2018). The city activities are dominantly releasing the greenhouse gases (GHGs) that lead to global warming and it can be directly coming from the fuel-based transport and indirectly through utilization of daily usage of the energy (e.g., electricity), manufacturing materials, and farming stuff. Furthermore, 80% of greenhouse gases (GHG) emissions are from the urban regions (Lamb et al., 2021).

The Western North Pacific (WNP), which generates over 30% of all tropical cyclones (TCs) worldwide, is the most active TC basin on the earth. WNP TCs, which are linked to powerful winds, a huge amount of rain, and large storm surges which are creating the vulnerable situation to people, and lot of people's lives in danger situation and harm the economies and societies along the Asian coastal cities (Chen et al., 2020). Preparedness of the vulnerable communities and evaluation of rigorous damage are crucial for efficient disaster risk management since cyclones are frequent phenomena that cause enormous costs, leading to budget deficits (George et al., 2021). Forecasts of the impact of human-caused climate change on tropical cyclones (TCs) typically focus on two issues: whether TCs have already been impacted by climate change, and if so, how much of that impact is caused by humans or natural factors, and how climate change may affect TCs in the future (Knutson et al., 2019; Walsh et al., 2016, 2019). Generally, fast-growing cities are the likely hot spots that are vulnerable to climate change impacts.

A. Krishnan (✉)
Department of Geography, Central University of Tamil Nadu, Thiruvarur, Tamil Nadu, India

J. Ramasamy
Department of Geography, University of Madras, Chennai, Tamil Nadu, India

A. L. Singh et al. (eds.), *Climate Change, Vulnerabilities and Adaptation*,
https://doi.org/10.1007/978-3-031-49642-4_3

Subsequently, impacts of climate change on the urban areas are rise in sea levels and a constant change in the intensity of temperature, rainfall and other phenomena such as wind speed and wind directions (Seneviratne et al., 2012; Thomas et al., 2017). Thus, local and regional climate change or variability is measured by different climate impacts from various cities. Temperature is a crucial climatic variable that has a direct impact on human and natural systems. As shown in numerous examination reports of the Intergovernmental Panel on Climate Change (IPCC), including the most comprehensive Fifth Evaluation Report, the average worldwide temperature at the surface is a critical indicator of climate change since it rises almost linearly with the total emission levels of GHGs (Abbs, 2012; Kaito et al., 2000; United, 2018).

The sea level rise increases the vulnerability of storm–surge flooding and rates of coastal erosion and accretion on the shoreline (Kantamaneni et al., 2022). In addition, the rising sea level threatens the coastal environments such as beaches, human settlements and wetland ecosystems, and through the saline water intrusion (Krishnan & Ramasamy, 2022a; Narmada & Annaidasan, 2019). Drinking water availability is gradually declining in many regions especially in cities with quality status including health and aquatic ecosystems (National & Pillars, 2008). Larger cities are unfortunately affected by Urban Heat Island (UHI) by storage of solar energy in the 'cities' it can be applicable for many coastal cities. Air pollution increases in cities, making the warm condition at night. Climate change severely impacts natural resources such as river, soils, and forest (Jaganathan et al., 2015; Krishnan & Ramasamy, 2022b). Many independent research works have shown that since Industrial Revolution, human activities have contributed considerably to climate change across the world. India should be concerned about climate change since a large portion of its population relies on industries that are vulnerable to it, such as agriculture, forestry, and fisheries. The severity of the nation's socioeconomic problems is increasing as a result of the negative effects of climate change, which include a decrease in rainfall and an increase in temperature. The ecological and socioeconomic systems, which are already under a great deal of stress from expanding industrialization, urbanization, and economic growth, will be further stressed by climate change (Sanjay et al., 2020; Balasubramanian, 2018). Tamil Nadu is categorized into seven agroclimatic zones, namely northeastern, northwestern, Cauvery delta, western, high-altitude, southern, and high precipitation zones. As rain-producing systems develop over the Bay of Bengal and move westward toward the coast of Tamil Nadu, rainfall is higher over coastal areas and decreases as one moves inland. Due to the creation or nonformation of rain-bearing systems and their spatial variance, the overall amount of rainfall during the season is not constant and has inter-seasonal and intra-seasonal variability (Bal et al., 2016; Krishnan et al., 2020).

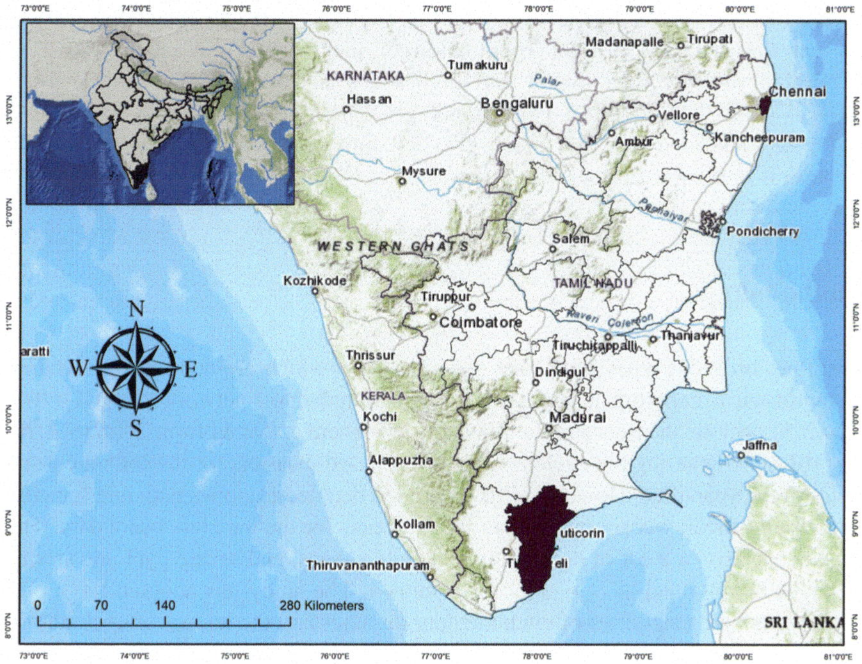

Fig. 3.1 Study area of Chennai and Thoothukudi

Study Area

The most significant cultural, economic, and educational hub of South India is Chennai, formerly known as Madras, which is the capital of Tamil Nadu and a most populous city. Chennai is situated on the Coromandel Coast of the Bay of Bengal. The city is divided into two for administrative purpose: Chennai Metropolitan Area (CMA) and Chennai City or Chennai Corporation. It is the fourth largest metropolitan city in India and holds 40th rank in the list of the largest inhabited cities in the world. The CMA will be completed as per the proposed urbanization by 2026 (Imam & Banerjee, 2016). The city extended upto 200 wards from North to South. Furthermore, Chennai is situated at 13.04°N and 80.17°E on the Southeast coast of India and in the Northeastern part of Tamil Nadu. The city's highest point is 60 m (200 feet) above sea level, with a usual elevation of 6 m (20 feet). The investigation area's average annual precipitation is 1382 mm, and the average maximum and minimum temperatures are 31.8 °C and 24.8 °C, respectively.

Thoothukudi, also known as Tuticorin, is also situated on the Coromandel Coast of the Bay of Bengal. It is geographically located at 8.906°N latitude and 78.019°E longitude (Fig. 3.1). This city is 590 km away from Chennai. The adjacent districts are Viruthunager and Ramanathapuram to the north; Thirunelveli to the South and West; and the Gulf of Mannar to the east. Furthermore, Tuticorin port is an emerging port in India and is the second largest port in Tamil Nadu (Rangarajan et al., 2019). As per the 2011 census data population of the district (not city) 1,750,176. According to IMD data lowest and highest average temperatures are 19.4 °C (January) and 38.3 °C (May), correspondingly. The humidity ranges between 52% and 90%. The area typically receives 625.8 mm of precipitation per year. The location is in seismic zone II.

Methodology

There are different types of dataset used in study which are aquired from indian meteorological department (IMD) and international flood network (IFNET). The historical cyclone datasets and the flood datasets, were collected from 2006 to 2020. The datasets regarding the cyclones and water-based disasters for the Indian coastal states in the year-wise. However, IMD provided detailed cyclones path and location in their an official website which can freely access for the past few years data. The dataset consists of a list of depression, cyclonic storms, and severe cyclonic storms throughout India and also in adjacent countries. The study has adopted the aforementioned three categories of cyclonic data for assessing the impact of cyclones on the Bay of Bengal in Tamil Nadu coastal region. In addition, the quantitative and qualitative datasets were adopted for evaluating the floods from 2006 to 2020, which are acquired yearly as printed report from the International Flood Network (http://www.internationalfloodnetwork.org/). However, detailed historical flood reports across the globe, concerning country, place, the number of deaths, missing, and the number of districts affected by the water-based disaster, are critical. This study has comprised a detailed account of cyclones which are crossed the at the Tamil Nadu coastline and to evaluate the impacts of cyclones and floods for sustainable management of the Tamil Nadu coastline.

Datasets to be used

The classification of the dataset adopted from cyclone warning in India. It can be a standard operating procedure at Indian Meteorological Department, and it was classified based on the intensity of the cyclone. This study considered parameters such as depression, cyclonic storm, and severe cyclonic storm, with a list of district and state data determined qualitatively. This study spatially assessed the cyclones using geospatial technology, and the spatial analysis representations exposed the location of the beginning point of cyclones with their classified parameters, i.e., depression, cyclonic storm, and severe cyclonic storm. The results showed the impact of climate change on Chennai and Thoothukudi cities (Table 3.1).

Table 3.1 Classification of cyclone

Sl. No.	Cyclone types	Sustained wind
1	Low-pressure area	Less than 31 Kmph (not exceeding 17 knots)
2	Depression	31–49 kmph (17–27 knots)
3	Deep depression	50–61 kmph (28–33 knots)
4	Cyclonic storm	62–88 kmph (34–47 knots)
5	Severe cyclonic storm	89–117 kmph (48–63 knots)
6	Very severe cyclonic storm	118–167 kmph (64–90 knots)
7	Extremely severe cyclonic storm	168–221 kmph (91–119 knots)
8	Super cyclonic storm	above 222 kmph (above 120 knots)

Source: Indian Meteorological Department

Result and Discussion

Climatic Parameters and Trends of Cyclones in Tamil Nadu

Observed Temperature Changes

The study was considered the environmental variables such as temperature, rainfall, and relative humidity, which were averaged out for the each year from 2006 to 2020, and with the help of the climatic parameters to evaluate the corresponding coastal cities of Tamil Nadu (Chennai and Thoothukudi). According to the results obtained, a highest average temperature of 28.39 °C in 2019 and a possible lowest average temperature of 27.74 °C in 2007 were recorded for Chennai. Similarly, the uppermost average temperature in Thoothukudi was 28.63 °C in 2019, while the lowest average temperature was 27.20 °C in 2006. However, the study revealed a slow and gradual increase in the average temperature from 2006 to 2020, the major cause being the increase in the number of vehicles and urbanization practices (Pal & Eltahir, 2016) (Fig. 3.2).

Observed Precipitation Changes

Appraisal of the average precipitation of Chennai and Thoothukudi coastal cities of Tamil Nadu and it was considered the, last 15 years of the precipitation from 2006 to 2020. The results showed the maximum and minimum rainfall of both cities with measurement unit of the precipitation in millimeter (mm) and is calculated by taking into account the total precipitation of both cities. In 2015, Chennai received a highest rainfall of 1929.72 mm throughout the year. Although a lowest precipitation of 788.54 mm was documented in 2006. Moreover, Thoothukudi has secured the highest precipitation in 2006, and the total precipitation is 1705.20 mm. Despite, lowest precipitation was registered in the year 2012, although overall assessment of

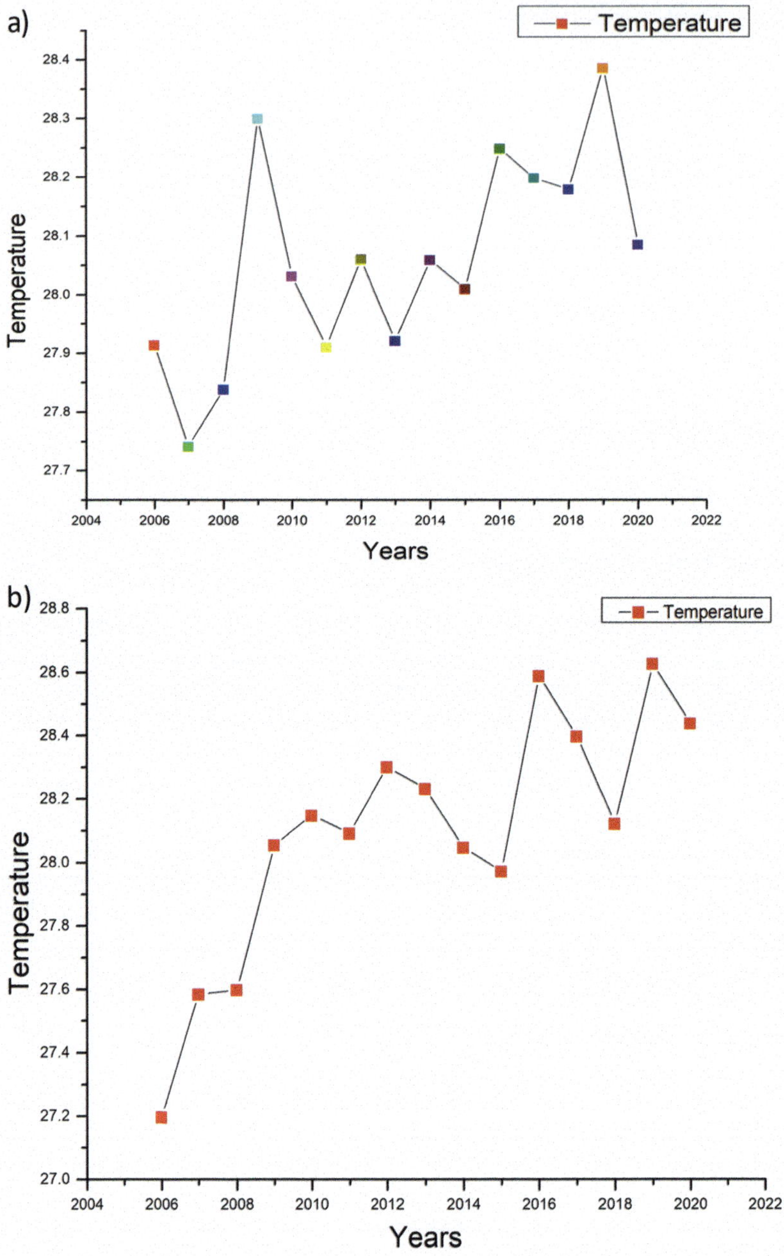

Fig. 3.2 Temperature of the coastal cities: (**a**) Chennai and (**b**) Thoothukudi

Table 3.2 Climatic data 2006–2020

	Chennai			Thoothukudi		
Years	Temperature	Precipitation	Relative humidity	Temperature	Precipitation	Relative humidity
2006	27.91	1232.45	72.46	27.20	1705.20	79.36
2007	27.74	1270.42	73.61	27.58	888.93	75.45
2008	27.84	1360.05	73.12	27.60	1041.70	75.69
2009	28.30	1159.28	70.99	28.06	804.65	72.82
2010	28.03	1378.34	75.29	28.15	890.29	73.52
2011	27.91	1345.79	72.42	28.09	653.70	72.33
2012	28.06	961.88	71.73	28.30	624.16	71.86
2013	27.92	1024.87	72.91	28.23	674.10	72.11
2014	28.06	1068.87	71.49	28.05	961.41	74.05
2015	28.01	1929.72	73.18	27.97	1244.21	76.13
2016	28.25	788.54	70.65	28.59	457.00	71.01
2017	28.20	1273.88	72.53	28.40	798.53	72.43
2018	28.18	798.51	71.09	28.12	671.40	73.28
2019	28.39	1186.17	72.97	28.63	1016.45	72.79
2020	28.09	1554.51	74.62	28.44	853.11	73.69

Source: NASA Power Data Access

the precipitation is 624.16 mm. Thus, Chennai secured the highest temperature and rainfall compared to Thoothukudi (Table 3.2).

Nevertheless, all monsoon seasons was documented some major deviations in annual precipitation., but the amount of precipitation is lesser than Chennai city (Rangarajan et al., 2019). The study has revealed that the Northeast Monsoon does not evenly distribute the precipitation to Tamil Nadu coastal cities (Fig. 3.3).

Observed Humidity Changes

The humidity is the natural asset of the atmosphere which created vast quantities of water on the earth's surface, such as rivers and lakes, evaporation reaches the atmosphere, tanks etc. However, the humidity of Chennai and Thoothukudi cities are comparatively quite different from each other. In Chennai city has a highest humidity 75.29% was recorded in 2010 and a lowest humidity of 71.09% which was documented in the 2018. Likewise, for Thoothukudi, a highest humidity of 79.36% was recorded in 2006 and a lowest humidity of 71.01% in 2016. Comparatively Chennai city has lesser than Thoothukudi city. The rest of the year deals with moderate percent of the humidity was recorded for both cities, and the influence of the temperature humidity is not stable (Fig. 3.4).

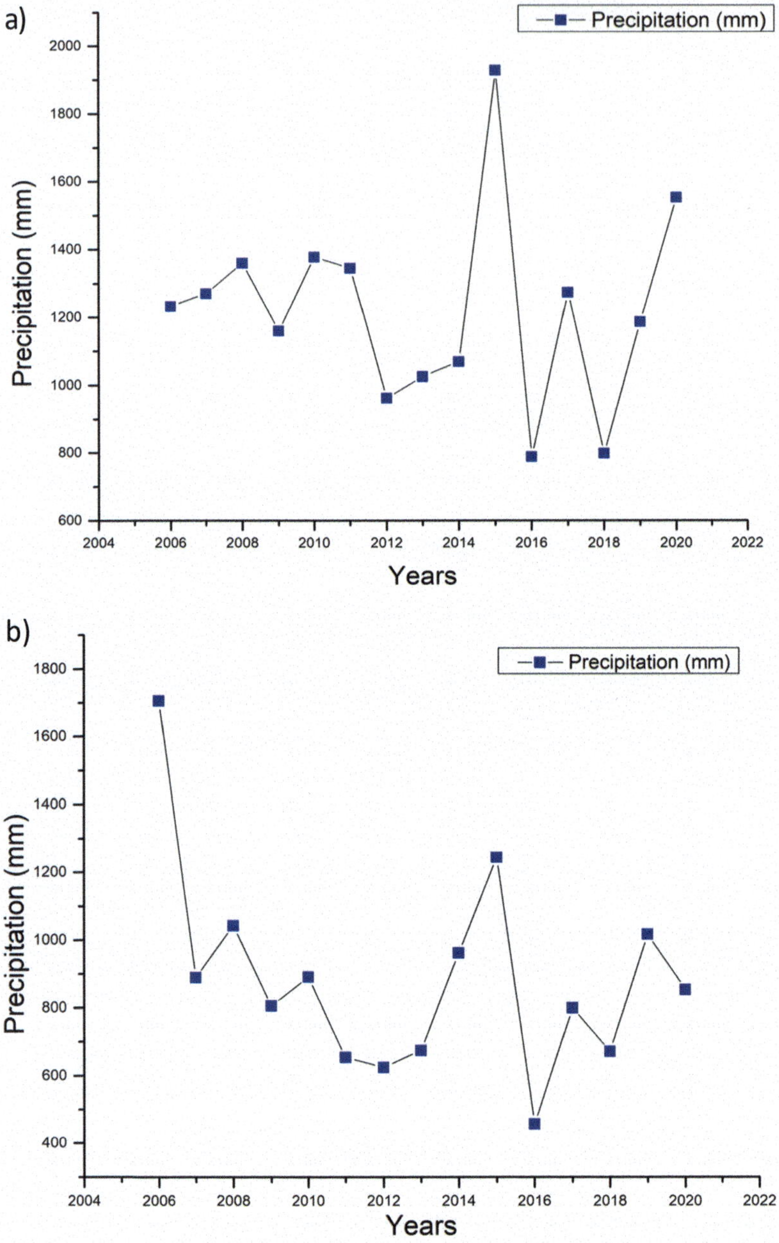

Fig. 3.3 Precipitation of the coastal cities: (**a**) Chennai and (**b**) Thoothukudi

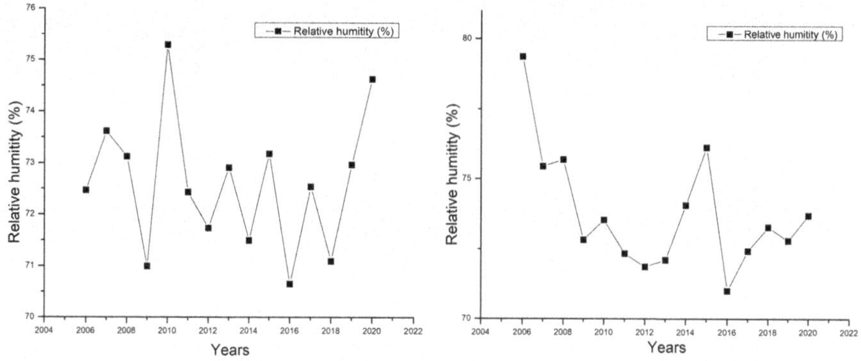

Fig. 3.4 Humidity at Chennai and Thoothukudi

Trends of the Cyclone Paths and Impacts

The tropical cyclones forms based upon the temperature in oceanic regions, and it is a relationship with the Greenhouse Gas (GHG) induced surface temperature increases, and tropical cyclones movement is an undecided than the surface temperature changes. Furthermore, the world meteorological organization (WMO) global level appraisal (Carter et al., 2015; Chen et al., 2020; Knutson et al., 2019, 2021; Wehner, 2021) which is concluded that the tropical cyclones are extremely unusual comparing with other natural causes. In addition, Tamil Nadu has 14 coastal districts, organized from North to South, including Thiruvallur, Chennai, Kancheepuram, Villupuram, Cuddalore, Mayiladuthurai, Nagapattinam, Thiruvarur, Thanjavur, Pudukkottai, Ramanathapuram, Thoothukudi, Thirunelveli, and Kanyakumari, these are the districts fall into the Tamil Nadu coastline, and Tamil Nadu state is secured more than 10 districts are located the coastline, therefore it is most vulnerable state for cyclones. Five depressions passed through Tamil Nadu, according to the Indian Meteorological Department (IMD), which brought about 11 cyclones upto 2020 by the northeast monsoon (Fig. 3.5). In addition, two cyclonic storms and four severe cyclonic storms crossed Tamil Nadu's coastline in the years 2011, 2016, 2018, and 2020. The severe cyclonic storm resulted in 120 fatalities and 1,09,238 people were displaced. During the Northeast Monsoon, severe cyclonic storms mostly struck the state, causing numerous losses to the environment, the human population, and habitats. Additionally, a tropical cyclone that crossed India's East coast resulted in two significant floods in 2014 and 2015, and low-pressure systems flooded Tamil Nadu. About 360 people lost their lives as a result of the storm in the districts of Chennai and Cuddalore (Table 3.3).

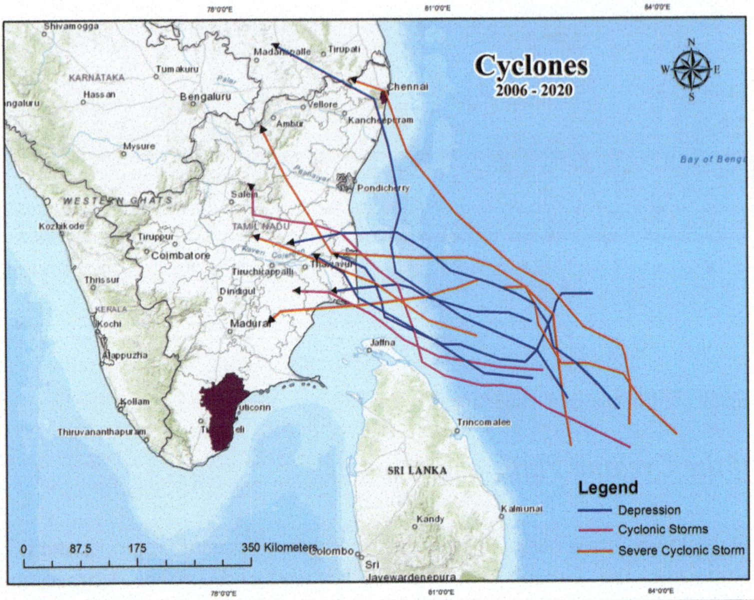

Years	Depressions	Cyclonic Storms	Severe Cyclonic Storms
25.11.2008	NIL	CS	NIL
04.11.2010	D	NIL	NIL
25.12.2011	NIL	NIL	SCS
28.10.2012	NIL	CS	NIL
13.11.2013	D	NIL	NIL
06.12.2013	D	NIL	NIL
08.11.2015	D	NIL	NIL
29.11.2016	D	NIL	NIL
06.12.2016	NIL	NIL	SCS
10.11.2018	NIL	NIL	SCS
22.11.2020	NIL	NIL	SCS

Fig. 3.5 List of cyclones that crossed the Tamil Nadu coastline from 2006 to 2020. (Source: Indian Meteorological Department E-Atlas)

Overall Assessment of Cyclone Trends and Climate Change

The state of Tamil Nadu comprises 14 coastal districts located near the Bay of Bengal, including two crucial cities, and study explored the comparative appraisal of climatic parameters such as precipitation, temperature, and relative humidity in Tamil Nadu coastal cities Chennai and Thoothukudi. The cyclones cross through Tamil Nadu during the Northeast monsoon (October–December) every year. Three types of cyclones have been crossing since 2006, and it is classified based upon the intensity of the cyclones (Table 3.1). The cyclone's path is tracked by the E-Atlas of the Indian Meteorological Department (IMD). Even though the study considered from 2006 to 2020, but cyclones are crossed in the study are in 2008. The study

Table 3.3 List of cyclones crossed near Chennai and Thoothukudi city from 2006 to 2020

Years	Depressions	Cyclonic storms	Severe cyclonic storms
25.11.2008	NIL	CS	NIL
04.11.2010	D	NIL	NIL
25.12.2011	NIL	NIL	SCS
28.10.2012	NIL	CS	NIL
13.11.2013	D	NIL	NIL
06.12.2013	D	NIL	NIL
08.11.2015	D	NIL	NIL
29.11.2016	D	NIL	NIL
06.12.2016	NIL	NIL	SCS
10.11.2018	NIL	NIL	SCS
22.11.2020	NIL	NIL	SCS

Source: Indian Meteorological Department
CS Cyclonic Storm, *D* Depression, *SCS* Severe Cyclonic Storm

revealed that most of the cyclones originated from the Bay of Bengal near the Andaman and Nicobar Islands, and Chennai city is consecutively affected by the disturbances from the cyclonic activities.

Conclusion and Recommendation

Among the most devastating problems in the world, cyclones and climate change (including changes in temperature, precipitation, and humidity) frequently cause havoc in South Asia, notably India. Tamil Nadu is a region where cyclones frequently occur. This study evaluated the cyclone and climate change trends in 14 coastal districts of Tamil Nadu from 2006 to 2020, particularly in the coastal cities of Chennai and Thoothukudi. For an evaluation of climatic parameters, GIS tools, NASA, and the IMD E-atlas, and the Indian Meteorological Data. According to this study, high-intensity cyclones are increasingly passing through the towns of Thoothukudi and Chennai. However, each year's monsoon cyclones frequently affect these two coastal cities. Further research into the connection between climate change and cyclone activity is urgently needed to study these trends further. The GIS maps that show cyclone patterns between two cities aid in determining the relationship between the severity of climate change and cyclone activity along the Tamil Nadu coast. The study's findings also assist policy- and decision-makers in developing and enhancing successful coastal management methods. Additionally, these outcomes contribute to the achievement of UN-SDG 13 (climate change), which enhances the sustainability of coastal management in Indian coastal regions.

References

Abbs, D. (2012). *The impact of climate change on the climatology of tropical cyclones in the Australian region* (CSIRO Climate Adaptation Flagship Working Paper Series, 24). http://www.csiro.au/Organiation-Structure/Flagships/Climate-Adaptation-Flagship/CAF-working-papers.aspx

Balasubramanian, M. (2018). Climate change, famine, and low-income communities challenge Sustainable Development Goals. The Lancet Planetary Health, 2(10), e421–e422. https://doi.org/10.1016/S2542-5196(18)30212-2

Bal, P. K., Ramachandran, A., Geetha, R., Bhaskaran, B., Thirumurugan, P., Indumathi, J., & Jayanthi, N. (2016). Climate change projections for Tamil Nadu, India: deriving high-resolution climate data by a downscaling approach using PRECIS. *Theoretical and Applied Climatology, 123*(3–4), 523–535. https://doi.org/10.1007/s00704-014-1367-9

Carter, J. G., Cavan, G., Connelly, A., Guy, S., Handley, J., & Kazmierczak, A. (2015). Climate change and the city: Building capacity for urban adaptation. *Progress in Planning, 95*, 1–66. https://doi.org/10.1016/j.progress.2013.08.001

Chen, J., Wang, Z., Tam, C. Y., Lau, N. C., Lau, D. S. D., & Mok, H. Y. (2020). Impacts of climate change on tropical cyclones and induced storm surges in the Pearl River Delta region using pseudo-global-warming method. *Scientific Reports, 10*(1), 1–10. https://doi.org/10.1038/s41598-020-58824-8

George, S. L., Balasubramani, K., Shekhar, S., Venkatesham, E., Prasad, K. A., Grover, A., Mahata, D., Kumar, A., Ashique, V. V., Libina, R. S., Swaminathan, D. R., Balasundareshwaran, A., & Annaidasan, K. (2021). A comprehensive study on preparedness, impacts, response and recovery from tropical severe cyclonic storm 'GAJA': Lessons for the future. *Journal of Coastal Conservation, 25*(6), 1–13. https://doi.org/10.1007/s11852-021-00842-3

Imam, A. U. K., & Banerjee, U. K. (2016). Urbanisation and greening of Indian cities: Problems, practices, and policies. *Ambio, 45*(4), 442–457. https://doi.org/10.1007/s13280-015-0763-4

Jaganathan, R., Annaidasan, K., Surendran, D., & Balakrishnan, P. (2015). Morphometric analysis for prioritization of watersheds in the Mullayar River Basin, South India. In *Environmental management of river basin ecosystems* (pp. 127–136). Springer. https://doi.org/10.1007/978-3-319-13425-3_7

Kaito, C., Ito, A., Kimura, S., Kimura, Y., Saito, Y., & Nakada, T. (2000). Topotactical growth of indium sulfide by evaporation of metal onto molybdenite. In. *Journal of Crystal Growth, 218*(2). https://doi.org/10.1016/S0022-0248(00)00575-3

Kantamaneni, K., Panneer, S., Krishnan, A., Shekhar, S., Bhat, L., & R, A. K., & Rice, L. (2022). Appraisal of climate change and cyclone trends in Indian coastal states: A systematic approach towards climate action. *Arabian Journal of Geosciences, 15*(9). https://doi.org/10.1007/s12517-022-10076-8

Knutson, T., Camargo, S. J., Chan, J. C. L., Emanuel, K., Ho, C. H., Kossin, J., Mohapatra, M., Satoh, M., Sugi, M., Walsh, K., & Wu, L. (2019). Tropical cyclones and climate change assessment. *Bulletin of the American Meteorological Society, 100*(10), 1987–2007. https://doi.org/10.1175/BAMS-D-18-0189.1

Knutson, T. R., Chung, M. V., Vecchi, G., Sun, J., Hsieh, T.-L., & Smith, A. J. P. (2021). ScienceBrief review: Climate change is probably increasing the intensity of tropical cyclones. *Critical Issues in Climate Change Science, 1*, 1–8. https://doi.org/10.5281/zenodo.4570334

Krishnan, A., & Ramasamy, J. (2022a). An assessment of land use and land cover changes in Muthupet Mangrove Forest, using time series analysis 1975–2015, Tamilnadu, India. *Geosfera Indonesia, 7*(2), 119. https://doi.org/10.19184/geosi.v7i2.28077

Krishnan, A., & Ramasamy, J. (2022b). Morphometric assessment and prioritization of the South India Moyar river basin sub-watersheds using a geo-computational approach. *Geology, Ecology, and Landscapes, 00*(00), 1–11. https://doi.org/10.1080/24749508.2022.2109819

Krishnan, R., Gnanaseelan, C., Sanjay, J., Swapna, P., Dhara, C., Sabin, T. P., Jadhav, J., Sandeep, N., Choudhury, A. D., Singh, M., Mujumdar, M., Parekh, A., Tewari, A., & Mehajan, R. (2020).

Introduction to climate change over the Indian region. In *Assessment of climate change over the Indian Region: A report of the Ministry of Earth Sciences (MoES), Government of India* (pp. 1–20). Springer. https://doi.org/10.1007/978-981-15-4327-2_1

Lamb, W. F., Wiedmann, T., Pongratz, J., Andrew, R., Crippa, M., Olivier, J. G. J., Wiedenhofer, D., Mattioli, G., Al Khourdajie, A., House, J., Pachauri, S., Figueroa, M., Saheb, Y., Slade, R., Hubacek, K., Sun, L., Ribeiro, S. K., Khennas, S., De La Rue, D., Can, S., et al. (2021). A review of trends and drivers of greenhouse gas emissions by sector from 1990 to 2018. *Environmental Research Letters, 16*(7). https://doi.org/10.1088/1748-9326/abee4e

Narmada, K., & Annaidasan, K. (2019). Estimation of the temporal change in carbon stock of Muthupet Mangroves in Tamil Nadu using remote sensing techniques. *Journal of Geography, Environment and Earth Science International, 19*(4), 1–7. https://doi.org/10.9734/jgeesi/2019/v19i430096

Pal, J. S., & Eltahir, E. A. B. (2016). Future temperature in southwest Asia projected to exceed a threshold for human adaptability. *Nature Climate Change, 6*(2), 197–200. https://doi.org/10.1038/nclimate2833

Rangarajan, S., Thattai, D., Cherukuri, A., Borah, T. A., Joseph, J. K., & Subbiah, A. (2019). A detailed statistical analysis of rainfall of Thoothukudi District in Tamil Nadu (India). In *Water resources and environmental engineering II*. Springer Singapore. https://doi.org/10.1007/978-981-13-2038-5

Sanjay, J., Revadekar, J. V., Ramarao, M. V. S., Borgaonkar, H., Sengupta, S., Kothawale, D. R., Patel, J., Mahesh, R., & Ingle, S. (2020). Temperature changes in India. In *Assessment of climate change over the Indian Region: A report of the Ministry of Earth Sciences (MoES), Government of India* (pp. 21–45). Springer. https://doi.org/10.1007/978-981-15-4327-2_2

Seneviratne, S. I., Nicholls, N., Easterling, D., Goodess, C. M., Kanae, S., Kossin, J., Luo, Y., Marengo, J., Mc Innes, K., Rahimi, M., Reichstein, M., Sorteberg, A., Vera, C., Zhang, X., Rusticucci, M., Semenov, V., Alexander, L. V., Allen, S., Benito, G., et al. (2012). Changes in climate extremes and their impacts on the natural physical environment. In *Managing the risks of extreme events and disasters to advance climate change adaptation: Special report of the Intergovernmental Panel on Climate Change, 9781107025* (pp. 109–230). Cambridge Univeristy Press. https://doi.org/10.1017/CBO9781139177245.006

Thomas, A., Pringle, P., Pfleiderer, P., & Schleussner, C.-F. (2017). *Briefing note: Tropical cyclones: Impacts, the link to climate change and adaptation*. 1–8. https://climateanalytics.org/media/tropical_cyclones_impacts_cc_adaptation.pdf

United, N. (2018). World Urbanization Prospects. *Demographic Research*. https://population.un.org/wup/publications/Files/WUP2018-Report.pdf

Walsh, K. J. E., Camargo, S. J., Knutson, T. R., Kossin, J., Lee, T.-C., Murakami, H., & Patricola, C. (2019). Tropical cyclones and climate change. *Tropical Cyclone Research and Review, 8*(4), 240–250. https://doi.org/10.1016/j.tcrr.2020.01.004

Walsh, K. J. E., Mcbride, J. L., Klotzbach, P. J., Balachandran, S., Camargo, S. J., Holland, G., Knutson, T. R., Kossin, J. P., Lee, T., & cheung, Sobel, A., & Sugi, M. (2016). Tropical cyclones and climate change. *Wiley Interdisciplinary Reviews: Climate Change, 7*(1), 65–89. https://doi.org/10.1002/wcc.371

Wehner, M. (2021). Simulated changes in tropical cyclone size, accumulated cyclone energy and power dissipation index in a warmer climate. *Oceans, 2*(4), 688–699. https://doi.org/10.3390/oceans2040039

Chapter 4
Climate Change and Weather Extremes in Tourist Towns: Challenges for Livelihood in Himachal Pradesh, India

Vishwa B. S. Chandel and Manu Raj Sharma

Introduction

Humans travel to experience and seek a better understanding of cultures, people and geographical locales (Kuus, 2016) fundamentally different to their usual settings. The purpose is not only to foster an insight into new areas but also to benefit in material and non-material sense. It is for this reason travelling was a vital pursuit in the ancient times (Kellerman, 2014). Gradually, it assumed a more structured shape thus leading to the emergence of tourism industry at a global scale. As amongst the largest contributors to global economy, travel and tourism industry contributes 3.6% to global gross domestic product (GDP) directly and an additional 10.4%indirectly (WTTC Methodology Report, 2021). The industry employs more than a quarter billion population worldwide and accounts for more than 11% global consumer spending (Kyrylov et al., 2020).

In globalized world, tourism offers a wide range of opportunities for development and growth. As tourism has established linkages with other major sectors of economy like agriculture, services industry, adventure sports and transportation, it also offers prospects for economic diversification (Germaine et al., 2022). In doing so, it provides a continuous flow of resources to create both direct and indirect employment. People visit different destinations for a variety of reasons; however, some locales such as the mountainous regions attract more people for a multitude of activities for which the guiding forces are enchanting landscapes, increasing connectivity and infrastructure, spiritual and centres of religious faiths, ethnic and cultural magnetism, extreme sports and adventure activities. Amongst all the

V. B. S. Chandel (✉)
Department of Geography, Panjab University, Chandigarh, India
e-mail: vishwa@pu.ac.in

M. R. Sharma
University Department of Geography, L. N. Mithila University, Darbhanga, Bihar, India

A. L. Singh et al. (eds.), *Climate Change, Vulnerabilities and Adaptation*,
https://doi.org/10.1007/978-3-031-49642-4_4

temptations that hill tourism offers, it remains largely sensitive to climatic stipulations and undercurrents.

India is a major hub for tourism activities; its geographical, cultural and climatic diversity offers a wide range of tourism opportunities. More than 1800 million domestic and ten million international tourists annually contribute to this growing industry (Menon et al., 2021). One of the key tourist destinations in India is the majestic Himalayan mountains; Himachal Pradesh is one such hill state that has seen the rise of tourism in the past four decades. Several hill towns of India established during the British rule developed into hill resorts in post-colonial independent India; lately these have emerged as centres of climate-based urban commercial tourism. Tourism in Himachal Pradesh is largely nature based, contributing to about 7% of the state's GDP. The domestic inflow of tourists was 16.8 million in addition to 3.8 lakh foreign tourists in 2019 (Government of Himachal Pradesh, 2019). The major influx of tourists comprises people from neighbouring states during summer seeking relief from scorching heat, remote location-based nature, culture and educational tourism. The winter tourism relies on snowfall and winter sports activities. This makes tourism in Himachal Pradesh highly dependent on climatic conditions and weather patterns.

Owing to pleasant weather conditions in hills during summer season and snowfall events during winter, the tourist influx in Himachal Pradesh is highest during summer, autumn and winter seasons. Climate change observations for India are raising serious concerns; the extreme events and multiple climate change drivers are maximizing the impacts of disasters (IPCC, 2021). Under the projected warming scenario, the attraction for snowfall-related winter tourism is likely to be affected. The warming of winters during the past century has been noticeable in Himachal Pradesh; the temperature rise is as high as 1 °C in the Greater Himalayan region. The magnitude of post-winter season warming is higher than the winters with a significant temperature upswing in the entire Himachal Pradesh (Sharma et al., 2018).

Globally, the frequency of heavy precipitation has increased (IPCC, 2021) since 1950s that has impacted the amount and seasonality of water availability, biodiversity, endemic species, ecotone shifts, global feedbacks and loss of human livelihood (Nijssen et al., 2001; Parmesan, 2006; Bates et al., 2008; Ma et al., 2009; IPCC, 2014). Mountainous regions experience notable variations in the accumulation of snow and the melting of ice that result in periodic excessive or insufficient water supply in many areas such as the Himalayan region (Sharma et al., 1997; Bahadur, 2003; Xu & Rana, 2005; Eriksson et al., 2009; Kulkarni et al., 2011; and Moore et al., 2013). The higher rates of glacial melting increase water discharge initially but its scarcity in the long run (Barry & Seimon, 2000; Vuille et al., 2008; Bradley et al., 2009) is likely to be even more serious. According to the IPCC reports (2007, 2014, 2021) alterations in hydrological systems will occur due to changes not only in the timing, intensity, magnitude and phase state of seasonal precipitation patterns but also in energy receipt, evaporation and transpiration (Sharma et al., 2022). Westerling et al. (2006) and van Mangtem et al. (2009) speculated that these changes will have serious implications for ecosystem sustainability.

In light of available literature and climate projections, it is obvious to contemplate marked impacts of weather extremes on tourist places in hilly areas. Moreover, the unplanned tourism-induced development in these areas is likely to trigger environmental issues in terms of sustainable tourism, declining carrying capacity and peoples' perception of safe tourism. It is in this context this chapter attempts to outline the direction of climate change by examining long-term precipitation and temperature extremes to understand the impending risk to life and livelihoods of people in select tourist towns of Himachal Pradesh.

Study Region

Himachal Pradesh is a part of the western Himalayas; the region has three distinct physiographic zones, viz. the Siwalik Hills, the Inner Himalayas (Dhauladhar and Pir-Panjal ranges) and the Great Himalayas. This state, which has an altitude ranging from 450 to 7000 meters above mean sea level, roughly forms a quadrilateral shape between latitudes 30°22′N to 33°12′N and longitudes 75°47′E to 79°04′E. Its landscape presents an assortment of hills, mountain ranges, snowy peaks and valleys. The study area is of significant climatic importance, with temperature fluctuations ranging from as high as 42 °C during summers in regions adjacent to the Punjab Plains to as low as −25 °C during winters in the cold-arid zones of the north and east. The spatial distribution of precipitation is highly diverse ranging between 350 and 3800 m and is largely received from southwest monsoons that account for 60% of total annual rainfall.

The study area is bestowed with natural bounties, pristine valleys and wilderness; places of worship of both Hindu and Buddhist faiths attract tourists. As a result, tourism provides numerous opportunities for numerous small-scale businesses and services. It has a wide range of multiplier spin-off effects that provide tangible and intangible benefits to local population depending on tourism for income. These geophysical and climatic attributes that support tourism in this hilly state often contribute negatively thus affecting local livelihoods. The area's physiographic and bioclimatic characteristics make it susceptible to climatic variability and extremes. The important tourist centres, viz. Dharamshala, Manali and Shimla examined in this study receive highly variable rainfall; the maximum rainfall is received during the monsoon season between July and September, whereas snowfall occurs during the winter months (December–February).

The temperature regime of the region is highly diverse, with an annual temperature range of over 40 °C, contributing to its climatic diversity. A cool weather with an average summer temperature around 25 °C provides soothing environment for tourists whereas snowfall attracts people during winters. It is this climatic regime that is central to tourism activities; trekking, hiking, paragliding and river rafting during summers, while skiing and other adventure sports in winter are the major attractions. As eco-tourism is gaining ground and creating a niche for economic growth, the far-flung corners of Himachal Pradesh's are gradually opening up for tourism development that essentially is driven by climate and weather stipulations. This would offer a source of income for local community in addition to the overall economic development of the state provided that the climatic regime and weather patterns remain stable.

Objectives

This research analysed the climatic regime of select towns of Himachal Pradesh that are placed on international map for tourism. By examining the trends and patterns of climatic extremes, it attempted to visualize the risk that such extremes may pose to tourism-dependent livelihood and economic activities. Moreover, the aim was to place emphasis on projecting tourism scenario and its sustainability in light of expected ecological sensitivity and vulnerability due to climatic changes.

Database and Methodology

The study is primarily based on secondary datasets on climate parameters; however, a great deal of interpretation of result is based on the researchers' prior knowledge and decade-long field experiences on local impacts of environmental change and climatic variability. The study is based on the analysis of long-term climate (rainfall and precipitation) data for three tourist towns, namely, Shimla, Dharamshala and Manali. The long-period datasets procured from the India Meteorological Department (IMD) were analysed for 12 climate extreme indices defined by WMO/CLIVAR.

Database

The access to long-term climate data for Himalayan mountains is challenging; datasets are either non-existent or have long gaps thus making task of trend analysis difficult. Instead of relying on station or gauge data, this study utilized daily gridded temperature data at a spatial resolution of 1°X1° and precipitation data at a spatial resolution of 0.25°X0.25° since 1951. The gridded datasets were analysed for six extreme temperature indices, viz. hottest day (TXx), warmest night (TNx), coldest day (TXn), coldest night (TNn), summer days (SU25) and tropical nights (TR20) and six precipitation indices, viz. consecutive dry days (CDD), consecutive wet days (CWD), very heavy precipitation days (R20), extremely wet days (R99p), max 1-day precipitation (RX1) and max 5-day precipitation (RX5). These 12 indicators were grouped into frequency indices and intensity indices (Table 4.1); the former indices measure the absolute number of days in which a specific weather extreme indicator

Table 4.1 Indicators for temperature and rainfall indices

ID		Indicator name	Definitions	Unit
Intensity indices				
1	TXx	Hottest day (max Tmax)	Monthly max. Value of daily max. Temperature	°C
2	TNx	Warmest night (max Tmin)	Monthly max. Value of daily min. Temperature	°C
3	TXn	Coldest day (Min Tmax)	Monthly min. Value of daily maximum temperature	°C
4	TNn	Coldest night (Min Tmin)	Monthly min value of daily minimum temperature	°C
5	CDD	Consecutive dry days	Max. No. of consecutive days when precipitation <1 mm	Days
6	CWD	Consecutive wet days	Max. No. of consecutive days when precipitation >=1 mm	Days
7	R20	Very heavy precipitation days	Annual count when precipitation >=20 mm	Days
Frequency indices				
8	SU25	Summer days	Annual count when TX (daily max. Temp.) is >25 °C	Days
9	TR20	Tropical nights	Annual count when TN (daily min. Temp.) is >20 °C	Days
10	R99p	Extremely wet days	Annual total precipitation from days >99th percentile	Mm
11	RX1 day	Max 1-day precipitation	Monthly maximum 1-day precipitation	Mm
12	RX5 day	Max 5-day precipitation	Monthly maximum 5-day precipitation	Mm

Source: Compiled from WMO/CLIVAR list for climate extreme indices

occurs over a given time frame (say, days per unit of time or days/months or year). Frequency indices define the behavioural changes of climate over a season and how these changes are manifested in terms of tangible length of different seasons. On the other hand, intensity indices describe the infrequency of events based on their magnitude, such as absolute temperature or precipitation values. It represents the dynamics of energy exchange within land–atmosphere system and defines threshold of extreme weather events.

Methodology

The daily gridded data utilized in this study is based on station-by-station interpolation. To generate gridded data, researchers have applied many objective analysis (OA) techniques (Barnes, 1973; Cressman, 1959; Gandin, 1965; Shepard 1968) on climatic gauge data to interpolate unevenly distributed data for conversion into a regular grid data. The modified Shepard's method is one of the best objective analysis techniques (New et al., 2000; Kiktev et al., 2003; Caesar et al., 2006) which IMD uses for preparation of gauge-based gridded datasets especially in the data sparse regions. This method is independent of data inhomogeneity and thus can be directly used for analysis. However, temperature and precipitation indices are sensitive to variations in location, changes in equipment, data collection and observation techniques. In order to eliminate the effects of these influences, this study employed the quality control (QC) module from the 'RClimDex 1.1' software developed by Haylock et al. (2006) to identify and eliminate errors and discrepancies in the data. In addition, presence of any outliers in the dataset was identified using the method suggested by Zhang et al. (2005). The mean values of daily temperature and precipitation variables falling outside the range of four standard deviations were treated as outliers. The least-square linear fit was used for trend analysis as it is simple to understand and estimate the uncertainty in fitted trends. Statistically significant trends were identified at a 95% confidence level. This study takes a different methodological approach from the majority of previous empirical studies by utilizing a rigorous climate analysis tool called RClimDex to examine climate extremes.

Results and Discussion

The precipitation and temperature conditions are variable; inter-annual fluctuations are common due to continuous energy exchange between land–ocean surface and atmosphere (Makarieva et al., 2021). However, a long-term directional distribution provides insight into trends in the changing nature of climate patterns. The analysis of precipitation and temperature conditions for over a century shows notable changes in climate patterns in the study area. The tourist towns selected for this study are situated in the Middle Himalayan region that receives plenty of rainfall during the monsoons and snowfall in the winter months. The annual rainfall in Dharamshala (above 2900 mm) is the highest amongst all stations due to strong orographic influence. The amount remains up to 1500 mm for other two stations. The rainfall is well spread in all stations throughout the year but a relatively higher variability exists at Manali. The trend analysis shows a decline in annual rainfall at Dharamshala (2.5 mm/year) and Shimla (1.5 mm/year) while Manali (1 mm/year) has recorded an increase since the early twentieth century (Chandel & Brar, 2013; Sharma et al., 2017). The decrease in rainy days at Manali and Shimla and the increase in Dharamshala is statistically non-significant. Similarly, rainfall intensity at Manali and Shimla has increased but declined slightly in Dharmshala. Interestingly, except for Manali, the other two stations show decline in monsoon rainfall but all three have experienced higher rainfall during summer months. Similarly, the winter precipitation has gone down except for Shimla. All these indicate a change in the patterns of precipitation in these towns. An increase in summer precipitation and its intensity may have negative impact on tourist flux.

The change in temperature regime is also evident; trends suggest an overall warming in the study area (Sharma, 2017a, b; Sharma et al., 2018) as evident from statistically significant increase (0.5 °C) in mean annual temperature. Similarly, the maximum mean and minimum mean annual temperature conditions show warming in all three stations. The warming of winter amounting to 0.8–1.0 °C while post-winter warming exceeding 1 °C is statistically significant. The trend is similar for summer and post-monsoon season; the change is however non-significant. On the contrary, the trend is statistically significant but declining for the monsoon season (Sharma et al., 2017). The overall precipitation and temperature trends reveal noteworthy changes; however, the impacts of such changes on tourism may be better visualized if patterns of extreme rainfall and temperature indices are taken into account. The following sections describe the key highlights of such climatic extremes.

Patterns of Precipitation Extremes

Precipitation extremes indicate the frequency of intense rainfall (Forestieri et al., 2017; Gebrechorkos et al., 2018) that may result in severe flood, cloudburst and landslide events leading to severe economic and societal implications (Trenberth & Hoar, 1997; Shrestha et al., 2000; Izrael & Anokhin, 2001; Mirza, 2002; Min et al., 2003; Gruza & Rankova, 2004). These may also include events related to prolonged deficiency of precipitation leading to drought-like conditions. This study mainly focuses on indices representing the wetter side of precipitation except for one indicator, i.e., Consecutive dry days (CDD) which indicate the number of consecutive days of daily precipitation <1 mm. It indicates the occurrence and length of dry spells that help to effectively measure probability of drought conditions. All three tourist stations show decline in dry days; however, this trend is statistically non-significant (Table 4.2 and Fig. 4.1a). The annual trends are fluctuating, thus indicating a high degree of variability in this dryness index. There is an overall decrease in CDD but high annual values imply presence of prolonged dry conditions. Such a scenario may translate into water shortage particularly during summers and other non-monsoon months.

The consecutive wet days (CWD) indicate continuous rainy spell (rainfall of more than 1 mm per day) during a given time period. Manali is the only town experiencing no significant increase, whereas Shimla and Dharamshala have witnessed decline in CWD with statistically significant decrease in the latter (Table 4.2 and Fig. 4.1b). For adequate water availability throughout the year and groundwater recharge, it is imperative to have longer wet periods with well-spread rainfall over consecutive days. Manali owing to its topography attracts monsoon winds in the valley and witnesses a longer wet period. But conditions are different and unbefitting for Shimla and Dharamshala where the declining rate of CWD threatens the water availability. Since there is no substantial change in total annual precipitation, reduction in wet days implies more short-duration intense rainfall.

Table 4.2 Trend analysis of extreme precipitation indices

Station	Indices	Equation	Significance	*p*-value	Trend
Shimla	CDD	$y = -0.0272x + 60.37$	$R^2 = 0.0013$	0.775	Non-significant decrease
	CWD	$y = -0.009x + 16.166$	$R^2 = 0.0015$	0.717	Non-significant decrease
	R20	$y = -0.0157x + 17.139$	$R^2 = 0.0078$	0.348	Non-significant decrease
	R99p	$y = 0.3797x + 63.798$	$R^2 = 0.0091$	0.312	Non-significant increase
	RX1	$y = 0.1089x + 63.137$	$R^2 = 0.0096$	0.297	Non-significant increase
	RX5	$y = 0.0453x + 147.82$	$R^2 = 0.0005$	0.812	Non-significant increase
Manali	CDD	$y = -0.0283x + 47.083$	$R^2 = 0.0025$	0.598	Non-significant decrease
	CWD	$y = 0.0177x + 15.877$	$R^2 = 0.0033$	0.540	Non-significant increase
	R20	$y = 0.0034x + 8.5913$	$R^2 = 0.0003$	0.851	Non-significant increase
	R99p	$y = -0.0867x + 87.349$	$R^2 = 0.0004$	0.832	Non-significant decrease
	RX1	$y = -0.0047x + 54.019$	$R^2 = 2E\text{-}05$	0.963	Non-significant decrease
	RX5	$y = -0.0272x + 118.36$	$R^2 = 0.0001$	0.902	Non-significant decrease
Dharamshala	CDD	$y = -0.0048x + 47.043$	$R^2 = 6E\text{-}05$	0.933	Non-significant decrease
	CWD	$y = -0.0633x + 17.873$	$R^2 = 0.0749$	0.003	**Significant decrease**
	R20	$y = 0.0429x + 7.8769$	$R^2 = 0.0594$	0.009	**Significant increase**
	R99p	$y = 0.8157x + 36.922$	$R^2 = 0.0367$	0.040	**Significant increase**
	RX1	$y = 0.2115x + 45.55$	$R^2 = 0.0404$	0.031	**Significant increase**
	RX5	$y = 0.3246x + 105.81$	$R^2 = 0.0223$	0.111	Non-significant increase

Thus, increased surface runoff results in less recharge of groundwater. Under such a scenario, the heavy tourist inflow in these destinations especially during summer may aggravate the problem of water scarcity.

The heavy precipitation poses a challenge for mountainous areas by triggering slope failure and flash floods. Very heavy precipitation days (R20) indices represent the annual count of days when precipitation exceeds 20 mm per day. Its patterns and trends are similar to other precipitation threshold indices like R10 and R25 but the magnitude of trend is less prominent. An increase in very heavy precipitation days at Manali and Dharamshala (statistically significant) is recorded, whereas non-significant but weak decreasing trend is observed for Shimla. Other indices of

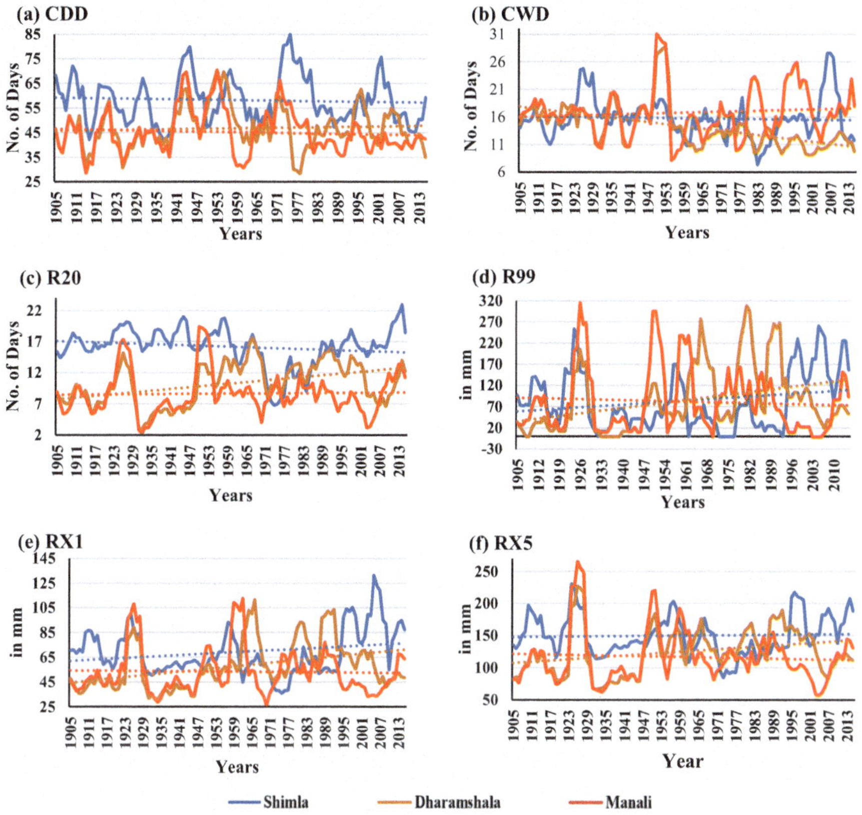

Fig. 4.1 Trends in precipitation indices

extreme precipitation, i.e., R99p, represent total annual precipitation from days greater than 99th percentile. Shimla and Dharamshala have witnessed increasing trend indicating receipt of more intense rainfall in these stations (Fig. 4.1d). A weak declining trend is however recorded for Manali. Most of such intense rainfall events have been recorded during monsoon and winter which suggests increasing risk of catastrophic events. In fact, the southern slopes of Dhauladhar range in Himachal Pradesh covering parts of Kangra, Dharamshala, Palampur and Jogindernagar area has maximum annual precipitation exceeding 2000 mm/year. Hence, the increasing trends of R20 and R99p (Fig. 4.1c, d) in these two areas might translate more intense rainfall activities leading to disasters. The two indices, viz. Rx1 and Rx5 representing the highest rainfall in a single day and five consecutive days, respectively show statistically significant increase over Dharamshala and non-significant increase over Shimla. However, a weak declining trend is observed for Manali. However, areas around Kullu-Mandi of Himachal Pradesh have become a hot spot for occasional high-intensity precipitation. Under such scenario, it is likely that vulnerability to cloudbursts and flash floods would be enhanced in these central Himalayan tourist circuits.

Patterns of Temperature Extremes

Hill tourism is fundamentally governed by conducive cool temperature conditions as these areas offer respite for tourists from scorching summer heat in the plain areas. The temperature conditions have significant impact on environmental conditions at different time scales. Long-term patterns, trends and variability indicate how thermal regime changes over the years (Table 4.3 and Fig. 4.2) and how such changes are likely to shape the future. Similarly, the nature and likelihood of temperature extremes are equally important. The hottest day (TXx) index represents the highest value of the daily maximum temperature for each month, thereby identifying the hottest day of the month and establishing the upper limit of the monthly temperature range. It depicts the daily march of temperature cycle, as well as the net amount of incoming solar radiation; hence, controls pan surface evaporation and relative humidity. An increase in TXx at Manali and Dharamshala (Fig. 4.2a) indicates warming of daytime temperatures. Such a situation is not welcoming for tourism activities.

On the contrary, the warmest night (TNx) index, i.e., monthly maximum value of the daily minimum temperature representing the warmest night temperature shows a decline across all tourist stations (Fig. 4.2b). Such trend indicates a dip in night-time temperatures; however, this change is statistically non-significant. TXn (coldest day) represents monthly minimum value of the daily maximum temperature (Min Tmax). It denotes the coldest daytime temperature during a month and thus establishes the lower limit for daytime temperatures during peak winters. All three stations show decrease (non-significant) in coldest day temperatures which shows a shift towards relatively colder daytime winter temperature conditions (Table 4.3 and Fig. 4.2c). On the contrary, the coldest night (TNn) index which represents the monthly minimum value of the daily minimum temperature, i.e., the coldest-night temperature shows increasing trends. The average coldest night temperature in Himachal Pradesh is higher (above 2 °C) in areas around Shimla that gradually decreases in the north towards Manali and Dharamshala. These stations have witnessed an overall increase in coldest-night temperature since 1950s but change is highly fluctuating as indicated by the trend line (Fig. 4.2d). It means that the coldest winter nights in these areas have become relatively warmer. However, the trends for last two decades show a reversal.

Summer days (SU25) indicates number of days in a year when daily maximum temperature remains above 25 °C and therefore defines the duration of warm spell. An increase in summer days in all stations suggests warming; however, the trend is not statistically significant (Table 4.3 and Fig. 4.2e). A distinct difference in directional change is observed; Dharamshala and Manali have recorded the highest rates of daytime warming compared to Shimla. Similarly, tropical night representing the

Table 4.3 Trend analysis of extreme temperature indices

Station	Indices	Equation	Significance	*p*-value	Trend
Shimla	TXx	$y = -0.0016x + 41.803$	$R^2 = 0.0005$	0.862	Non-significant decrease
	TNx	$y = -0.0106x + 27.849$	$R^2 = 0.0517$	0.073	Non-significant decrease
	TXn	$y = -0.009x + 13.891$	$R^2 = 0.0179$	0.296	Non-significant decrease
	TNn	$y = 0.0055x + 1.9624$	$R^2 = 0.0112$	0.41	Non-significant increase
	SU25	$y = 0.0522x + 256.38$	$R^2 = 0.0085$	0.472	Non-significant increase
	TR20	$y = 0.076x + 143.28$	$R^2 = 0.0171$	0.307	Non-significant increase
Manali	TXx	$y = 0.0016x + 39.382$	$R^2 = 0.0005$	0.855	Non-significant increase
	TNx	$y = -0.0091x + 25.648$	$R^2 = 0.0344$	0.146	Non-significant decrease
	TXn	$y = -0.0025x + 11.641$	$R^2 = 0.0013$	0.779	Non-significant decrease
	TNn	$y = 0.0076x + 0.7066$	$R^2 = 0.0204$	0.264	Non-significant increase
	SU25	$y = 0.0799x + 229.62$	$R^2 = 0.0176$	0.3	Non-significant increase
	TR20	$y = 0.0921x + 105.27$	$R^2 = 0.016$	0.323	Non-significant increase
Dharamshala	TXx	$y = 0.0053x + 43.056$	$R^2 = 0.0059$	0.55	Non-significant increase
	TNx	$y = -0.0114x + 28.719$	$R^2 = 0.0578$	0.199	Non-significant decrease
	TXn	$y = -0.01x + 14.115$	$R^2 = 0.0206$	0.262	Non-significant decrease
	TNn	$y = 0.0083x + 1.7557$	$R^2 = 0.0269$	0.058	Non-significant increase
	SU25	$y = 0.0805x + 258.92$	$R^2 = 0.0168$	0.311	Non-significant increase
	TR20	$y = 0.1072x + 146.62$	$R^2 = 0.0345$	0.145	Non-significant increase

total number of days with daily minimum temperature exceeding 20 °C (indicating the duration of cool nights) shows warming due to shortening of cool-night conditions in all three tourist stations. Overall, the increase in both day and night temperatures during the peak summer season provides evidence of warming in these popular tourist destinations.

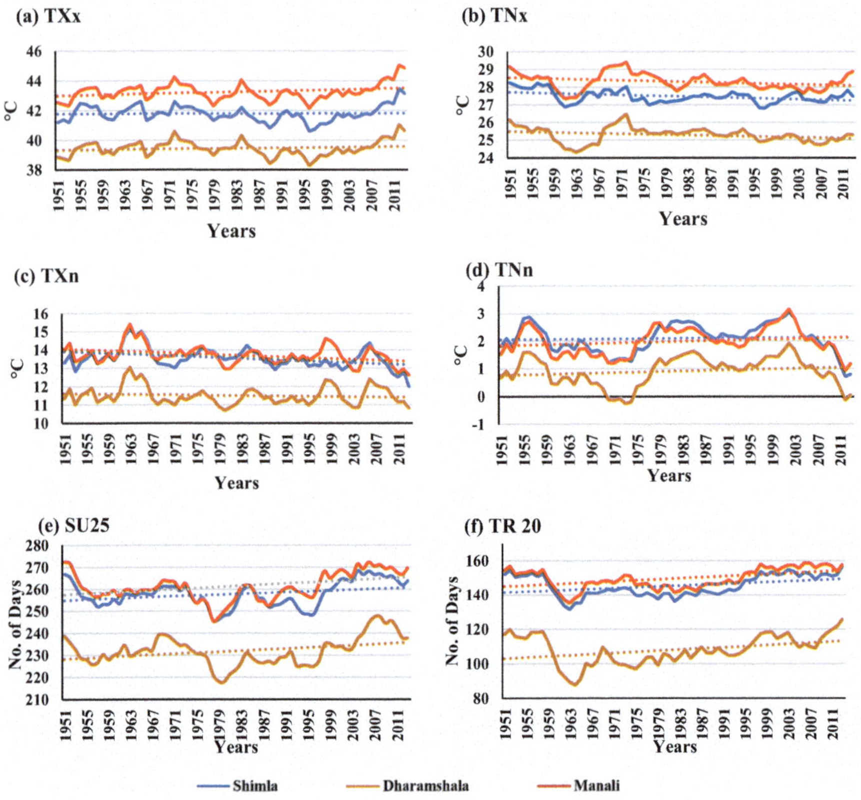

Fig. 4.2 Trends in temperature indices

Climate Extremes and Implications for Tourism and Livelihood

Humans have always been adapting to changing climates; however, as scientific knowledge on this issue grows, it is apparent that massive setbacks to various economic activities are likely to happen. Such changes have serious repercussions on areas where people rely on climate for livelihood. A bidirectional relationship exists between climate change and tourism; both affect and get affected by each other. Climate is a key driver and an important attribute that facilitates tourism and thus offers livelihood to a large population in the tourist town selected for this study. This mountainous landscape offers enchanting cool mountain climate, aesthetically appealing snowclad and green landscapes, adventure sports activities, pilgrimage, spiritual experiences, and historical places. Hence, the implications of climate extremes on tourism are manifold and might manifest through different ways to impact the natural processes and functions as well as human endeavours for development.

The observed increase in temperature, particularly the extreme hot and cold conditions, might affect the timing of peak tourist season and inflow of tourists. The warmer temperature will affect tourism via altered seasonality, incidences of heat-related stress, uneconomic tourist ventures, lack of winter-sport destinations. As a result, these areas are likely to face increased pressure on existing resources. Climate controls hydrological cycle; a substantial proportion of insolation drives hydrological cycle, and any change in these dynamics alters the amount and circulation of water. The warming of winter and post-winter seasons might lead to a reduction in the extent and duration of snowfall due to less winter precipitation in the middle Himalayas (Bhutiyani et al., 2010) and higher ablation rates. The significant temperature increase in post-winter and early summer seasons is likely to trigger the problem of water excess in early spring season due to rapid melting of snow, and relatively lower discharge during summers.

As observed in the analysis, it is also possible that a slight decline in seasonal rainfall may cause less river discharge while frequent episodes of short-duration extreme rainfall may cause flash floods. Similarly, the fast snow-melt and high runoff due to winter and post-winter warming might exhaust the water sources before the arrival of monsoon thus creating water scarcity in summers when the tourist flux is at peak. Consequently, the projected warming scenario is likely to cause a shift in the preferred time and destinations and hence will cause pressure on livelihood options. Moreover, the inter-annual and inter-seasonal flux of river discharges might become highly variable. More importantly, these changes will directly impact adventure sports and related tourism activities. As a result, the growing urban population and tourists in these hilly areas will be the first to face the brunt of such changes. This scenario is likely to be more serious for areas in and around tourist centres. Since development in these areas is dependent on water, such a stress may aggravate local conflicts related to water sharing and access. As a result of extremely variable rainfall coupled with growing urbanization and tourism, it is likely to cause hydrological stress and water shortage, the unavailability of which would force people in these tourist destinations to alter their long-established livelihood options.

Another aspect of economy that might face serious setback is hydropower generation. The lack of winter snow and higher summer ablation rates will lead to insufficient water for power generation throughout the year. The episodes of high-intensity rainfall-induced flash floods causing high turbidity in river water may halt power generation. Consequently, lower production and higher demand will exert pressure on energy systems resulting in higher maintenance costs of power grids and transmission. It is but obvious that tourism industry will be severely affected that might cause livelihood loss. Thus, temperature and rainfall unpredictability particularly during the winter and summer seasons may have negative impact on urban energy systems. More importantly, the traditional irrigation practices might be rendered ineffective thus increasing the production costs leading to economic vulnerability.

Climate extremes and disasters are inseparable. Increase in rainfall intensity and extreme precipitation events in the temperate zones may impact tourism due to enhanced risk of cloudbursts and floods during summer and monsoon seasons. The districts of Kullu, Kangra and Shimla are already experiencing higher incidences of extreme rainfall-driven disasters in these seasons; Dharamshala and Kullu-Manali area have been identified as cloudburst hot spots (Chandel & Brar, 2010) whereas Kullu and Kangra are most flood-prone areas (Chandel et al., 2014). Such climate extreme may result in fewer tourists that would cause loss of livelihood and increase in economic vulnerability of population dependent on tourism. Such scenario is already showing signs in many parts of Himachal Pradesh where extreme rainfall in pre-monsoon and monsoon seasons has been affecting tourism.

Conclusion

Climate change implications on tourism and livelihood are undeniable. This work explored trends and potential changes in climatic extremes in select tourist stations of Himachal Pradesh. The higher rainfall regions of northern and southern Himachal Pradesh have become less rainy and seasonal intensity appears to have moderated over the time. However, the episodes of extreme events have increased; the daily maximum and 5-day maximum rainfall patterns show an increasing trend. Therefore, it would not be premature to envisage that precipitation extremes in the study area are likely to be more dramatic and unpredictable in times to come especially in the central parts of Himachal Pradesh. The selected tourist stations under study shows significant increase in temperature; the warming is evident with an increase in both daytime and night-time temperature conditions. The rise in the number of summer days and tropical nights implies an increase in the temperature during the summer season. Furthermore, the noticeable increase in the frequency of summer days in hilly areas indicates a relatively higher rate of warming during the summer season. Moreover, due to increasing overall temperature, winters and post-winter seasons might witness less snowfall and early melt leading to higher water discharge in early spring season. It is also plausible that dry season may have insufficient water in rivers leading to hydrological stress and thus the water scarcity in the long run. The entire discussion on climate change and variability is concluded with the fact that there are clear markers of significant escalation in the intensity of precipitation and temperature extremes. Moreover, there are indications of significant shift in season precipitation patterns, rise of summer season rainfall and intensity, and winter snowfall in the study area.

In light of above observation, it is expected that the projected warming of climate will affect the tourism through loss of income, change in tourist destinations, scarcity of tourism resources and increasing risk of hazards. Increase in extreme precipitation events especially in the temperate zones of Himachal Pradesh may impact tourism due to more frequent disasters and enhanced risk during tourist seasons. The markers of such problem have already been seen in many parts of Himachal Pradesh where effects of extreme events like droughts, flash floods and snow-free winters have wreaked havoc on tourism activities in the recent past. For instance, the year 2018 recorded massive impact on horticulture wherein lack of rainfall impacted fruit production; the pre-monsoon shower caused massive economic loss to farmers and tourism industry while high-intense short-duration rainfall in the months of August and September led to massive scale damage to various sectors of economy due to floods.

It is vital to acknowledge the need for translating scientific knowledge of climate change and weather extremes for appropriate solutions and adaptive measures to counter its implications for the future of tourism. This research outlined the climate extremes and its potential impact on water resources, hydrological regimes, and ecosystem well-being which collectively affect tourism and many other aspects of human development. This invokes an idea of major paradigm shift in tourism development based on clean and green energy. More importantly, this study calls upon the preparation of climate action plans at regional and administrative levels for aiding tourism-based economy in the long run to ensure sustainability of livelihoods in the study area.

References

Bahadur, J. (2003). *Indian Himalayas – An integrated view*. Vigyan Prasar.

Barnes, S. L. (1973). *Mesoscale objective map analysis using weighted timeseries observations*. NOAA Tech. Memo, ERL NSSL-62, National Severe Storms Laboratory, Norman: Oklahoma. https://repository.library.noaa.gov/view/noaa/17647

Barry, R., & Seimon, A. (2000). Research for mountain area development: Climatic fluctuations in the mountains of the Americas and their significance. *Ambio, 29*(7), 364.

Bates, B. C., Kundzewicz, Z. W., Wu, S., & Palutikof, J. P. (2008). Climate change and water. In *Technical paper of the intergovernmental panel on climate change*. IPCC Secretariat.

Bhutiyani, M. R., Kale, V. S., & Pawar, N. J. (2010). Climate change and the precipitation variations in the northwestern Himalaya:1866–2006. *International Journal of Climatology, 30*(4), 535–548. https://doi.org/10.1002/joc.192

Bradley, S. L., Milne, G. A., Teferle, F. N., Bingley, R. M., & Orliac, E. J. (2009). Glacial isostatic adjustment of the British Isles: New constraints from GPS measurements of crustal motion. *Geophysical Journal International, 178*(1), 14–22. https://doi.org/10.1111/j.1365-246X.2008.04033.x

Caesar, J., Alexandar, L., & Russell, V. (2006). Large-scale changes in observed daily maximum and minimum temperatures: Creation and analysis of a new gridded data set. *Journal of Geophysical Research, 111*, DO5101. https://doi.org/10.1029/2005JD006280

Chandel, V. B. S., & Brar, K. K. (2013). Drought in Himachal Pradesh, India: A historical-geographical perspective, 1901–2009. *Transactions of the Institute of Indian Geographers, 35*(2), 259–273.

Chandel, V. B. S., Kahlon, S., & Brar, K. K. (2014). Flood disaster in mountain environment: A study of Himachal Pradesh, India. In D. D. Sharma & B. R. Thakur (Eds.), *Managing our resources: Perspectives and planning* (pp. 9–19). Bharti Publications. ISBN: 978-93-81212-85-1.

Chandel, V. B. S., & Brar, K. K. (2010). Climatic extremes and changing climate in Western Himalayas: A study of cloudburst incidences in Himachal Pradesh. *Punjab Geographer, 6*, 29–40.

Cressman, G. P. (1959). An operational objective analysis system. *Monthly Weather Review, 87*(10), 367–374.

Eriksson, M., Xu, J., Shrestha, A. B., Vaidya, R. A., Nepal, S., & Sandstrom, K. (2009). *The changing Himalayas – Impact of climate change on water resources and livelihoods in the greater Himalayas*. ICIMOD.

Forestieri, A., Arnone, E., Blenkinsop, S., Candela, A., Fowler, H., & Noto, L. V. (2017). The impact of climate change on extreme precipitation in Sicily, Italy. *Hydrological Processes, 32*, 332–348.

Gandin, L. S. (1965). *Objective analysis of meteorological fields* (Translated from Russian by R. Hardin.) (p. 242). Israel Program for Scientific Translations.

Gebrechorkos, S. H., Hülsmann, S., & Bernhofer, C. (2018). Changes in temperature and precipitation extremes in East Africa. *Geophysical Research Abstracts, 20*, EGU2018-12310.

Germaine, M.-A., Ducourtieux, O., Stone, J. O., & Cournet, X. (2022). Landscapes as drivers of ecotourism development: A case study in Northern Laos. *Canadian Journal of Development Studies/Revue Canadienne d'études Du Développement, 0*(0), 1–23. https://doi.org/10.1080/02255189.2022.2029734

Government of Himachal Pradesh. (2019). *The Himachal Pradesh tourism policy*. Department of Tourism & Civil Aviation, HP Government (n.d.).

Gruza, G., & Rankova, E. (2004). Detection of changes in climate state, climate variability and climate extremity. In Y. Izrael, G. Gruza, S. Semenov, & I. Nazarov (Eds.), *Proceedings of the world climate change conference* (pp. 90–93). Institute of Global Climate and Ecology.

Haylock, M. R., Peterson, T. C., Alves, L. M., Ambrizzi, T., Anunciação, Y. M. T., Baez, J., Barros, V. R., Berlato, M. A., Bidegain, M., Coronel, G., Corradi, V., Garcia, V. J., Grimm, A. M., Karoly, D., Marengo, J. A., Marino, M. B., Moncunill, D. F., Nechet, D., Quintana, J., Rebello, E., Rusticucci, M., Santos, J. L., Trebejo, I., & Vincent, L. A. (2006). Trends in total and extreme south American rainfall in 1960–2000 and links with sea surface temperature. *Journal of Climate, 19*(8), 1490–1512.

Intergovernmental Panel on Climate Change (IPCC). (2007). *Fourth assessment report. Summary for policymakers*. Cambridge University Press.

Intergovernmental Panel on Climate Change (IPCC). (2014). *Climate change – Impacts, adaptations, and vulnerability. Summary for policy makers*. Cambridge University Press.

Intergovernmental Panel on Climate Change (IPCC). (2021). Climate Change 2021: The physical science basis. In V. Masson-Delmotte, P. Zhai, A. Pirani, S. L. Connors, C. Péan, S. Berger, N. Caud, Y. Chen, L. Goldfarb, M. I. Gomis, M. Huang, K. Leitzell, E. Lonnoy, J. B. R. Matthews, T. K. Maycock, T. Waterfield, O. Yelekçi, R. Yu, & B. Zhou (Eds.), *Contribution of Working Group I to the sixth assessment report of the Intergovernmental Panel on Climate Change*. Cambridge University Press. In Press.

Izrael, Y. A., & Anokhin, Y. A. (2001). Climate change impacts on Russia. *Integrated Environmental Monitoring*, 112–117.

Kellerman, A. (2014). The satisfaction of human needs in physical and virtual spaces. *The Professional Geographer, 66*(4), 538–546. https://doi.org/10.1080/00330124.2013.848760

Kiktev, D., Sexton, D. M. H., Alexander, L., & Folland, C. K. (2003). Comparison of modeled and observed trends in indices of daily climate extremes. *Journal of Climate, 16*, 3560–3571.

Kulkarni, A. V., Rathore, B. P., Singh, S. K., & Bahuguna, I. M. (2011). Understanding changes in the Himalayan cryosphere using remote sensing techniques. *International Journal of Remote Sensing, 32*, 601–615.

Kuus, M. (2016). To understand the place: Geographical knowledge and diplomatic practice. *The Professional Geographer, 68*(4), 546–553. https://doi.org/10.1080/00330124.2015.1099450

Kyrylov, Y., Hranovska, V., Boiko, V., Kwilinski, A., & Boiko, L. (2020). International tourism development in the context of increasing globalization risks: On the example of Ukraine's integration into the global tourism industry. *Journal of Risk and Financial Management, 13*, 303. https://doi.org/10.3390/jrfm13120303

Ma, X., Xu, J. C., Luo, Y., Aggarwal, S. P., & Li, J. T. (2009). Response of hydrological processes to land cover and climate change in Kejie Watershed, Southwest China. *Hydrological Processes, 23*(8), 1179–1191. https://doi.org/10.1002/hyp.7233

Makarieva, A. M., Nefiodov, A. v., Nobre, A. D., Sheil, D., Nobre, P., Pokorný, J., Hesslerová, P., & Li, B.-L. (2021). Vegetation impact on atmospheric moisture transport under increasing land-ocean temperature contrasts. http://arxiv.org/abs/2112.12880

Menon, S., Bhatt, S., & Sharma, S. (2021). A study on envisioning Indian tourism – Through cultural tourism and sustainable digitalization. *Cogent Social Sciences, 7*(1). https://doi.org/10.1080/23311886.2021.1903149

Min, S. K., Kwon, W. T., Park, E. H., & Choi, Y. (2003). Spatial and temporal comparisons of droughts over Korea with East Asia. *International Journal of Climatology, 23*, 223–233.

Mirza, M. Q. (2002). Global warming and the changes in the probability of occurrence of floods in Bangladesh and implications. *Global Environment Change, 12*, 127–138.

Moore, W., Meyer, W. M., Eble, J. A., Franklin, K., Wiens, J. F., & Brusca, R. C. (2013). Introduction to the Arizona Sky Island Arthropod Project (ASAP): Systematics, biogeography, ecology and population genetics of arthropods of the Madrean Sky Islands. In *Biodiversity and management of the Madrean Archipelago III* (Vol. 67, pp. 140–164). U.S. Department of Agriculture RMRS-P.

New, M. G., Hulme, M., & Jones, P. D. (2000). Representing twentieth-century space-time climate variability part II: Development of 1901–96 monthly grids of terrestrial surface climate. *Journal of Climate, 13*, 2217–2238.

Nijssen, B., O'Donnell, G. M., Hamlet, A. F., & Lettenmaier, D. P. (2001). Hydrologic sensitivity of global rivers to climate change. *Climatic Change, 5*, 143–175.

Parmesan, C. (2006). Ecological and evolutionary response to recent climate change. *Annual Review of Ecology, Evolution, and Systematics, 37*, 637–669.

Sharma, M. R., Chandel, V. B. S., & Brar, K. K. (2022). Markers of climate change: Analysing extreme temperature indices over the Himalayan Mountains and adjoining Punjab Plains. In U. Schickhoff, R. Singh, & S. Mal (Eds.), *Mountain landscapes in transition. Sustainable development goals series*. Springer. https://doi.org/10.1007/978-3-030-70238-0_2

Sharma, M. R. (2017a). Long term analysis of temperature extremes over Punjab and Himachal Pradesh, India (1951–2013). *Online International Interdisciplinary Research Journal, 07*(5), 29–40. (ISSN 2249-9598).

Sharma, M. R. (2017b). Trends and variability of rainfall in Punjab: A statistical analysis, 1981-2010. *Research Guru, 11*(3), 73–81. (ISSN 2349-266X).

Sharma, M. R., Brar, K. K., & Chandel, V. B. S. (2017). Long-term analysis of trend and variability in annual rainfall in Punjab and Himachal Pradesh. *The Deccan Geographer, 55*(1 & 2), 1–16.

Sharma, M., Chandel, V. B. S., & Brar, K. K. (2018). Punjab and Himachal Pradesh: Temperature variability and trends since 1900 AD. *Punjab Geographer, 14*, 5–32.

Sharma, S., Jackson, D. A., & Minns, C. K. (1997). Will northern fish populations be in hot water because of climate change? *Global Change Biology, 13*, 2052–2064.

Shepard, D. (1968). A two-dimensional interpolation function for irregularly-spaced data. In *Proceedings of the 1968 23rd ACM national conference* (pp. 517–524). ACM.

Shrestha, A. B., Wake, C. P., Dibb, J. E., & Mayewski, P. A. (2000). Precipitation fluctuations in the Nepal Himalaya and its vicinity and relationship with some large-scale climatology parameters. *International Journal of Climatology, 20*, 317–327.

Trenberth, K. E., & Hoar, T. J. (1997). El Niño and climate change. *Geophysical Research Letter, 24*(23), 3057–3060.

Van Mangtem, P. J., Stephenson, N. L., Bryne, J. C., Daniels, L. D., Franklin, J. F., Fule, P. Z., Harmon, M. E., Larson, A. J., Smith, J. M., Taylor, A. H., & Veblen, T. T. (2009). Widespread increase of tree mortality rates in the Western United States. *Science, 323*, 521–524.

Vuille, M., Francou, B., Wagnon, P., Juen, I., Kaser, G., Mark, B. G., & Bradley, R. S. (2008). Climate change and tropical Andean glaciers: Past, present and future. *Earth Science, 89*, 79–96.

Westerling, A. L., Hidalgo, H. G., Cayan, D. R., & Swetnam, T. W. (2006). Warming and earlier spring increases western U.S. Forest wildfire activity. *Science, 313*, 940–943.

WTTC. (2021). *Travel & tourism economic impact research methodology report 2021*. Oxford Economics.

Xu, J., & Rana, G. M. (2005). Living in the mountains. In T. Jeggle (Ed.), *Know risk* (pp. 196–199). UN Inter-agency Secretariat of the International Strategy for Disaster Reduction.

Zhang, X., Hegerl, G., Zwiers, F. W., & Kenyon, J. (2005). Avoiding inhomogeneity in percentile-based indices of temperature extremes. *Journal of Climate, 18*, 1641–1651.

Chapter 5
Evaluation of the Effect of Urban Heat Island on Local Climatic Zones by Using Remote Sensing and Statistical Analysis in Khartoum Sudan

Elhadi K. Mustafa, Hanan E. Osman, and Hazem T. Abd El-Hamid

Introduction

Due to the city's environment warming up at an unprecedented rate, urban heat islands (UHI) are an occurrence. Urban heat island (UHI) could be defined as the difference in temperature between an urban area and the nearby rural area. Uncertainty still exists regarding the contribution of various local urbanization trends to UHI intensity and persistence in regions with ongoing growth (Sun et al., 2019). It happens when the surrounding rural area is cooler than the urban area. Urbanization has altered the surface and atmosphere, resulting in a changed thermal environment that is hotter than the surrounding areas (Voogt & Oke, 2003). The purpose of local climate zones (LCZs) framework is to resolve this contradiction by epitomizing climatic differences across the sites and organizing them into distinct sectors that may be converted into accurate assessable parameters of urban environments (Bechtel et al., 2019). UHI phenomenon has been observed globally (Clinton & Gong, 2013; Zhou et al., 2014). The primary elements influencing the spatiotemporal variability of urban climate are well known and include, for instance, the thermal features of the substances used to construct surfaces and building structures, their

E. K. Mustafa
Department of Surveying Engineering, College of Engineering, Karary University, Omdurman, Sudan

H. E. Osman
Biology Department, Faculty of Science, Umm-Al-Qura University, Makkah, Saudi Arabia

Botany and Microbiology Department, Faculty of Science, Al-Azhar University, Cairo, Egypt

H. T. Abd El-Hamid (✉)
Department of Marine Pollution, National Institute of Oceanography and Fisheries (NIOF), Cairo, Egypt

A. L. Singh et al. (eds.), *Climate Change, Vulnerabilities and Adaptation*,
https://doi.org/10.1007/978-3-031-49642-4_5

layout, mass, height, and land-use patterns (Seto et al., 2015). However, inadequate quantification of these factors and their non-standard distributions have long limited research on urban climate (Stewart, 2011). Over the last ten years, the concept of local climatic zones (LCZs) has evolved to address these difficulties in urban climatology (Stewart & Oke, 2012). Stewart and Oke (2012) categorized local climate sectors as zones with uniform land cover, building components, and anthropogenic activities ranging from a few meters to a few kilometers on a flattened level. Each LCZ‘s warm climate is affected by the features of its surface, including the structure (height/density of buildings and trees), texture (albedo, warm admittance), land cover (penetrability), and metabolism (heat from transportation and space warming/cooling). Each LCZ class designates either a natural land cover sort or a built type. Moreover, each LCZ type differs from the others due to the LCZ classification systems' consideration of geometric, radioactive, thermal, surface cover, and metabolic characteristics. The LCZ categorization seeks to standardize the exchange of worldwide urban temperature monitoring data. Concerning the critical role of the LCZ and its origin in the UHL investigation database (Stewart et al., 2014), the LCZ classification scheme has been used in investigations to categorize the thermal features of towns using weather station data, mobile measurements, land use and land cover (LULC) data, remotely sensed pictures, and urban morphology data attained from various sources (Bechtel et al., 2016; Unger et al., 2014; Lelovics et al., 2014). Up to date, there is a severe shortage of research maps of LCZs in African urban areas, particularly in Sudan. Only the LCZ maps for Harare, a metropolitan city in Zimbabwe, are done to date. For the cool and hot seasons in Harare, LCZs were linked with Landsat 8 derived land surface temperature (LST). Most studies of the African countries utilized LULC classes that were not standardized to contribute to the global urban metadata call (Mushore et al., 2019). For instance, in the work of Tarawally et al. (2018), Sierra Leone's Freetown and Bo towns were recently mapped into LULC classes and coupled to dry season surface temperature patterns. Numerous investigations in South Africa have interconnected single date and temporal changes in general LULC to land surface temperature. Odindi et al. (2015) demonstrated the significant value of urban vegetation in reducing heat. The results of this research are difficult to compare between cities as, already said, the classification names were region-specific and had subjective connotations. Most of the current LCZ literature were about Europe, North America, and Asia. For instance, Wang et al. (2018) assess the local region climate in North America's dry towns of Las Vegas, Nevada, Phoenix, and Arizona. Lehnert et al. (2015) investigated the local climatic district characteristics in a metropolitan station network in Olomouc (Czech Republic), and compared the temperature features of the stations, to identify the key factors which can encourage the development of an urban heat island (UHI) in the town. Moreover, Kaloustian and Bechtel (2016) utilized local climatic zoning mapping in Beirut, a highly populated town along the Mediterranean Sea. They give insights into the city's dominant urban form and materials. The development of remote sensing technology has persuaded large-scale research on UHI for small and extensive urban communities globally (Weng et al., 2004). Using satellite-derived data, the UHI investigation is conducted by separating the land surface temperature (LST). A wide scope of sensors, such as Landsat 4, 5 Thematic Mapper (TM), 7 Enhanced Thematic Mapper Plus (ETM+), 8 Thermal Infrared Sensor (TIRS) 1 and 2), advanced spaceborne thermal emission and reflection (ASTER), moderate resolution imaging spectroradiometer (MODIS), advanced very high-resolution radiometer (AVHRR), and others are used for the extraction of LST. Accurate and speedy information acquisition concerning the magnitude of the changes is required for monitoring and forecasting future changes (Mustafa et al., 2019). The fundamental objective of this research is the LCZ changes and the associated LST modifications. The results showed that the LST and LCZ have a strong link, and the development of UHI is evaluated using the shift in the LCZ and LST. Remote sensing also fails to completely capture radiant emission from vertical surfaces, such as building walls, because sensors predominantly detect energy radiated from flat surfaces, such as roadways, rooftops, and trees. LST data must be corrected for radiometric and atmospheric effects because the environment is so densely populated. On the other hand, satellite images provide fine-scale warm data that is difficult to obtain through transect estimating campaigns or climate station systems, allowing investigation of the LST mark of LCZs. This study connected LCZ changes and future climate forecast to recognize the modification in a pattern on land surface temperature for sustainable urban expansion and planning. For the monitoring and forecasting future conditions and changes in land surface temperature, the accurate and transparent gathering of fundamental data indicating the extent of the changes is necessary. To forecast land surface temperature and gather essential data about the climate system, prediction algorithms are used. With this knowledge, it will be simpler to discover unique methods and legislation in town planning and designs that lessen the effects of UHI. Remote sensing data can be used to collect this information. In light of this, the study attempted to investigate the following issues: temporal mapping of LCZs in Khartoum town, evaluation of the impact of LCZ change on urban heat islands, simulation of LST for various seasons, and analysis of the simulated LST in various LCZs.

Materials and Methods

Description of the Study Area

The study area is conducted in Khartoum state, the capital of Sudan. The study area sited between latitudes 15.384°N to 15.854°N and longitudes 32.295° to 33.795°′E covers about 2785.62 km^2. Figure 5.1 shows the general location of the study area. It is located in the middle of Sudan and is semi-arid, with a hot weather season lasting from March to mid-June and a cold, dry winter from November to February. April is the hottest month. The highest annual rainfall is mostly recorded from July to September at about 157 mm (Tucker 1979). A flat plain at an elevation of between 380 and 400 m makes up the majority of the area. The Blue Nile, White Nile, and River Nile are the main waterways, along with a number of seasonal streams. The

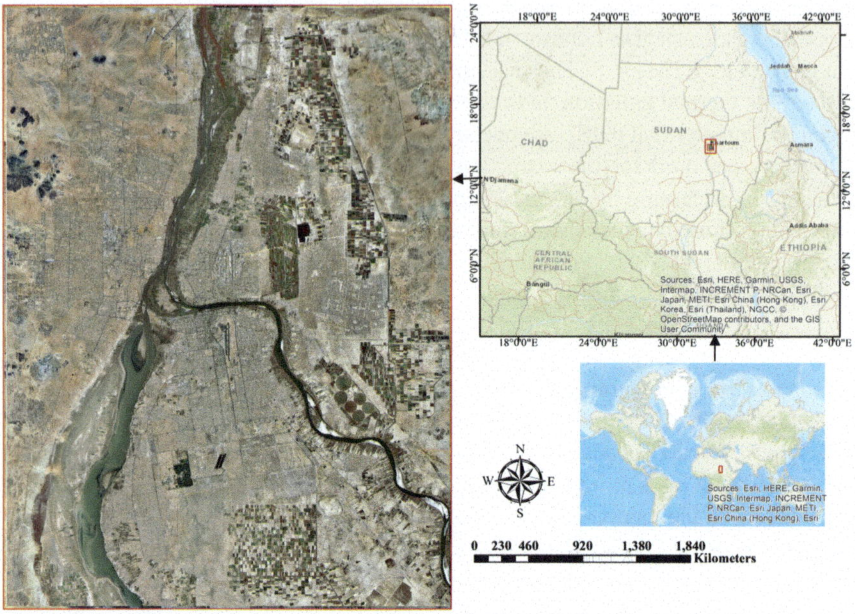

Fig. 5.1 Location of the area under study

population of the developed Khartoum state ranges from 7.6 million to 15.7 million. Generally speaking, this kind of development has detrimental repercussions and may cause climatological issues for urbanized areas, including urban heat island. As there is less vegetation in the urban area and more concrete and asphalt are used in the built environment, an excess of heat builds up there. This phenomenon consequently causes extreme temperatures in the majority of the urban areas covered by this investigation. Particularly in the summer, these extreme temperatures happen during the day, which makes the locals feel even worse. This study aims to ascertain the relationship between Khartoum City's land surface temperature and the local climate zones.

Datasets

The Landsat imagery utilized in this study was made available through the Earth Resources Observation and Science (EROS) center. EROS played a crucial role in providing high-quality images that were both free of clouds and geometrically corrected, ensuring accurate and reliable data for our investigation, which was available via the USGS Global Visualization Viewer. The images used were cloud-free Landsat images of path/row 173/49 from the hot, cool, and rainy seasons as depicted in Table 5.1. For mapping the LCZs, images from the hot season (April 1984, April 2000, and April 2016 Operational Land Imager (OLI)), the cold season (January 1986, January 2001, and January 2016 [OLI]), and the rainy season

Table 5.1 Cloud-free Landsat 8 imageries used for LCZ and LST retrieval in this study

No.	Sensor	Date	Season	Path/Row	Bands	Source
1	Landsat 4/5 TM	20.4.2000	Hot	173/049	3, 5, 7, 6	usgs.gov
2	Landsat 4/5 TM	08.4.1984	Hot	173/049	3, 5, 7, 6	usgs.gov
3	Landsat 4/5 TM	08.1.1986	Cool	173/049	3, 5, 7, 6	usgs.gov
4	Landsat 4/5 TM	17.1.2001	Cool	173/049	3, 5, 7, 6	usgs.gov
5	Landsat 4/5 TM	12.9.1986	Rainy	173/049	3, 5, 7, 6	usgs.gov
6	Landsat 4/5 TM	30.9.2001	Rainy	173/049	3, 5, 7, 6	usgs.gov
7	Landsat 8 OLI_TIR	16.4.2016	Hot	173/049	4, 6, 7, 10	usgs.gov
8	Landsat 8 OLI_TIR	18.4.2016	Hot	173/049	4, 6, 7, 10	usgs.gov
9	Landsat 8OLI_TIR	11.1.2016	Cool	173/049	4, 6, 7, 10	usgs.gov
10	Landsat 8 OLI_TIR	27.1.2016	Cool	173/049	4, 6, 7, 10	usgs.gov
11	Landsat 8 OLI_TIR	23.9.2016	Rainy	173/049	4, 6, 7, 10	usgs.gov
12	Landsat 8 OLI_TIR	22.8.2016	Rainy	173/049	4, 6, 7, 10	usgs.gov

(September 1986, September 2001, and September, August 2016 [OLI]) were chosen. Bands 2, 3, 4, 5, and 6 of OLI (the 30-m reflective bands of TM5) and their corresponding spatial and spectral ranges were used in this experiment. The first four bands in each of the aforementioned categories span the visual near-infrared (VNIR) portion of the electromagnetic spectrum. The last two bands in each category, however, are those that correspond to the short-wave infrared (SWIR) spectrum.

Image Preprocessing

Regional climate zones and their impact on land surface temperature were evaluated using Landsat satellite pictures (Landsat 5 TM and Landsat 8 OLI/TIRS) with metadata (MTL) file (LST). Several preprocessing steps were performed on these satellite photos. The preprocessing consists of geometrical correction, atmospheric adjustment (dark-object subtraction) with Envi 5.3, and cloudy radiometric calibration. After that, the photos underwent a second resampling with pixels that were 30 × 30 m in size throughout all bands, including the thermal band. As was previously mentioned, a single multiband record for the classification was created by combining 30-m goal-resolution reflecting bands from Landsat images taken over a number of seasons.

Local Climate Zones Classification

These satellite photos underwent a number of preprocessing steps. Full radiometric calibration, atmospheric correction (dark-object subtraction) using Envi 5.3, and correction of geometrical aberrations are included in preprocessing. Following that,

Fig. 5.2 Visual interpretation of local climate zones

the photos underwent a second resampling with pixel sizes of 30 × 30 m for all bands, including the thermal band. For the classification, 30-m goal-resolution reflective bands from Landsat images taken over different seasons were stacked into a single multiband record, as previously mentioned. Figure 5.2 shows the Khartoum-recognized LCZ, which generates a description of each class based on a standard. Using a stratified random technique, at least 150 points for each LCZ type were acquired during field evaluations, as indicated in Table 5.2.

LST Retrieval from Thermal Infrared Data

Based on the methods outlined by Weng et al. (2019), we were able to obtain LST in this study. For each year, an average thermal infrared digital numbers (DN) image was calculated using two images that were captured each month. In the process, digital numbers (DN) from Landsat thermal infrared data were converted to radiance, brightness (blackbody) temperature, and land surface temperature, respectively (emissivity correction). The map of land surface emissivity was created using normalized difference vegetation index (NDVI), which was borrowed from the work of Sobrino et al. (2004).

Conversion of the Digital Number to Spectral Radiance

The DNs of the ETM+ and TM5 images of the TIR bands for each year are converted to spectral radiance utilizing the formula proposed by Markham and Barker (1986)

Table 5.2 Description of local climate zones

Id	LCZ	Description	Training on Google Earth
a	LCZ 5	Open mix of buildings; few trees; land cover mostly paved; concrete, steel, and glass construction material	Training polygons digitized: 11 Total area of digitized polygons: 1.4 km^2
b	LCZ 2	Dense mix of material of <8 stories; few trees; land cover mostly paved; stone, brick, tile, and concrete construction material	Training polygons digitized: 21 Total area of digitized polygons: 8.3 km^2
c	LCZ 10	Dense arrangement of low-rise buildings (1–3 stories); abundance of pervious land cover; scattered trees; stone, tile, and concrete construction material	Training polygons digitized: 21 Total area of digitized polygons: 35.9 km^2
d	LCZ 8	Open arrangement of low-rise buildings (1–3 stories); abundance of pervious land cover; scattered trees; stone, tile, concrete, and steel construction material	Training polygons digitized: 21 Total area of digitized polygons: 6.9 km^2
e	LCZ D	Featureless landscape of grass or herbaceous plants/ crops; few or no trees; zone function is natural grassland, agriculture, or urban park; these were also observed in sporting field	Training polygons digitized: 26 Total area of digitized polygons: 25.2 km^2
f	LCZ A	Heavily wooded landscape of deciduous and evergreen trees; land cover mostly pervious "low plants"; zone function is natural "forest"; tree cultivation	Training polygons digitized: 12 Total area of digitized polygons: 2.2 km^2
g	LCZ F	Bare soil	Training polygons digitized: 22 Total area of digitized polygons: 75 km^2
h	LCZ H	Large open "water" bodies such as seas and lakes or small water bodies such as rivers, reservoirs, and lagoons	Training polygons digitized: 16 Total area of digitized polygons: 3 km^2
i	LCZ E	Sand land	Training polygons digitized: 6 Total area of digitized polygons: 1.2 km^2

as shown by Eq. (5.1). Landsat 8 thermal infrared images were converted using the United States Geological Survey (USGS) standard, followed by Eq. (5.2):

$$L_\lambda = L_{\min} + \frac{L_{\max} - L_{\min}}{\mathrm{QCAL}_{\max} - \mathrm{QCAL}_{\min}}(\mathrm{DN} - \mathrm{QCAL}_{\min}) \tag{5.1}$$

$$L_\lambda = M_L \times Q_{\mathrm{cal}} + A_L \tag{5.2}$$

In the provided equations, L_λ represents the spectral radiance received by the sensor from each pixel in the image, measured in W/(m^{2}srμm). The values M_L and A_L correspond to specific multiplicative and additive rescaling factors for each band, which are obtained from the image MTL file. Q_{cal} refers to the digital number (DN) of each image pixel, while $QCAL_{max}$ represents the maximum DN value (65,535 for 16-bit Landsat 8 and 255 for other Landsat missions). Similarly, $QCAL_{min}$ represents the minimum DN value (0). L_{max} and L_{min} denote the top-of-atmospheric (TOA) radiances, which are scaled to $QCAL_{max}$ and $QCAL_{min}$, respectively, and are measured in W/(m^{2}srμm).

Conversion of Spectral Radiance to Brightness Temperature

The radiant pictures were also transformed to the blackbody temperature using DNs after being converted to spectral radiance, using Eq. (5.3):

$$T_b = \frac{K_2}{\ln\left\{\left(\frac{K_1}{L_\lambda}\right) + 1\right\}} \tag{5.3}$$

where K_1 and K_2 are prelaunch calibration constants in Kelvin units acquired from the image MTL file, T_b is the effective at-sensor brightness temperature in Kelvin units, and L is the spectral radiance in W/(m^2 srm). However, the brightness temperatures make the mistake of assuming that the earth is a blackbody, which it is not, leading to surface temperature inaccuracies. Emissivity correction is required to reduce these inaccuracies, and the equation was used to finally derive land surface temperature (LST).

Surface Emissivity (ε) Retrieval

Using the normalized difference vegetation index (NDVI) threshold approach, the land surface emissivity was determined (Sobrino et al., 1990). The emissivity was obtained from the red spectral area, and the approach states that pixels are deemed to be bare lands when the NDVI is less than 0.2. When the NDVI is greater than 0, the pixels are deemed to have full vegetation coverage, and an emissivity value of 0.99 is assumed (Sobrino et al., 2004). When the NDVI is between 0.2 and 0.5, the pixels are regarded as having a mix of vegetation and soil. In this instance, emissivity is obtained by Eq. (5.4) as:

$$\varepsilon = \varepsilon_v P_v + \varepsilon_s(1 - P_v) + \Delta\varepsilon \tag{5.4}$$

where $\varepsilon_v$the emissivity of vegetation coverage, ε_s is the emissivity of soil surface and, P_v is the proportion of vegetation calculated from using Eq. (5.5):

$$Pv = \left[\frac{\mathrm{NDVI} - \mathrm{NDVI}_s}{\mathrm{NDVI}_v - \mathrm{NDVI}_s}\right]^2 \tag{5.5}$$

where NDVI_s and NDVI_v are the NDVI values of pure soil and vegetation, respectively that were derived from an NDVI image. The internal reflection, whose value is regarded as negligible for the plain and homogeneous surfaces, is shown by the quantity in Eq. (5.7) as well as the geometrical distribution of the natural surface (Sobrino et al., 2004). But in the case of rough and heterogeneous surface the value is assumed as 2% and is expressed by the following Eq. (5.6):

$$\Delta\varepsilon = (1 - \varepsilon_s)(1 - P_v)F\varepsilon_v \tag{5.6}$$

where F is the shape factor whose mean value for different geometrical distributions is assumed as 0.5.

Summarizing Eqs. (5.5) and (5.6), the final equation for emissivity estimation is obtained as Eq. (5.7):

$$\varepsilon = mP_v + n \tag{5.7}$$

where m and n co-efficient are calculated as given in Eq. (5.8):

$$m = \varepsilon_v - \varepsilon_s - (1 - \varepsilon_s)F\varepsilon_v, \text{ and } n = \varepsilon_s + (1 - \varepsilon_s)F\varepsilon_v \tag{5.8}$$

Brightness Temperature to LST

$$\mathrm{LST} = \frac{T_b}{1 + \left\{\lambda T_b\left(\frac{K}{\rho}\right) \times \ln \varepsilon\right\}} \tag{5.9}$$

In Eq. (5.9), λ is the wavelength of emitted radiance (11.5μm) (Markham & Barker, 1986), $\rho = hc/\sigma$ (Stone et al., 2010), K is the Stefan–Boltzmann's constant (1.38×10^{-23} JK^{-1}), h is the Planck's constant (6.26×10^{-34} Js), c is the velocity of light (2.998×108 ms^{-1}), and ε is the surface emissivity. Finally, the derived LST values were converted to the conventional degree Celsius (°C) unit by adding the absolute zero, approximately minus 273.5 °C.

Surface Urban Heat Island Intensity Classification

LST divides the surface UHI in Khartoum into five levels based on the standard deviation from the mean value (Zhang et al., 2013). The advantage of using the mean and standard deviation is that they may be used regardless of the temporal volatility in the actual LST data, allowing the exposure of spatial differentials of LST (Table 5.3).

Simulation of Land Surface Temperature Distribution in Khartoum Using Land Cover Indices

Computation of Urban and Vegetation Indices

Table 5.4 contains urban and vegetation indices that were calculated using the total digital bands that were measured. As previously said, a few indices were tested with the end goal of examining the variations in the characteristics of relationship with surface temperature and to identify indices with the most firmly established capacity to measure urban surface temperature. Simulating LST using polynomial curve fitting is shown in Fig. 5.3.

Correlation Analyses between LST and Indices

To be evaluated using multiple factors, surface temperatures must have a strong correlation with the predictor variables and no inter-factor interactions. We evaluated how well the values in Table 5.5 are associated with LST. As seen in Fig. 5.6, a linear regression model was developed to examine the projected future land surface temperatures utilizing the variables that have an amazing relationship with LST. Multiple linear regression models were developed for each period (different seasons) to evaluate the relationship between land cover indices (such as UI, MNDWI, NDVI,

Table 5.3 Threshold values for LST classification

LST category	Range
Very low	$TS < \mu - 1$ SD
Low	$\mu - 1$ SD $\leq TS < \mu -$ SD/3
Medium	$\mu -$ SD/3 $\leq TS < \mu +$ SD/3
High	$\mu +$ SD/3 $< TS < \mu + 1$ SD
Very high	TS $\geq \mu + 1$ SD

Note: μ and SD are the mean value and the standard deviation of LST, respectively

Table 5.4 Derivation of urban and vegetation indices from Landsat data

Index	Computation	Reference
Normalized difference built-up index	$\text{NDBI} = \frac{\text{SWIR1} - \text{NIR}}{\text{SWIR1} + \text{NIR}}$	Zha et al. (2003)
Urban index	$\text{UI} = \frac{\text{SWIR2} - \text{NIR}}{\text{SWIR2} + \text{NIR}}$	Kawamura et al. (1996)
Normalized difference vegetation index	$\text{NDVI} = \frac{\text{NIR} - \text{RED}}{\text{NIR} + \text{RED}}$	Tucker (1979)
Modified normalized difference water index	$\text{MNDWI} \frac{\text{GREEN} - \text{NIR}}{\text{GREEN} + \text{NIR}}$	Gutman and Ignatov (1998)

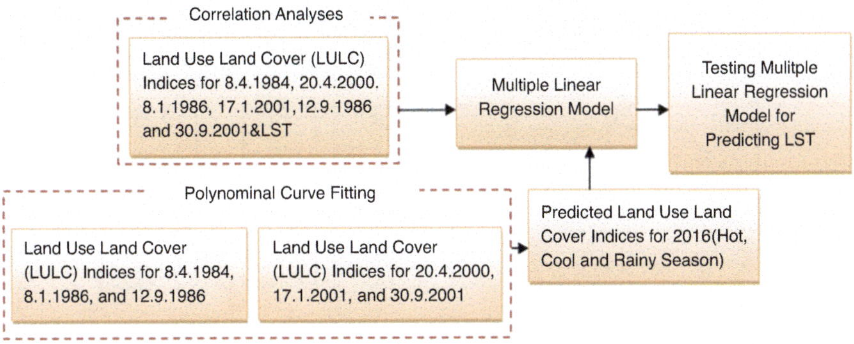

Fig. 5.3 Detailed flowchart for simulating LST using polynomial curve fitting

Table 5.5 LCZ category statistics in Khartoum in 1984, 2000, and 2016

Local climate zones	1984 (km^2)	%	2000 (km^2)	%	2016 (km^2)	%
Compact mid-rise	76.38	2.74	136.26	4.89	171.90	6.17
Open mid-rise	5.13	0.18	11.59	0.42	17.91	0.64
Large low-rise	382.82	13.74	479.63	17.22	626.47	22.49
Heavy industry	25.07	0.90	44.25	1.59	62.06	2.23
Dense trees (woodland)	56.65	2.03	20.15	0.72	13.30	0.48
Low plants	259.69	9.32	343.72	12.34	439.78	15.79
Sand land	8.30	0.30	7.80	0.28	6.21	0.22
Bare soil	1903.96	68.35	1663.00	59.70	1361.20	48.87
Wastewater	9.61	0.35	10.01	0.36	13.38	0.48
Water	58.00	2.08	69.21	2.48	73.42	2.64
Total area	2785.62	100.00	2785.62	100.00	2785.62	100.00

and NDBI) and the LST. The results of the analysis demonstrated that the correlations are statistically significant and long-lasting. For each type of point data, LST values and land cover indices were mined for each pixel in the research zone in order to accomplish this. These points calibrate the linear regression model (Kumar & Shekhar, 2015). The model provides a broad overview of the connection and relationship between the LST and LULC indexes. This view is very consistent with the findings of earlier studies (Liu & Zhang, 2011).

Polynomial Curve Fitting for Land Use and Land Cover Indices Prediction

In order to map the future status of the indices for 2016 in various seasons, the land use land cover indices for 1984, 1986, 2000, and 2001 during various seasons were used as inputs. To anticipate the situation of the land use land cover indices in 2032, the land use land cover indices of 2000, 2001, and 2016 were employed in a polynomial curve fitting study, as shown below:

$$\begin{bmatrix} m & \sum x_i & \sum x_i^2 & \cdots & \sum x_i^n \\ \sum x_i & \sum x_i^2 & \sum x_i^3 & \cdots & \sum x_i^{n+1} \\ \sum x_i^2 & \sum x_i^3 & \sum x_i^4 & \cdots & \sum x_i^{n+2} \\ \vdots & \vdots & \vdots & \ddots & \vdots \\ \sum x_i^n & \sum x_i^{n+1} & \sum x_i^{n+2} & \cdots & \sum x_i^{2n} \end{bmatrix} \begin{bmatrix} P_1 \\ P_2 \\ P_3 \\ \vdots \\ P_{n+1} \end{bmatrix} = \begin{bmatrix} \sum y_i \\ \sum x_i y_i \\ \sum x_i^2 y_i \\ \vdots \\ \sum x_i^n y_i \end{bmatrix}$$

Curve fitting over m pairs of data (x_1,y_1), (x_2,y_2), . . ., (x_m, y_m) is a method to get a polynomial regression between the two pairs as follows: $p(x) = P_1x^n + P_2x^{n-1} + \ldots + P_nx + P_{n+1}$; where, $p(x)$ is a curve fitting of data pairs; $p_1, p_2, \ldots, p_{n+1}$ are the model parameters; and the $x, \ldots, x_n$ is the n inputs variables; and this depends on the possible data available across. The least-square method is used in this study to find the unknown parameters. The $n + 1$ coefficient can be calculated to provide polynomials of n degree parameters across m points. Finally, the distributions of land surface temperatures for 2016 and 2032 over various seasons were obtained by converting the projections of the land use land cover indices into multiple linear regression analysis functions.

Accuracy Assessment of Prediction of Land Surface Temperature

To evaluate the model's performance, we utilized it to forecast the established land surface temperature for the year 2016. We measured the accuracy of the predictions by calculating the mean absolute percentage error (MAPE) given by Eq. (5.10) (Owen et al., 1998):

$$\text{MAPE\%} = \frac{1}{N}\sum_{i=1}^{N}\left(\left|T_{\text{predicted}} - T_{\text{observed}}\right)/T_{\text{observed}}\right|\right) * 100 \tag{5.10}$$

where $T_{\text{predicted}}$ is the modeled surface temperature and T_{observed} is the actual land surface temperature recorded from Landsat data for the ith pixel. The mean absolute percentage is a measurement of prediction accuracy that expresses error as a percentage. The root mean square error (RMSE) and the ratio RMSE/STD. were also used to evaluate the model's accuracy in forecasting temperature. The model was then used to forecast the distribution of land surface temperatures starting in 2032 after an accuracy assessment.

Statistical Analyses of LST in LCZs

The research is predicated on the idea that particular LCZs should exhibit characteristics typical of a specific LST regime. LSTs were computed for each zone, and LST fields were overlaid with LCZs. The one-way analysis of variance was used to assess variations in mean LCZ temperatures (ANOVA). A thorough analysis of the application of this method's assumptions was conducted before using ANOVA. This included comparing LCZ temperature variability, Q-Q graphs, and checking for normalcy using the Kolmogorov-Smirnov test. When the ANOVA *F*-test indicated statistically significant differences in LST, the Tukey Honestly Significant Difference (HSD) test was employed to identify which LCZs varied significantly in their mean LST (Kumar & Shekhar, 2015). It also helped to take the results of repeated comparisons into account. Each of the 16 produced temperature fields underwent a separate analytical run (two study areas with eight temperature fields each). Finally, the score for each LCZ was generated by tallying the number of experiments that showed significant temperature differences. This score's meaning is simple: LCZs with higher scores can better distinguish them from other zones in terms of their typical LST. LCZs with lower scores, on the other hand, are harder to tell apart from different zones in terms of their normal LST.

Results and Discussion

LCZs Mapping

Figure 5.4 displays the spatial distribution of land cover zones in Khartoum using the random forest classification method. Across the entire study area, the LCZ categories, such as open mid-rise, compact mid-rise, and significant low-rise "built-up cover" types, are typically surrounded by "natural cover" types like vegetation and water. To assess the accuracy of our mapping, we overlaid the LCZ classification image on Google Earth imagery and visually inspected each pixel. Most of our LCZs aligned well with the corresponding features in Google Earth. We employed a random sampling method and analyzed a total of 250 pixels. The LCZ statistics are presented in Table 5.5 and Fig. 5.4. In 1984, LCZ A (dense trees) covered an area of 56.6 km^2, accounting for 2.03% of the study area. In 2000, the coverage decreased to 20.15 km^2 (0.72%), and by 2015, it further reduced to 13.30 km^2 (0.48%). This indicates a decline in dense tree coverage in Khartoum, potentially attributed to deforestation activities. Conversely, LCZ 2 (compact mid-rise visually inspected) covered an area of 76.38 km^2 (2.74%) in 1984, which increased to 136.26 km^2 (4.895%) in 2000 and reached 171.90 km^2 in 2015. This signifies an increase in compact mid-rise structures due to rapid construction and urbanization in the study area. The proportion of compact low-rise buildings increased from 13.74% in 1984 to 17.22% in 2000 and 22.49% in 2015. Overall

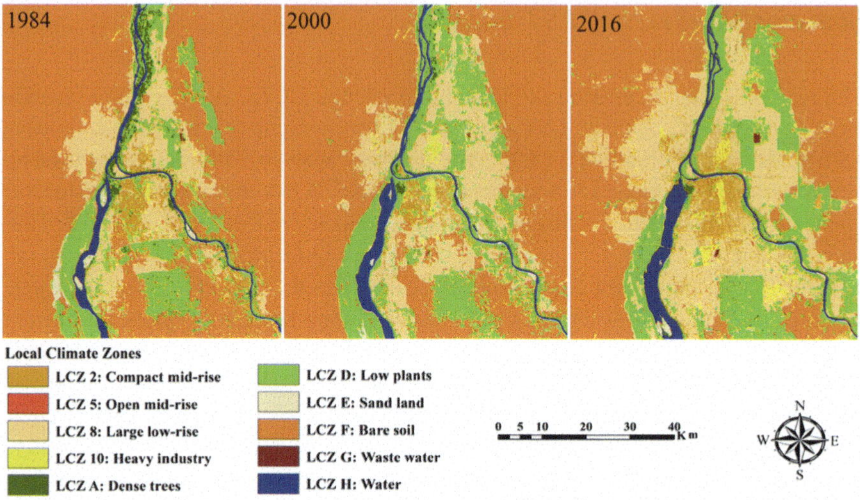

Fig. 5.4 Spatial distribution of local climate zones for Khartoum

accuracies for the study area in all years exceeded 85%. Table 5.6 provides the user accuracy (UA) and producer accuracy (PA) for each individual LCZ class, with accuracies of 95.63%, 96.3%, and 94% for 1984, 2000, and 2016, respectively. Incorporating Lidar data enabled the inclusion of detailed urban and vegetation morphology information in LCZ mapping, contrasting with the limitations of direct object-based image analysis or supervised pixel-based classification techniques. To date, only one study has employed Lidar data for LCZ mapping, conducted in a small suburban area of Sydney, Australia, primarily due to data acquisition constraints (Koc et al., 2017). The temporal variability of urban climate is influenced by various factors, including the thermal characteristics, mass, height, and design of surfaces and structures, as well as land-use patterns, which are well understood.

Land Surface Temperature

Older buildings in industrial locations and city centers traditionally have higher surface temperatures than the surrounding environment. Aside from the densely inhabited areas, the warmest regions of the city also had near-large lengths of relatively flat, impervious surfaces. Hotspots were commonly connected to significant economic zones, distribution hubs, and grounds for international trade fairs wherever they were in the city (which are rather expansive). But the coolest places were near water and in predicted zones (Figs. 5.5 and 5.6). These findings are consistent with the research area's climate's temporal variability. The geography, urbanization, and pollution may all be the contributing causes to this difference. The urban areas are the most urbanized of the regions, and because they are so closely

Table 5.6 Accuracy of multitemporal LULC classification maps

Year	Local climate zones	Producer accuracy (%)	User accuracy (%)	
1984	Compact mid-rise	86.36	72.43	
	Open mid-rise	100.00	97.08	
	Large low-rise	98.62	99.40	
	Heavy industry	98.76	99.80	
	Dense trees (woodland)	88.30	78.64	
	Low plants	83.73	79.61	
	Sand land	98.2	91.05	
	Bare soil	94.47	99.51	
	Wastewater	93.01	90.2	
	Water	98.11	99.40	
2000	Compact mid-rise	87.36	79.03	
	Open mid-rise	100.00	96.01	
	Large low-rise	98.02	98.11	
	Heavy industry	98.70	91.01	
	Dense trees (woodland)	90.30	81.64	
	Low plants	88.71	81.61	
	Sand land	91.21	88.9	
	Bare soil	95.47	90.7	
	Wastewater	99.2	90.01	
	Water	99.40	91.23	
2016	Compact mid-rise	96.55	93.33	
	Open mid-rise	96.55	93.33	
	Large low-rise	95.88	91.67	
	Heavy industry	95.65	73.33	
	Dense trees (woodland)	91.95	78.00	
	Low plants	85.71	80.00	
	Sand land	93.21	88.9	
	Bare soil	100.00	99.01	
	Wastewater	92.31	89.81	
	Water	95.99	91.31	
	Overall accuracy (%)	95.63 (1984)	96.33 (2000)	94 (2016)
	Kappa	0.89 (1984)	0.88 (2000)	0.90(2016)

knit together and concentrated, they have the largest spatial agglomerations. In this region, industrial and commercial areas make up the majority of the developed areas. This region is highly urbanizing, and it is likely it will develop into a megalopolis one day.

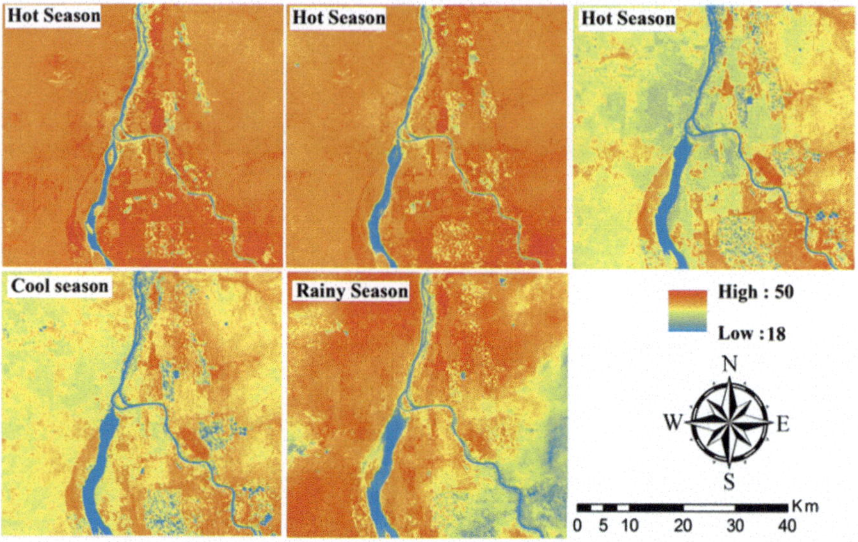

Fig. 5.5 Distribution of land surface temperature for different seasons

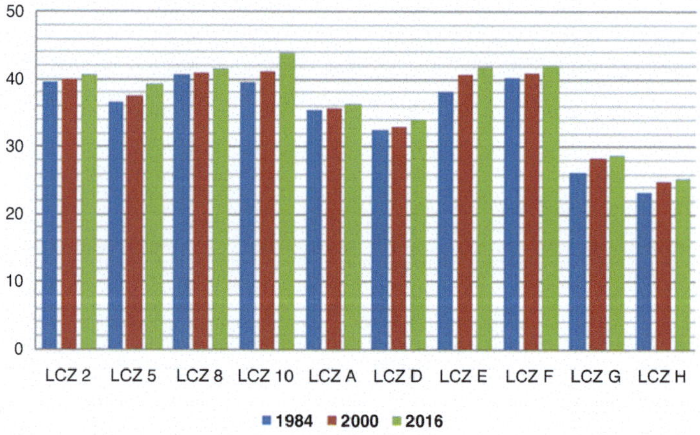

Fig. 5.6 Distribution of land surface temperature for LCZs

Local Climate Zones and Land Surface Temperatures

As shown in Fig. 5.6, box plots were generated, showing the similarities and differences between LST and LCZ. The line within the box indicates the median; the bottom of the box is the first quartile, and the top is the third quartile of LST values. The results show a clear variation of LST with respect to LCZs. LCZ 2 (compact mid-rise) and LCZ 5 (open mid-rise), followed by large low-rise LCZ 8, were the first and second hottest zones, respectively, in all the study periods. On

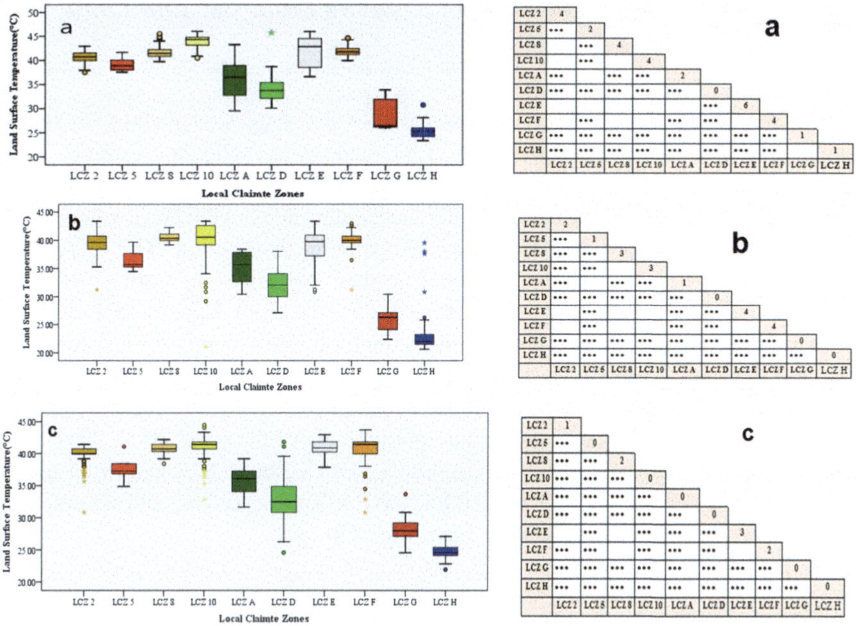

Fig. 5.7 Box-plots summary and Tukey Honestly Significant Difference (HSD) test of LST values for individual LCZs in Khartoum at different time span: (**a**) 8 April 1984; (**b**) 20 April 2000 and (**c**) 16 April 2016 (Significance at 0.001 level for that pair of LCZs; Empty space: no significant difference of the corresponding LCZs pairs. The count of empty spaces for the respective LCZs is represented by the number mentioned at the diagonal. The bottom of the box represents the first quartile of LST values, the top represents the third quartile, and the line inside the box represents the median (Fig. 5.7). [LCZ H, the coldest zone, and LCZ 10, the warmest zone])

the other hand, as shown in the LST map, LCZ 8 (water), LCZ_G (wastewater), and bare land LCZ F had the lowest surface temperature, respectively. The two low plants and dense trees region show a moderate temperature compared to other classes. For each LCZ, LST values were also analyzed in relation to the distribution density to verify normalization and validate the underlying premise of the ensuing ANOVA study. Although further research is needed, it is interesting that open urban areas with trees nearby displayed lower LST levels than regions with low plants. This suggests that trees have a greater cooling effect than low plants, such as grass. According to our research, it is important to understand how the microclimate of LCZs responds to changes in the size and structure of urban neighborhoods and regions to make informed decisions about how to apply LCZs to assessments of urban transformation in the future. In the interim, some types of water bodies (such as streams and rivers) are more susceptible to the environment and climate. It suggests that by using LCZs mapping to identify homogenous zones based on surface temperature, these zones can be distinguished from one another (Fig. 5.7).

Retrieval of Surface Temperature from the Land Use and Land Cover Indices

One of the major contributions of this work is the suggested method of using temperature forecast to determine future LST patterns to model and estimate UHI according to various LCZs. In the Polynomial curve fitting analysis, the urban index (UI) and modification of normalized difference water index (MNDWI) were shown to be the best predictors of land surface temperature distribution. The regression model was tested on independent Landsat data collected in April 2015 and September 2017. The temperature obtained directly from Landsat 8's thermal infrared data (Band 10) was compared to that obtained from the urban index (UI) and the modification of normalized difference water index (MNDWI). Urban index (UI) and MNDWI (based on 400 points samples over the investigation region) made highly accurate surface temperature predictions for the years 2000, 2001, and 2016 (mean relative rate percentages of 4.88%, 4.11%, and 5.88% with root mean square errors of 0.78, 0.83, 0.77, 0.78, 0.85, and 0.81). (Hot season, cold season, and rainy seasons, respectively) (Table 5.7 and Fig. 5.8).

Simulated and Predicted LST of LCZ

Table 5.8 displays anticipated and simulated LST for various LCZ. The expected peak LST values for LCZ 10, LCZ 5, and LCZ E in 2032 were 43.93, 39.18, and 32.77, respectively. The anticipated levels for 2032 were higher than the simulated values for 2016 based on LST values. (LCZ H, the coldest zone, and LCZ 10, the warmest zone) Sand, industrial, and huge low-rise buildings are predicted to have the highest future value based on LST intensity. This study advises decision-makers and the government to avoid building new cities close to these high LST zones. Research into how the environment is impacted and how climate change is measured may increase as a result of the correlation between environmental indices (Abd El-Hamid et al., 2022) (Fig. 5.9).

Table 5.7 Correlation analysis of LST and difference indices in different seasons

No	Year	Season	Model	R	R^2
1	2000	Hot season	$16.6 \times UI - 11.1 \times MNDWI + 38.6$	0.89	0.78
2	2001	Cool season	$11.6 \times UI - 5.32 \times MNDWI + 30.9$	0.91	0.83
3	2001	Rainy season	$17.9 \times UI - 4.51 \times MNDWI + 35.7$	0.88	0.77
4	2016	Hot season	$17.1 \times UI - 13.78 \times MNDWI + 39.1$	0.89	0.78
5	2016	Cool season	$10.9 \times UI - 8.53 \times MNDWI + 29.8$	0.92	0.85
6	2016	Rainy season	$15.8 \times UI - 7.43 \times MNDWI + 36.6$	0.90	0.81

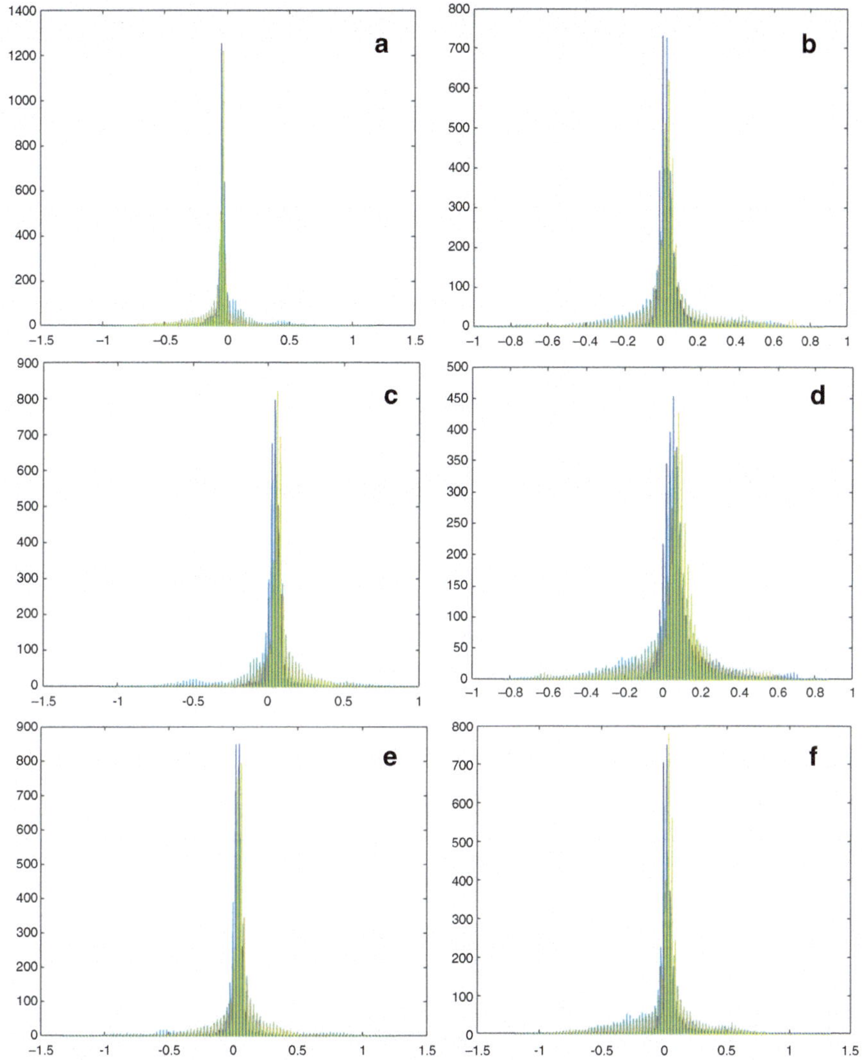

Fig. 5.8 Frequency of occurrence versus MNDWI (**a**), UI (**b**), MNDWI (**c**), UI (**d**) MNDWI (**e**) and UI (**f**) – residuals for 2016 (cool, hot, and rainy seasons, respectively)

Table 5.8 Simulated and predicted LST of different local climate zones

Mean of LST for LCZ (y/m)						
LCZ	2016/4	2016/1	2016/9	2032/4	2032/1	2032/9
LCZ 2	40.69	28.93	35.38	41.73	30.24	35.96
LCZ 5	39.19	28.23	34.54	39.42	28.79	34.86
LCZ 8	41.59	30.15	36.87	42.06	31.25	36.84
LCZ 10	43.51	32.37	38.67	43.93	32.96	39.18
LCZ A	36.24	29.56	30.86	37.23	29.58	32.25
LCZ D	33.84	28.31	34.57	34.63	28.46	34.83
LCZ E	41.81	32.46	33.45	43.07	32.77	33.80
LCZ F	41.86	31.81	37.20	42.84	32.58	38.12
LCZ G	28.76	22.71	29.10	29.91	23.33	29.40
LCZ H	25.18	19.66	26.80	25.61	20.57	26.80

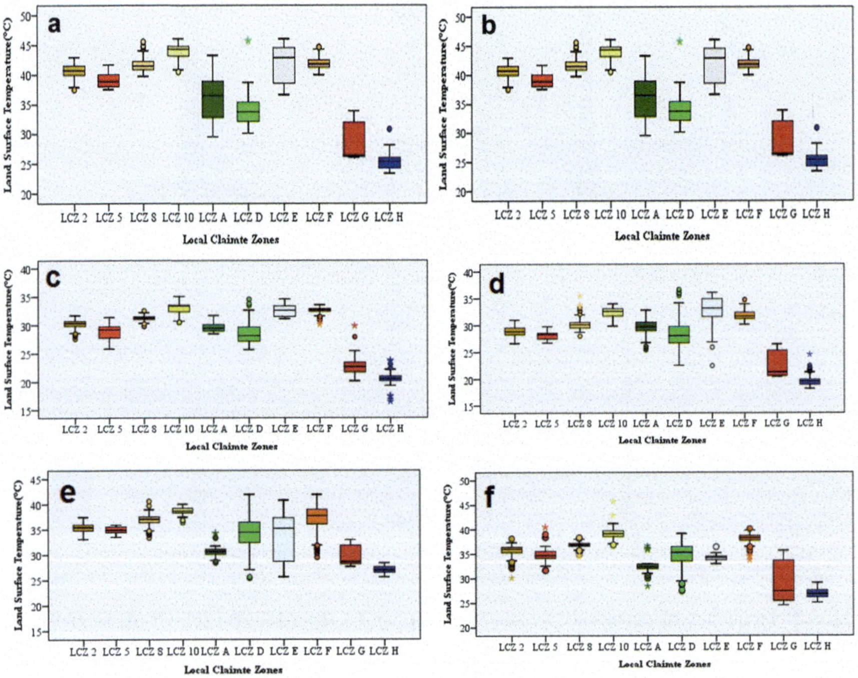

Fig. 5.9 Box-plots summary of LST values for individual LCZs in Khartoum on (**a**) April 2016; (**b**) January 2016; (**c**) September 2016; (**d**) April 2032; (**e**) January 2032; and (**f**) September (Fig. 5.9)

Conclusion

Using information from two Landsat photos, this study contrasted the geographical distribution of land surface temperatures with regional climate zones in Khartoum City, Sudan. Since our investigation looks at temperature variations across LCZ classes rather than the more traditional "urban" and "rural" classes, it shows the relevance of LCZ mapping for understanding UHI. The LCZs and LST linkage demonstrated the use of the LCZ mapping for comparing and analyzing the UHI. The majority of LCZ pair comparisons reveal variations in average LSTs that are highly significant. Comparing intra-urban temperatures within various urban classes' helps researchers better understand how diverse urban morphology affects regional climate change. Water, on the other hand, makes up the least LST. According to LSTs' satellite data, normal surface temperatures vary greatly between zones. Dense trees covered an area of 56.6 km^2 in 1984, accounting for 2.03%, while in the year 2000, the area cover is 20.15 km^2, which represents 0.72%. In 2015, the area cover is 13.30 km^2 with a percentage of 0.48% of the study area, which means that the coverage of the dense trees in Khartoum is reducing, perhaps due to deforestation activities. In addition, the area of LCZ 2 (compact mid-rise visually inspected) in 1984 made up to76.38 km^2, accounting for 2.74% of the built types classes, while in 2000 the area cover was 136.26 km^2, accounting for 4.895%, and in 2015, the area cover was 171.90 km^2.This means that there is an increase in compact mid-rise as a result of rapid construction and urbanization in the study area. The proportion of the compact low rise in 1984 is 13.74%, in 2000 is 17.22%, and in 2015 is 22.490%. It demonstrates that the majority of the study area's structures are mid-low rises, the majority of which are compact. The total accuracy for the study area was more than 85% in all years. Regarding surface temperatures, large low-rise, heavy industry, and low plant zones were distinguished; however, both cities' compact mid-rise, open high-rise, and sparsely built-up areas are less easily distinguished. Water bodies, areas with effluent, and other such locations were the coldest. Heavy industrial, crowded low-rise buildings, and tightly packed mid-rise structures were the warmest locations. This LCZ–LST comparison might be considered the most significant finding of this contribution when coupled with applying the GIS-based technique. Some issues need further research, like the seasonality of LST discrepancies and thermal anisotropy (differences in complete surface temperatures).

References

Abd El-Hamid, H. T., Arshad, M., & Eid, E. M. (2022). The effects of coastal development on the urban heat Island in the mangrove ecosystem along the Jazan coast, KSA. *Journal of Coastal Conservation, 26*, 70. https://doi.org/10.1007/s11852-022-00915-x

Bechtel, B., See, L., Mills, G., & Foley, M. (2016). Classification of local climate zones using SAR and multispectral data in an arid environment. *IEEE Journal of Selected Topics in Applied Earth*

Observations and Remote Sensing, 9(7), 3097–3105. https://doi.org/10.1109/JSTARS.2016.2531420
Bechtel, B., Alexander, P. J., Beck, C., Böhner, J., Brousse, O., Ching, J., Demuzere, M., & Fonte, C. (2019). Generating WUDAPT Level 0 data – Current status of production and evaluation. *Urban Climate, 27*, 24–45. https://doi.org/10.1016/j.uclim.2018.10.001
Clinton, N., & Gong, P. (2013). MODIS detected surface urban heat islands and sinks: Global locations and controls. *Remote Sensing of Environment, 134*, 294–304. https://doi.org/10.1016/j.rse.2013.03.008
Gutman, G., & Ignatov, A. (1998). The derivation of the green vegetation fraction from NOAA/AVHRR data for use in numerical weather prediction models. *International Journal of Remote Sensing, 19*(8), 1533–1543. https://doi.org/10.1080/014311698215333
Kaloustian, N., & Bechtel, B. (2016). Local climatic zoning and urban heat Island in Beirut. *Procedia Engineering, 169*, 216–223. https://doi.org/10.1016/j.proeng.2016.10.026
Kawamura, M., Jayamana, S., & Tsujiko, Y. (1996). Relation between social and environmental conditions in Colombo Sri Lanka and the urban index estimated by satellite remote sensing data. *The International Archives of the Photogrammetry, Remote Sensing, 31*, 321–326.
Koc, C. B., Osmond, P., Peters, A., & Irger, M. (2017). Mapping local climate zones for urban morphology classification based on airborne remote sensing data. In *Urban remote sensing event (JURSE), joint* (pp. 1–4). IEEE.
Kumar, D., & Shekhar, S. (2015). Statistical analysis of land surface temperature–vegetation indexes relationship through thermal remote sensing. *Ecotoxicology and Environmental Safety, 121*, 39–44. https://doi.org/10.1016/j.ecoenv.2015.07.004
Lehnert, M., Geletič, J., & Husák, J. (2015). Urban field classification by "local climate zones" in a medium-sized Central European city: The case of Olomouc (Czech Republic). *Theoretical and Applied Climatology, 122*, 531–541. https://doi.org/10.1007/s00704-014-1309-6
Lelovics, E., Unger, J., Gál, T., & Gál, C. V. (2014). Design of an urban monitoring network based on local climate zone mapping and temperature pattern modelling. *Climate Research, 60*(1), 51–62. https://www.jstor.org/stable/24896173
Liu, L., & Zhang, Y. (2011). Urban heat Island analysis using the Landsat TM data and ASTER data: A case study in Hong Kong. *Remote Sensing, 3*(7), 1535–1552. https://doi.org/10.3390/rs3071535
Markham, B. L., & Barker, J. L. (1986). *Landsat MSS and TM post-calibration dynamic rangers, exoatmospheric reflectance and at-satellite temperatures* (pp. 3–8). EOSAT Landsat Technical Notes.
Mushore, T. D., Dube, T., Manjowe, M., Gumindoga, W., Chemura, A., Rousta, I., Odindi, J., & Mutanga, O. (2019). Remotely sensed retrieval of Local Climate Zones and their linkages to land surface temperature in Harare metropolitan city, Zimbabwe. *Urban Climate, 27*, 259.
Mustafa, E. K., Liu, G., El-Hamid, H. T. A., & Kaloop, M. R. (2019). Simulation of land use dynamics and its impact on land surface temperature in Beijing using satellite data. *GeoJournal, 86*, 1089–1107. https://doi.org/10.1007/s10708-019-10115
Odindi, J., Bangamwabo, V., & Mutanga, O. (2015). Assessing the value of urban green spaces in mitigating multi-seasonal urban heat using MODISL and surface temperature (LST) and Landsat 8 data. *International Journal of Environmental Research, 9*(1), 9–18. https://doi.org/10.22059/ijer.2015.868
Owen, T. W., Carlson, T. N., & Gillies, R. (1998). An assessment of satellite remotely-sensed land cover parameters in quantitatively describing the climatic effect of urbanization. *International Journal of Remote Sensing, 19*(9), 1663–1681. https://doi.org/10.1080/014311698215171
Seto, K. C., Solecki, W. D., & Griffith, C. A. (2015). *The Routledge handbook of urbanization and global environmental change*. Routledge. https://doi.org/10.4324/9781315849256
Sobrino, J., Caselles, V., & Becker, F. (1990). Significance of the remotely sensed thermal infrared measurements obtained over a citrus orchard. *ISPRS Journal of Photogrammetry and Remote Sensing, 44*(6), 343–354. https://doi.org/10.1016/0924-2716(90)90077

Sobrino, J. A., Jiménez-Muñoz, J. C., & Paolini, L. (2004). Land surface temperature retrieval from LANDSAT TM 5. *Remote Sensing of Environment, 90*(4), 434–440. https://doi.org/10.1016/j.rse.2004.02.003

Stewart, I. D. (2011). A systematic review and scientific critique of methodology in modern urban heat Island literature. *International Journal of Climatology, 31*(2), 200–217. https://doi.org/10.1002/joc.2141

Stewart, I. D., & Oke, T. R. (2012). Local climate zones for urban temperature studies. *Bulletin of the American Meteorological Society, 93*(12), 1879–1900. https://doi.org/10.1175/BAMS-D-11-00019

Stewart, I. D., Oke, T. R., & Krayenhoff, E. S. (2014). Evaluation of the 'local climate zone' scheme using temperature observations and model simulations. *International Journal of Climatology, 34*(4), 1062–1080. https://doi.org/10.1002/joc.3746

Stone, B., Hess, J. J., & Frumkin, H. (2010). Urban form and extreme heat events: Are sprawling cities more vulnerable to climate change than compact cities? *Environmental Health Perspectives, 118*(10), 1425–1428. https://doi.org/10.1289/ehp.0901879

Sun, R., Lü, Y., Yang, X., & Chen, L. (2019). Understanding the variability of urban heat islands from local background climate and urbanization. *Journal of Cleaner Production, 208*, 743–752. https://doi.org/10.1016/j.jclepro.2018.10.178

Tarawally, M., Xu, W., Hou, W., & Mushore, T. D. (2018). Comparative analysis of responses of land surface temperature to long-term land use/cover changes between a coastal and inland city: A case of Freetown and Bo Town in Sierra Leone. *Remote Sensing, 10*(1), 112. https://doi.org/10.3390/rs10010112

Tucker, C. J. (1979). Red and photographic infrared linear combinations for monitoring vegetation. *Remote Sensing of Environment, 8*(2), 127–150. https://doi.org/10.1016/0034-4257(79)90013-0

Unger, J., Lelovics, E., & Gál, T. (2014). Local Climate Zone mapping using GIS methods in Szeged. *Hungarian Geographical Bulletin, 63*(1), 29–41. https://doi.org/10.15201/hungeobull.63.1.3

Voogt, J. A., & Oke, T. R. (2003). Thermal remote sensing of urban climates. *Remote Sensing of Environment, 86*(3), 370–384. https://doi.org/10.1016/S0034-4257(03)00079

Wang, C., Middel, A., Myint, S. W., Kaplan, S., Anthony, J., Brazel, A. J., & Lukasczyk, J. (2018). Assessing local climate zones in arid cities: The case of Phoenix, Arizona and Las Vegas, Nevada. *ISPRS Journal of Photogrammetry and Remote Sensing, 141*, 59–71. https://doi.org/10.1016/j.isprsjprs.2018.04.009

Weng, Q., Lu, D., & Schubring, J. (2004). Estimation of land surface temperature–vegetation abundance relationship for urban heat Island studies. *Remote Sensing of Environment, 89*(4), 467–483. https://doi.org/10.1016/j.rse.2003.11.005

Zha, Y., Gao, J., & Ni, S. (2003). Use of normalized difference built-up index in automatically mapping urban areas from TM imagery. *International Journal of Remote Sensing, 24*(3), 583–594. https://doi.org/10.1080/01431160304987

Zhang, Q., Schaaf, C., & Seto, K. C. (2013). The vegetation adjusted NTL Urban Index: A new approach to reduce saturation and increase variation in nighttime luminosity. *Remote Sensing of Environment, 129*, 32–41.

Zhou, D., Zhao, S., Liu, S., Zhang, L., & Zhu, C. (2014). Surface urban heat Island in China's 32 major cities: Spatial patterns and drivers. *Remote Sensing of Environment, 152*, 51–61. https://doi.org/10.1016/j.rse.2014.05.017

Weng, Q., Firozjaei, M. K., Kiavarz, M., Alavipanah, S. K., & Hamzeh, S. (2019). Normalizing land surface temperature for environmental parameters in mountainous and urban areas of a cold semi-arid climate. Science of the Total Environment, 650, 515-529.

Chapter 6
A Preliminary Investigation into the Social Perceptions of Urban Residents Exposed to River Floods

Gowhar Farooq Wani, Rayees Ahmed, Syed Towseef Ahmad, Sumaira Javaid, Ajinder Walia, and Pervez Ahmed

Introduction

The general idea behind conducting a preliminary or a pilot study is to determine feasibility of selected methods, tools (a questionnaire or an interview guide) and techniques to perform a detailed or substantive study. It gives researchers enough freedom to test the instrument (an interview schedule) for the collection of data, interact with the respondents to know about the difficulties of gaining access to more respondents and conducting interviews in a manner that meets the pre-set criteria or research plan, resolving alterable ethical and field issues and make modifications wherever necessary. Any preliminary study (although not many researchers mention it in the text) proves to be a value addition to a substantive study. It is usually carried out on a smaller scale (a distinctive feature of a pilot study) to improve the quality and effectiveness of the main study and obtain quality results (Malmqvist et al., 2019). Commonly known as 'mini or feasibility studies', pilot studies can be crucial for both quantitative and qualitative studies. Highly recommended in clinical and social research, such studies are considered an important element of a good research design, and rightly so for immensely benefitting the substantive or full-scale studies (In, 2017; Van Teijlingen & Hundley, 2001). Prior to beginning such a study, the researchers not only fully acquaint themselves with the study problem, its objectives and main purpose but also with the different procedures, materials and methods incorporated at the later stage. Although the pilot studies have significantly evolved

G. F. Wani (✉) · R. Ahmed · S. T. Ahmad · S. Javaid · P. Ahmed
Department of Geography and Disaster Management, School of Earth and Environmental Sciences, University of Kashmir, Srinagar, India

A. Walia
Department of Governance and Inclusive Disaster Risk Reduction, National Institute of Disaster Management (NIDM), Ministry of Home Affairs, Government of India, New Delhi, India

A. L. Singh et al. (eds.), *Climate Change, Vulnerabilities and Adaptation*,
https://doi.org/10.1007/978-3-031-49642-4_6

in the recent past, benefitting many beginners and clinical and social researchers, not many researchers find it interesting due to different reasons, such as publication bias. There is no denying the fact that pilot studies can be much informative for both the researchers conducting them and those leading similar works (Thabane et al., 2010), regardless of the problems under discussion. However, this and a lot of information about pilot studies never becomes part of final publications, often due to their wrong implementation and presentation of results (Van Teijlingen & Hundley, 2001), while placing emphasis on the statistics involved and not the feasibility thing – which is unarguably the main highlight of a pilot study (Thabane et al., 2010). Nevertheless, performing a pilot study does not guarantee a major success in the substantive study, but it does increase the possibility of generating valuable and reliable outcomes (Van Teijlingen & Hundley, 2001). As we discuss more about the process and outcome of pilot studies (by adopting a qualitative approach for this study) in the next sections, we sincerely believe that there is a need for more positive deliberations in the scientific community on the application/implementation of the pilot studies. The researchers need to reflect on both the processes and results of pilot studies and report more details/findings in the text, so as to offer guidance to others on addressing their problems.

This study focuses on understanding residents perspectives living in the low-lying areas of Srinagar city in the aftermath of a major flood, namely Kashmir Flood 2014. The historic flood of 2014 was an overwhelming situation as it brought the entire populace of Kashmir Valley to its knees (although not affecting everyone in the same way) and left huge devastation in its wake, while exposing/failing all the existing mechanisms of flood risk reduction/management. The main contributory causes for this abnormal situation were unprecedented rains, mismanagement and intense pressures from unrelenting anthropogenic activity in the region. This study provides valuable insights into the ground realities around the flooding problem in the region learned from the first-hand accounts of people but at a smaller scale. More insights into the usefulness, appropriateness and applicability of implementing such a study prior to studying the problem in depth are reflected in Fig. 6.1. Nevertheless, pilot studies serve great purpose, but we find certain inconsistencies, which we

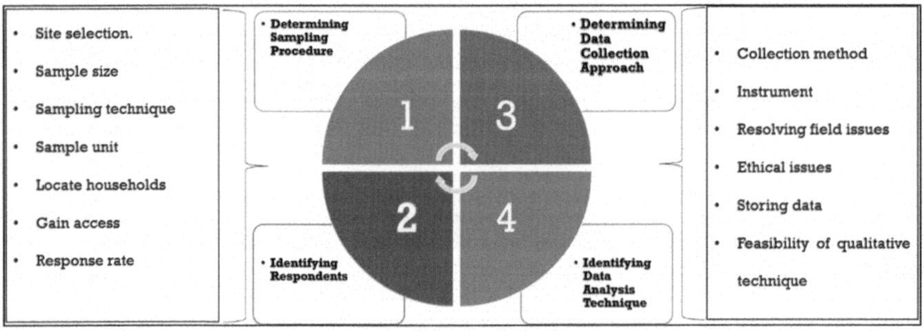

Fig. 6.1 A schematic diagram underlining the purpose and scope of conducting this preliminary study

deliberate upon later in this chapter. There can be many reasons behind performing such a study but our main objective was to 'gather and analyse useful data to provide guidance for conducting a substantive study'. The purpose was very much straightforward, keeping in mind the following: firstly, the findings/first-hand experiences would indicate if the study's existing design needs any improvements; secondly, designing and testing the adequacy of an interview guide and see if it was workable and having a realistic appeal and, finally, help assess beforehand the feasibility of conducting a comprehensive study to save time and energy. The idea behind this was to uncover the realities about the local residents' continued presence at their current location and their returning to the original places after the devastation of the 2014 flood. Nevertheless, the present mini-study exploring residents' perceptions of the flooding problem proved immensely beneficial for Wani et al. (2022) in developing a nuanced discourse through a substantive study.

Study Site

The research was conducted in the flood-prone areas of Srinagar city, including Dal Lake interiors and some low-lying areas along the Jhelum River, which have a troubled flood history (see Fig. 6.2). The city is home to around 1.2 million people, spread over an area of about 278 sq. km and situated at an average elevation of 1600 m a.s.l. (Census of India, 2011). Out of this population, around 0.34 million are urban poor, who live in several unnotified slum pockets within the city limits (Srinagar Master Plan (SMP) J&K, Srinagar Metropolitan Region – 2035, 2019). Srinagar City, in the heart of breath-taking Kashmir Valley, is not just about mountains, mosques, temples, gurdwaras, hospitals, lakes and gardens. It has lifted itself to global prominence time and again, through its rich cultural heritage, local traditions, hospitality, generosity, food, art, music and architecture.

Methodology

Design and Sampling Plan

Taking into account the scope of this small but important preliminary investigation, the qualitative research design was applied to unearth the underlying reasons or ground realities about the flooding problem from a social perspective. Simple random sampling (SRS) technique was used to achieve the desired sample. The sample size was 20 with household as a sampling unit. SRS typically follows a sequential process, wherein a population of interest is well defined, numbers are assigned, a reasonable sample size is chosen and finally, a list of potential respondents is generated for survey implementation (see Fig. 6.3). In this process, every household has an equal chance of getting selected. However, before applying SRS,

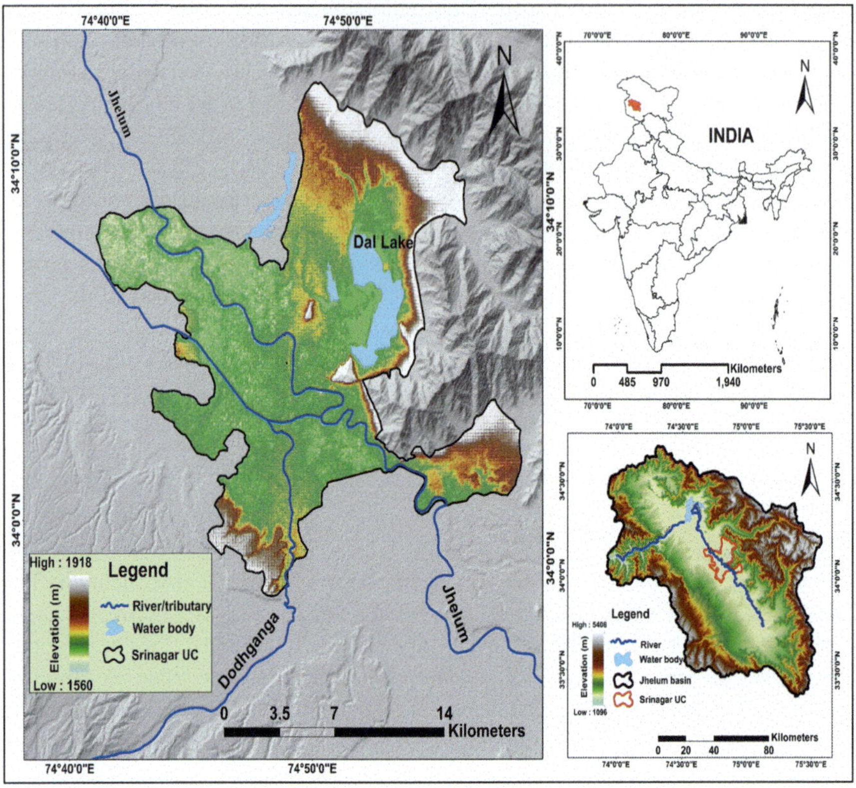

Fig. 6.2 Map of Srinagar City, the largest urban center in the Himalayan region and situated on the banks of River Jhelum

Fig. 6.3 A graphical representation of simple random sampling technique

two pre-conditions were considered: households along the Jhelum River with direct flood experience were preferred and household heads or members of age above 18 (and irrespective of gender) were chosen as interviewees.

Interview Guide

Part 1: Past Experience

- Have you ever experienced a flood?

Part 2: Kashmir Flood of 2014

- What is your story?

Part 3: Place of Residence

- What is the first thing that comes to your mind about your place?

Part 4: Level of Awareness

- How well do you know what to do before, during and after a flood?

Fig. 6.4 A sketch of the interview guide

Instrument

In research studies of qualitative nature and implementing a survey method, the interview protocol plays an important role in collecting meaningful information/data from the study participants concerning a research problem.

In this research, a simple and flexible interview guide (see Fig. 6.4 for more details) was carefully followed for research enquiry, consisting of four parts: (1) past experience, (2) the Kashmir Flood of 2014, (3) place of residence, and (4) level of awareness. The questions were intentionally (and as the study demanded) kept open-ended and unstructured to make respondents feel much more comfortable while providing unconventional answers. The sequential order was not meant for strict implementation but to provide essential guidance on different aspects meant for the effective and successful interview process.

Data Collection and Analysis Process

The collection of information/data was started right after conceptualizing the idea and detailing the study plan in black and white. However, the highly exposed stretches to riverine flooding were identified using an inundation map of the Kashmir Flood of 2014 and households were visited in April and May 2021 during daytime. Prior to beginning the interviews of participants subject to their oral consent, they

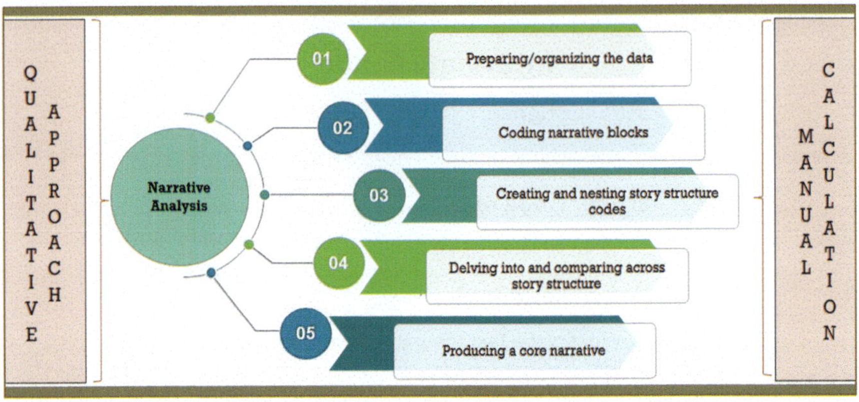

Fig. 6.5 A schematic presentation of the narrative analysis technique (Riessman, 1993; Creswell et al., 2007; Merriam & Tisdell, 2015; Herman & Vervaeck, 2019)

were informed about the purpose of the interview. The interviews which were conducted at the interviewees' home lasted for 70 minutes on an average.

It was a difficult but good experience, sticking to the topical agenda at a time when the entire world was fighting COVID-19. The pandemic upended our lives to a great extent, and together with it, people in Kashmir were also absorbing an overwhelming change (the abrogation of Artcle 370 and bifurcation of J & K State into two Union Territories). Among the 20 study participants, 13 were male and 7 female, preferably the household heads (and in a few cases eldest person available at the survey time) residing at current location for a long time, typically more than 20 years.

Nevertheless, after the culmination of an effective interview process, the preparation, organization and transcription of the recorded narrations into meaningful chunks/bits of information were carried out and subjected to narrative analysis (a qualitative data analysis technique) to obtain the productive outcomes (Fig. 6.5), as deliberated by Riessmen (1993), Creswell et al. (2007), Merriam and Tisdell (2015) and Herman and Vervaeck (2019), well grounded in the theory of substantiating the qualitative approaches in sociological research (Collins & Stockton, 2018). The narrative analysis is generally a five-step process: (1) preparing/organizing the data; (2) coding of narrative blocks; (3) coding and nesting the story structure; (4) delving into and comparing across the story structure, and (5) producing a report or core narrative. The calculations in terms of organization of narrative blocks (straight from the transcripts), development of initial codes and arrangement of the same in the corresponding categories were done manually with utmost care and while exercising great caution to avoid personal bias.

Findings

The outcome of this research was achieved after applying case-specific methods and techniques for data collection and analysis. Initially, after the conceptualization of this particular research problem and some pre-planning to achieve the main objectives, the idea of going for a pilot study did not appeal much. However, with new details emerging during the data collection process to which the researchers were not familiar with, it really proved to be a beneficial exercise. The researchers got an opportunity to familiarize themselves with the study area and get a first-hand experience of the flooding problem, while having an in-person conversation with the affected people. It helped them to identify the actual study participants, know more about the gender-diverse groups and uncover peoples' stories of great courage shown during the flood event of 2014, as well as many ordeals filled with a sense of hopelessness and helplessness. Nevertheless, it was important to stick to the topical agenda and ask appropriate questions to the respondents, so that the study objectives are met. Although the answers were interesting, very relevant and consistent with the interview protocol, the exercise proved to be exhausting. It took some effort to make the respondents understand about the scope of this research and conclude in agreed time, while fully respecting their point of view/opinions/stories and showing empathy. The flexibility in the interview guide (having both advantages and disadvantages), allowed the respondents to narrate more details about their past flood experience, uncaring of time. The singling out of authorities (un)intentionally for not doing enough during the 2014 floods was to register a statement that Kashmiri people showed great courage during the response, recovery and rehabilitation stages, as the devastation shattered their dreams and hopes.

The findings of this study generated from the narrative blocks have been discussed in detail under the standalone five categories or themes after the rigorous coding and revision process supported by participant (direct) quotations across the dataset. Since we adopted a qualitative approach to achieve the objectives of this study, the data was analysed using the narrative analysis technique, which is a five-step process: (1) preparing/organizing the data; (2) coding of narrative blocks; (3) coding and nesting the story structure; (4) exploring into and comparing across the story structure and (5) producing a report or core narrative. Since the sample size was small, every study participant was given ample time and great deal of freedom to narrate his/her story/past flood experience. This in return uncovered a lot of relevant information necessary to understand the important aspects of the flooding problem.

The following sections (4.1, 4.2, 4.3, 4.4, and 4.5) give further details on the five main themes generated in the narrative analysis process, namely sense of place, understanding threat, flood experience, coping abilities and social cohesion. The emergent themes in this process and subsequent thick descriptions around each of them give valuable insights into the flooding problem and peoples' response and precautionary behaviour towards it. It further provides us with some convincing answers supported with evidence as narrative descriptions to our initial enquiry that 'why did the occupants return to their original place of residence after the 2014 flood waters receded. How and why they continue to live there despite the risk of future floods in the study region.' In addition to this, the specifics on each theme are provided separately (see Table 6.1).

Table 6.1 The description of outcomes after analysing recorded narrations of study participants by applying narrative analysis technique and culminating into five main themes/categories

Narrative Blocks	Initial Code	Corresponding Categories
Participant 1 (Male-50): "I have experienced flooding a few times…umm but remember 2014 only. Floods are real. It was a narrow escape from death; all chaos…and it eat my precious stuff too. A little I knew till 2014. Now I know a lot. Home is heaven & am lucky to have my neighbours"	1) Childhood memories 2) Birthplace 3) Homeplace affection	1), 2), 3), Sense of place
Participant 3 (Female-70): "2014 experience was a happening that caught us unaware. Flooding is a looming threat here. My garden was gone and so were the essential items on ground floor. I know I am being exposed to it but never thought it can actually come. Ah! My home, my small world that I rule feels safe. And Yes, I have my dear ones around me to look after me"	4) Loss of belongings 5) Worry	4), 5), Understanding threat
Participant 8 (Female-39) "There has been a few instances of heavy rains but September 2014 was huge. I know I am exposed but I have some experience to avoid loss of life. Normal life was disturbed & remained same for months because of temporary relocation. I can't imagine a life away from waters. Born and brought up here."	6) Terrifying ordeal 7) Personal accounts 8) Increased awareness	6), 7), Flood experience
Participant 11 (Male-32) "All of us have fresh memories of 2014, I guess. It can flood in future too. My home library was destroyed and books rendered useless. I am a bit cautious now and have a library at the Second floor. A place that has given wings to your dreams means a lot"	9) Capabilities 10) Liking for community	8), 9), Coping abilities
Participant 19 (Male-63) "May Allah save us all, 2014 Flood seems just a story of yesterday. It has happened and can happen again. Our young men made sure I am evacuated to safety. We learn from first-hand experiences and prepare. Our locality is safe and welcoming. We live together peacefully"	11) Caring relatives 12) Peace loving	10), 11), 12), Social cohesion

Our understanding and impressions of the outcomes/findings related to this study reveal that the benefits at the present site outweigh the threats in light of the past flood events severely impacting the region. The motivations (especially socially driven) have a good influence on people's lives and their decision-making process irrespective of gender, social or economic status and having disproportionate bearings on the individuals within same community in the past. There is a common understanding that if you face a constant threat in a particular place and can do a little about it while being aware, you might be an innocent person in the wrong place. However, there is a misconception that a hazard, even if it has turned into a disaster before, will not unleash havoc. This leads to underestimating the existing threats to one's life and property. This state of self-denial is unarguably distant from the ground realities that keeps haunting locals, time and again, with many negative consequences on its realization. Something very popular among the masses living near water bodies and exposed to flooding.

The outcomes/findings in terms of rich narrative descriptions around each theme (a total of five themes in our case) are as under:

Sense of Place

The study participants described their current place of living as a safe base, the land of their forefathers and having uncountable memories attached to it. A male respondent in his fifties beautifully put it as 'My home is my heaven in the world, you know.' The respondents know their place of residence, as their place of birth, which they celebrate every year with great pride and enthusiasm. Many of them associate several childhood memories with the place, with one of the middle-aged female participants sharing her motivation as 'This is the place where I was born, bred and brought up and it means a lot,' while the other participant goes a bit emotionally exhaustive about it and shares that 'I can't imagine a life away from waters. Water is everything to me and more.' These narrations of unheard masses undoubtedly provide a better reflection of the idea of people–place bonding and a deep sense of attachment to things. It also gives us an indication that their current place of living does not only hold something visible to a naked eye but there is a lot more (e.g., people–place attachment) that we would not be able to count until we make serious efforts in this direction.

Understanding Threat

Although the residents do not face an imminent threat of flooding when we study the flood history of this region, it has been established time and again that the low-lying areas and those areas adjacent to water bodies are exposed to a future flooding. The majority of study respondents understand that there is some kind of flood threat at the current place which has the potential to disturb their normal life to a great extent. The threat levels can vary as we move away from the water body, as one of the participants laments as ‘flooding is a real problem here but my house is still located at some distance than others’, while another participant states that ‘floods have struck this place before and same can repeat in the future as well’. This gives an impression that the residents to a certain degree do recognize the impending flood threat. Understanding threat is a good thing; however, there is every chance that the residents may underestimate or overestimate these threats and put themselves and their assets at risk. Their understanding of risk or threat perception will more or less depend on the consequences of their previous encounters with the similar (abnormal) situations, as well as their knowledge of the existing flood risk management mechanism. The residents do remember the 2014 flood event as a thing of yesterday and do worry about a similar situation unfolding in the future. At the same time, they remain watchful about the present and hopeful about the future by not letting such a situation (unfortunately if that arises ever) go worse or out of the control, the next time.

Flood Experience

The Kashmir Flood 2014, in the first week of September that caught the entire region unawares and leaving none untouched, was one of worst things in the residents living memory. ‘I remember everything (the good, the bad and the ugly) about 2014 flood. It made me realise that the water I treat as my friend can turn to be a foe too),’ narrated a female respondent when requested to share her ordeal. Everyone remembers the heroic efforts of young men during the search, rescue and immediate response phase while risking their own lives. According to a respondent who runs a business in the locality that ‘I have had heard many stories of courage and goodwill until 2014. It was the time when I became myself a witness to it. The youth of our area extended extraordinary help to us during that tough time. Ah! Something I can never forget.’ Many participants accept that whenever it rains for a few days they start worrying for their safety. However, this and some incidents of water logging/ storm surges make them believe that the past learnings will help them tackle similar situations (Allah forbids) in future.

Coping Abilities

Floods did not just bring destruction to the region but it also exposed the human-created protective/safety systems, which, in turn, the local residents recognize as a good learning experience. In addition to this, the traditional knowledge gained over time and especially the latest face-off in 2014 (knowingly or unknowingly) with some success while giving them a tough time made residents believe that they are able people. However, if provided with financial support, capacity building and training for flood risk reduction, they can make serious attempts to reduce their risk to flood hazard and protect themselves from future floods.

Social Cohesion

The study participants share a deep sense of belonging and togetherness at their current place of living. An adult male participant in his mid-thirties describes this feeling as 'a comfortable and heart-warming place, my area (without any doubt), where neighbours asking for your well-being and whereabouts of your other family members is a routine, a feeling that can't be described in words.' According to another participant (who has recently retired after serving for decades in the education department), 'the idea of "one community-progressive community" is only achievable when we come together under one roof for the sake of our own people, set aside trivial matters and denounce individual differences that pop-up over small things for the greater pursuit of building a strong and resilient community.' He explains further by maintaining that 'The strength of any community lies in the fact that members live peacefully, help and care about each other and develop as one community while independently growing as nicer individuals.'

Concluding Remarks

The preliminary study was successfully conducted after effectively implementing the study plan, while considering different aspects of performing mini-studies. At this stage, there was more clarity in terms of research problem under observation, objective, study design, area, participants, sampling plan, data collection and analysis process. Although the process was successful but it was not so smooth. The main conclusions are:

1. The residents perspectives towards living in flood-prone were useful to understand the people's psyche about existing and future flooding problem in the region.
2. The thick descriptions on themes, namely sense of place, understating threat, flood experience, coping abilities and social cohesion, after applying the narrative analysis technique gave some idea about residents' motivations for occupying exposed areas.
3. The narrative analysis technique was good to go with at this stage but a more reflexive qualitative data technique like thematic analysis could have been a better alternative to generate comprehensive discourse.
4. The interview process was time-consuming. Because the correct length of interview was not fixed in advance.

5. The interview guide was found to be somewhat insufficient, recognizing that more follow-up and probing questions were needed for better implementation.
6. The interviewer felt the need to follow the respondent, who brought more new information, while struggling to stick to the time schedule.
7. The need was felt to treat the interview as an interpersonal encounter, showing empathy towards the interviewee, adding humour (wherever appropriate) for the successful interview process.
8. The lack of stratification element prevented the interviewer from gaining a deeper understanding of (and access to) the small but vulnerable population of interest (e.g., urban poor).
9. The manual technique for coding was found laborious and very time-consuming. It made organization, coding and analysis processes more challenging.
10. The research inquiry, field observations and study findings after applying an emergent research design were interesting and, therefore, initiating a substantive study was feasible. Besides this, it gives an impression that 'pilot-studies should be a permanent feature of all similar/relevant studies'.

References

Census of India, Provisional Population Totals, Office of the Registrar General and Census Commissioner, New Delhi, 2011 Paper 1 of 2011 India, Series-1.

Collins, C. S., & Stockton, C. M. (2018). The central role of theory in qualitative research. *International Journal of Qualitative Methods, 17*(1), 1609406918797475.

Creswell, J. W., Hanson, W. E., Clark Plano, V. L., & Morales, A. (2007). Qualitative research designs: Selection and implementation. *The Counseling Psychologist, 35*(2), 236–264.

Herman, L., & Vervaeck, B. (2019). *Handbook of narrative analysis*. U of Nebraska Press.

In, J. (2017). Introduction of a pilot study. *Korean Journal of Anesthesiology, 70*(6), 601–605.

Malmqvist, J., Hellberg, K., Möllås, G., Rose, R., & Shevlin, M. (2019). Conducting the pilot study: A neglected part of the research process? Methodological findings supporting the importance of piloting in qualitative research studies. *International Journal of Qualitative Methods, 18*, 1609406919878341.

Merriam, S. B., & Tisdell, E. J. (2015). *Qualitative research: A guide to design and implementation*. John Wiley & Sons.

Riessman, C. K. (1993). *Narrative analysis* (Vol. 30). Sage.

Srinagar Master Plan (SMP) J&K, Srinagar Metropolitan Region – 2035. A publication of Town Planning Organization Kashmir, Srinagar Development Authority and approved by State Administrative Council (SAC), Government of Jammu and Kashmir in 2019.

Thabane, L., Ma, J., Chu, R., Cheng, J., Ismaila, A., Rios, L. P., et al. (2010). A tutorial on pilot studies: The what, why and how. *BMC Medical Research Methodology, 10*, 1–10.

Van Teijlingen, E., & Hundley, V. (2001). The importance of pilot studies. *Social Research Update, 35*, 1–4.

Wani, G. F., Ahmed, R., Ahmad, S. T., Singh, A., Walia, A., Ahmed, P., ... & Mir, R. A. (2022). Local perspectives and motivations of people living in flood-prone areas of Srinagar city, India. *International Journal of Disaster Risk Reduction, 82*, 103354.

Chapter 7
Determining the Impact of Land Use and Land Cover on Microclimate with Reference to Thermal Variability in Srinagar Municipal Corporation

Mohd Saqib, Saleha Jamal, Manal Ahmad, Md Ashif Ali, Aakib Yaqoob Mir, and Md Babor Ali

Introduction

Urbanization is a complex and multifaceted process that involves the growth and expansion of cities and towns, driven by social, economic, and environmental factors. It is a global phenomenon that has accelerated in recent decades, especially in developing nations like India, where the urban population rose from 27.8% in 2001 to 34.0% in 2011 (Census of India, 2011a). This rapid urbanization has profound implications on local and regional climates, land use and land cover (LULC) patterns, and the well-being of urban residents. The urban heat island (UHI) phenomenon, which sees urban regions endure hotter temperatures than their rural surroundings, is one of the most significant climatic effects of urbanization. This effect has significant implications for energy consumption, air quality, and public health in urban areas. Higher temperatures in cities increase the demand for air conditioning, leading to higher energy consumption and greenhouse gas emissions (EPA, 2008). Further, warmer temperatures can make air pollution worse by encouraging the production of ground-level ozone and other secondary pollutants (Jacob & Winner, 2009). Also, this has direct and indirect impacts on public health, including heat-related illnesses, respiratory and cardiovascular diseases, and reduced overall well-being (Hajat et al., 2010). It is primarily attributed to changes in LULC, including the replacement of natural vegetation with built-up areas, that absorb and re-emit solar radiation more efficiently than natural landscapes, which alters the energy balance and increases the overall land surface temperature (LST) (Weng, 2009; Jamal et al., 2022). In addition, the reduction of vegetation in urban areas

M. Saqib · S. Jamal (✉) · M. Ahmad · M. A. Ali · A. Y. Mir · M. B. Ali
Department of Geography, Aligarh Muslim University, Aligarh, India

A. L. Singh et al. (eds.), *Climate Change, Vulnerabilities and Adaptation*,
https://doi.org/10.1007/978-3-031-49642-4_7

decreases evapotranspiration, a process that cools the surface by releasing water vapor into the atmosphere (Grimmond, 2007).

The assessment of LST and UHI effects in urban areas is crucial for understanding the climatic impacts of urbanization, urban planning, and environmental management. Remote sensing data provides a valuable tool for this purpose, offering large-scale and high-resolution data on LST and LULC (Weng, 2009). Several studies have used remote sensing data to assess the effect of LST and UHI on Indian cities, revealing a strong correlation between LST and LULC (Singh et al., 2018a, b, c; Kumar et al., 2018a, b). For instance, a study by Singh et al. (2018a, b, c) analyzed the UHI effect in Jaipur, India, using remote sensing data and found a strong relationship between LST and the normalized difference vegetation index (NDVI) and the normalized difference built-up index (NDBI). The study concluded that increasing vegetation cover and reducing built-up areas could help mitigate the UHI effect in Jaipur. Similarly, Kumar et al. (2018a, b) conducted a comparative study of the UHI effect in Delhi and Mumbai, two of India's largest cities, using remote sensing data. The study found that both cities experienced a significant UHI effect, with higher LST values in built-up areas compared to vegetated areas. The authors suggested that urban planning strategies, such as increasing green spaces and promoting sustainable building practices, could help reduce the UHI effect in these cities. Also, these studies highlight the importance of understanding the relationship between urbanization, LST, and the UHI effect in the context of rapidly urbanizing countries like India. By assessing the spatial and temporal variations of LST and its relationship with LULC, researchers and policymakers can develop targeted strategies to mitigate the UHI effect and its implications for energy consumption, air quality, and public health.

High Mountain Asia has undergone substantial warming recently, with the most pronounced increase in temperature observed in the Kashmir Himalayas over the past several years. This warming trend is anticipated to persist across the Himalayas (Bhutiyani et al., 2007). Changes in temperature will impact all facets of the Himalayan alpine ecosystems, comprised of hydrology and the cryosphere (Meraj et al., 2014; Murtaza & Romshoo, 2017; Marazi & Romshoo, 2018). Srinagar, located in the Kashmir Himalaya, is notably characterized by urban sprawl. The temperature in the city of Srinagar is 2–5 °C higher than in the countryside (Ackerman, 1985; Jiang & Tian, 2010). Because of the increased work prospects brought about by its socioeconomic center status, more individuals from rural areas are moving to the city in search of better amenities. The city's structure has been permanently altered as a result of this urban flux and unrestrained growth, which has also caused serious environmental and ecological damage (Jamal et al., 2023). Using ArcGIS, this study aims to assess how these land system changes affect the city's (especially in it is a municipal corporation) land surface temperatures. This tool allows for the analysis of large-scale geospatial data, providing a comprehensive view of the changes in land surface temperatures over time. In addition to this, the study also aims to establish a statistical relationship between land surface temperature (LST) and biophysical indices such as the normalized difference vegetation index (NDVI), normalized difference built-up index (NDBI), modified normalized difference water index (MNDWI), normalized difference bareness index (NDBaI). These indices provide valuable insights into vegetation health, built-up, bareness and water content on the land surface. By correlating these indices with LST, the study hopes to gain a deeper understanding of how changes in vegetation and water content may influence the temperature variations in the city.

Database

This research has been conducted through geospatial satellite data. The satellite data were obtained from the USGS (United States Geological Survey) Earth Explorer (Table 7.1).

Methodology

Land Use and Land Cover Indices

Land use and land cover (LULC) indices are essential tools for monitoring and assessing changes in the Earth's surface, particularly in the context of urbanization, deforestation, and climate change. Among the widely used LULC indices are the normalized difference vegetation index (NDVI), modified normalized difference water index (MNDWI), normalized difference built-up index (NDBI), and normalized difference bareness index (NDBaI). These indices have been found to correlate with land surface temperature (LST), providing valuable insights into the relationships between land cover types and microclimatic conditions.

Normalized Difference Vegetation Index (NDVI)

The NDVI is a widely used index for assessing vegetation cover and monitoring vegetation dynamics (Rouse et al., 1974). It is calculated using Eq. (7.1):

Table 7.1 Geospatial data

Satellite	Sensor	Date	Resolution (m)	Path	Row	UTM zone
Landsat 7	Enhanced thematic mapper plus (ETM+)	29-08-2001	30	149	036	43
Landsat 5	Thematic mapper (TM)	01-08-2011	30	149	036	43
Landsat 8	Operational land imager (OLI)	12-08-2021	30	149	036	43

$$\mathrm{NDVI} = \frac{(\mathrm{NIR} - \mathrm{RED})}{(\mathrm{NIR} + \mathrm{RED})} \tag{7.1}$$

Lower NDVI values suggest sparse or non-vegetated environments, while greater values indicate richer vegetation. Studies have shown a negative correlation between NDVI and LST, suggesting that increased vegetation cover can help mitigate urban heat island effects (Weng et al., 2004).

Modified Normalized Difference Water Index (MNDWI)

The MNDWI is an index used for mapping and monitoring water bodies (Xu, 2006). It is calculated using the difference between green (GREEN) and mid-infrared (MIR) reflectance values, normalized by their sum, which is given by Eq. (7.2):

$$\mathrm{MNDWI} = \frac{(\mathrm{GREEN} - \mathrm{MIR})}{(\mathrm{GREEN} + \mathrm{MIR})} \tag{7.2}$$

Higher MNDWI values indicate the presence of water, while lower values represent non-water areas. The MNDWI has been found to correlate with LST, with water bodies generally exhibiting lower surface temperatures compared to surrounding land surfaces (Zhang et al., 2013).

Normalized Difference Built-Up Index (NDBI)

The NDBI is an index used for mapping and monitoring built-up areas (Zha et al., 2003). It is calculated using the difference between mid-infrared (MIR) and near-infrared (NIR) reflectance values, normalized by their sum, given by Eq. (7.3):

$$\mathrm{NDBI} = \frac{(\mathrm{SWIR} - \mathrm{NIR})}{(\mathrm{SWIR} + \mathrm{NIR})} \tag{7.3}$$

Built-up areas are represented by higher NDBI values, whereas undeveloped regions are represented by lower values. The urban heat island effect causes built-up regions to typically have higher surface temperatures, and studies have found a positive association between NDBI and LST (Weng et al., 2004).

Normalized Difference Bareness Index (NDBaI)

The NDBaI is an index used for mapping and monitoring bare soil and sparsely vegetated areas (Boschetti et al., 2008). It is calculated using Eq. (7.4):

$$\mathrm{NDBaI} = \frac{(\mathrm{SWIR} - \mathrm{NIR})}{(\mathrm{SWIR} + \mathrm{NIR})} \tag{7.4}$$

Higher NDBaI values indicate bare soil or sparsely vegetated areas, while lower values represent non-bare areas. The NDBaI has been found to correlate with LST, with bare soil and sparsely vegetated areas generally exhibiting higher surface temperatures compared to vegetated areas (Boschetti et al., 2008).

The calculation of LULC indices, such as NDVI, MNDWI, NDBI, and NDBaI, provides valuable information on land cover types and their dynamics. These indices have been found to correlate with LST, offering insights into the relationships between land cover types and microclimatic conditions. Understanding these relationships is essential for creating effective land management strategies and mitigating the impacts of urbanization and climate change.

Retrieving of Land Surface Temperature from Landsat Data

Land surface temperature (LST) refers to the temperature of the surface of the Earth's land area (Abbas et al., 2021). It differs from air temperature and can be affected by factors such as vegetation, soil type, and urbanization. LST can be measured using remote sensing techniques, such as thermal infrared (TIR) imaging. It is an important variable in understanding and predicting weather patterns, studying land-atmosphere interactions, and exploring the effects of land-use changes on the climate.

In this study, the mono-window technique described by Qin et al. (2001) was used to obtain land surface temperature (LST) data from multiple Landsat satellite sensors over time. This method requires specific parameters, including ground emissivity, atmospheric transmittance, and effective mean atmospheric temperature. Initially, the original thermal infrared (TIR) bands, with resolutions of 100 meters for Landsat 8 OLI data and 120 meters for Landsat 5 TM data, were resampled to a 30-meter resolution by the USGS data center for subsequent analysis.

Calculation of LST from Thermal Band of Landsat 5 TM

Conversion of Digital Number (DN) to Top-of-Atmosphere (TOA) Spectral Radiance

To convert the spectral radiance of each thermal band into brightness temperature, Eq. (7.5) was used:

$$L_\lambda = \left(\frac{L\_\max\lambda - L\_\min\lambda}{\text{Qcal\_max} - \text{Qcal\_min}}\right) \times (\text{Qcal} - \text{Qcal\_min}) + L\_\min\lambda \tag{7.5}$$

Here, L_λ represents spectral radiance, while Qcal represents the quantized calibrated pixel value in DN. L_maxλ refers to spectral radiance scaled to Qcal_max (in Watts/m^2 × sr × μm), and L_minλ refers to spectral radiance scaled to Qcal_min.

Conversion of Radiance into Brightness Temperature and Land Surface Temperature

$$T = \frac{K2}{ln\left(\frac{K1}{L_\lambda} + 1\right)} - 273.15 \tag{7.6}$$

Here, T represents the at-satellite brightness temperature; L_λ represents the top-of-atmosphere spectral radiance; and $K1$ and $K2$ are the constant values for the band being analyzed.

Calculation of LST from Thermal Band of Landsat 8 OLI

Conversion of Digital Number (DN) to Top-of-Atmosphere (TOA) Spectral Radiance

Initially, the Thermal Infrared Digital Numbers were converted to top-of-atmosphere (TOA) spectral radiance using the "radiance rescaling factor" proposed by Cohen (1960). This is calculated using Eq. (7.7) as:

$$L\lambda = \text{ML} \times \text{Qcal} + \text{AL} - Qi \tag{7.7}$$

Here, $L\lambda$ represents TOA spectral radiance (in Watts/m^2 × sr × μm), ML represents the radiance multiplicative band, AL represents the radiance add band, Qcal represents the quantized and calibrated standard product pixel value (in DN), and Qi represents the correction value for Band 10, which is 0.29.

Conversion of Top-of-Atmosphere (TOA)/Spectral Radiance to Brightness Temperature

$$T = \frac{K2}{ln\left(\frac{K1}{L_\lambda} + 1\right)} \tag{7.8}$$

In Eq. (7.8), T refers to the at-satellite brightness temperature, L_λ refers to the top-of-atmosphere spectral radiance, and $K1$ and $K2$ are the constant values specific to the band being analyzed.

Proportion of Vegetation (PV)

It is calculated using Eq. (7.9):

$$\mathrm{PV} = \left(\frac{\mathrm{NDVI} - \mathrm{NDVI_{min}}}{\mathrm{NDVI_{max}} - \mathrm{NDVI_{min}}}\right)^2 \tag{7.9}$$

Land Surface Emissivity

NDVI (normalized difference vegetation index) is a characteristic of natural objects and is a crucial surface parameter derived from the radiance emitted by a specific location. The NDVI values are utilized to evaluate the average emissivity of an element present on the Earth's surface. The mathematical representation of this is given by Eq. (7.10):

$$E = 0.004 \times \mathrm{PV} + 0.986 \tag{7.10}$$

Land Surface Temperature (LST)

The result of performing all the computations indicated above is the average temperature calculated from the object's observed brightness on Earth's surface. This is represented by Eq. (7.11):

$$LST = (BT/1) + W(BT/14280) * 1n(E) \tag{7.11}$$

The formula to compute the land surface temperature (LST) is given by Eq. (7.11), where LST is determined using the at-satellite brightness temperatures (BT) and the wavelength of emitted radiance (W). The equation also involves the natural logarithm of emissivity (*E*) and a constant value of 14,280 (Fig. 7.1).

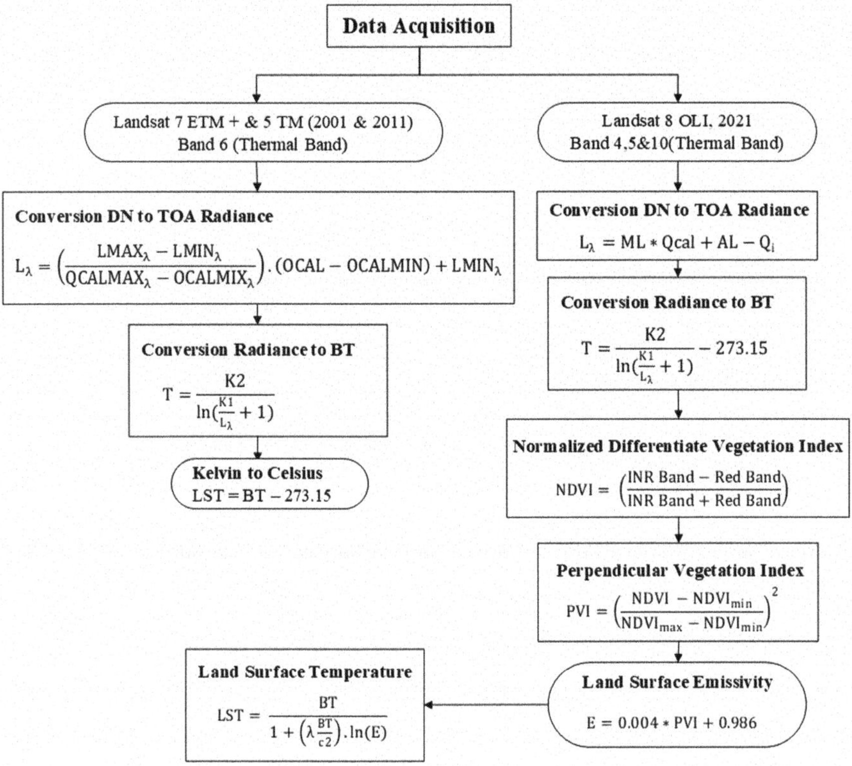

Fig. 7.1 Methodological flowchart for retrieving LST

Conversion of LST Data from Kelvin to Degree Celsius

Once all the five steps were completed, the LST data was obtained in Kelvin. Since this unit is not commonly used, the data was converted to the Celsius scale (°C) for ease of understanding. To get the result of LST in degree celcius, the formula i.e., C=(K-273.15) has been applied.

Statistical Tools

To explore the link between the indices of land use, land cover, and land surface temperature (LST), this study employs two statistical methods: Karl Pearson's correlation and simple linear regression. The methodology for each method is described below.

Karl Pearson's Correlation

Karl Pearson's correlation coefficient, also known as Pearson's *r*, is a measure of the linear relationship between two continuous variables (Pearson, 1895). In this study, Pearson's correlation is used to assess the strength and direction of the association between LST and LULC indices.

To calculate Pearson's correlation coefficient, Eq. (7.12) will be used:

$$r = \frac{\sum (X - \overline{X})(Y - \overline{Y})}{\sqrt{\sum (X - \overline{X})^2}\sqrt{(Y - \overline{Y})^2}} \tag{7.12}$$

Where, $\overline{X}$ = Mean of X Variables

$\overline{Y}$ = Mean of Y Variables

Simple Linear Regression

Simple linear regression is used to simulate the relationship between a dependent variable (in this case, LST) and an independent variable (in this case, LULC indices) (Montgomery et al., 2012). It seeks to identify the straight line that best captures the relationship between the two variables. The simple linear regression model can be represented as:

$$y = \beta 0 + \beta 1x + \varepsilon \tag{7.13}$$

where y is the dependent variable (LST), x is the independent variable (LULC indices), β_0 is the intercept, β_1 is the slope, and ε is the error term (Montgomery et al., 2012).

To estimate the parameters β_0 and β_1, the method of least squares will be used, which minimizes the sum of the squared differences between the observed values and the predicted values of the dependent variable (Montgomery et al., 2012).

Once the regression model is estimated, the coefficient of determination (R^2) will be calculated to assess the proportion of the variance in LST that can be explained by the LULC indices. In addition, hypothesis tests will be conducted to determine the statistical significance of the estimated parameters.

Study Area: Srinagar Municipal Corporation

One of the largest cities in the Himalayas, Srinagar is situated in the Kashmir valley between 34°5′23″ and 34°89′72″ north latitude and 74°47′24″ to 74°79′ east longitude at an altitude of around 1580–1595 meters (amsl) (Jamal & Ahmad, 2020). It is the summer capital of the Union Territory of Jammu and Kashmir, India, famous for its natural beauty, gardens, waterfronts, and houseboats (Kaul, 2011).

Srinagar Municipal Corporation (SMC) is the urban local body responsible for providing essential services and infrastructure to the city's residents. Established in 2000, the SMC is governed by the Jammu and Kashmir Municipal Corporation Act, 2000 (Government of Jammu and Kashmir, 2000). The corporation's jurisdiction covers an area of approximately 294 square kilometers, encompassing 74 municipal wards (SMC, 2021) (Fig. 7.2). The weather here is typically moderate throughout the summer, cool in the spring and fall, and frigid in the winter. The average mean minimum temperature is 7.29 °C, while the average mean maximum temperature is 19.27 °C. The wintertime lows are below zero, and the summertime highs reach 38 °C. The Valley receives 84 cm of rain annually on average (Shafiq et al., 2022). The city's population, as per the 2011 Census, was 1,273,312, with a population density of 4313 persons per square kilometer (Census of India, 2011a, b). It has a diverse population, comprising various ethnic and religious groups, including Kashmiri Muslims, Kashmiri Pandits, and Sikhs. The city's economy is primarily driven by tourism, handicrafts, and agriculture, particularly horticulture (Kaul, 2011). Due to the considerable expansion of transport, tourism, economic, and other networked amenities, the size of the Srinagar Municipal Corporation as well as its population have increased. The main issues that have been explored in this area include urban planning and development, solid waste management, sanitation, public health, and disaster management due to which the city is facing several challenges such as unplanned urbanization, inadequate infrastructure, environmental degradation, and temperature rise which have implications on the quality of life of its residents (Bhat & Rather, 2017).

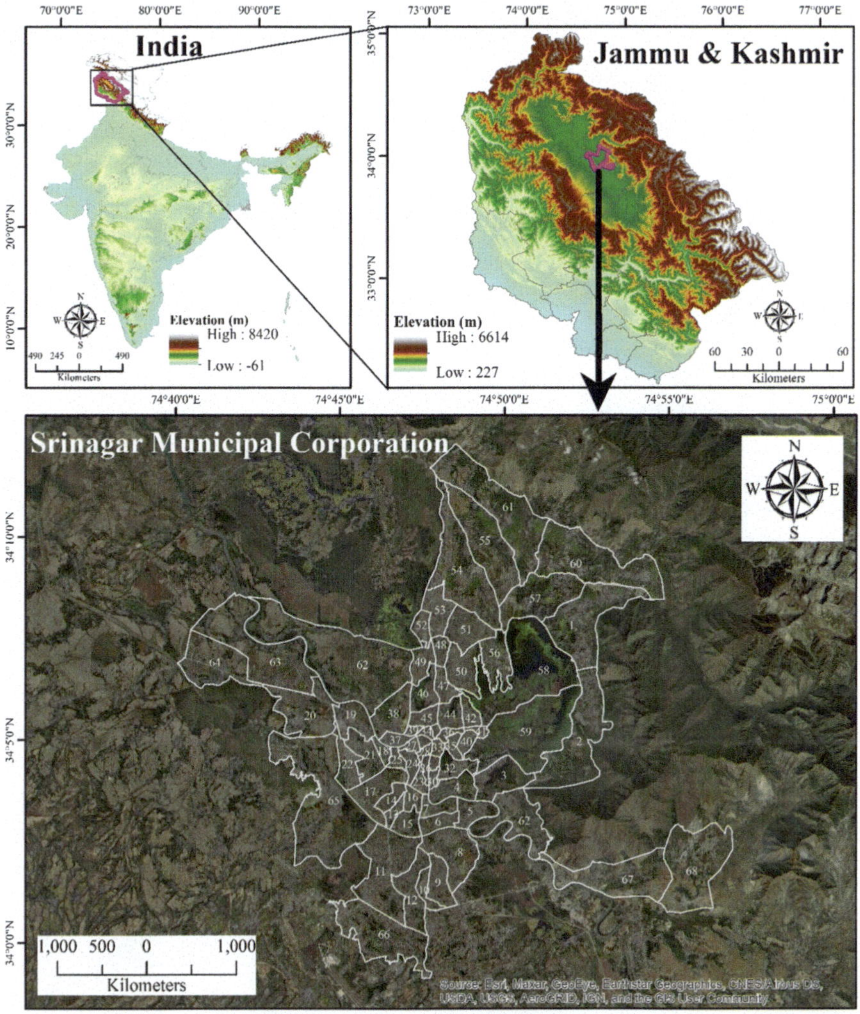

Wd. No.	Ward Name	Wd. No.	Ward Name	Wd. No.	Ward Name	Wd. No.	Ward Name
1	HARWAN	18	QAMARWARI	35	AQILMIR KHANYAR	52	SOURA
2	NISHAT	19	PARIMPORA	36	KHAWJA BAZAR	53	BUCHPORA
3	DALGATE	20	ZAINAKOTE	37	SAFA KADAL	54	AHMAD NAGAR
4	LAL CHOWK	21	BEMINA EAST	38	IDDGAH	55	ZAKORA
5	RAJBAGH	22	BEMINA WEST	39	TARABAL	56	HAZRATBAL
6	JAWAHAR NAGAR	23	SHAHEED GUNJ	40	JOGI LANKAR	57	TAILBAL
7	WAZIR BAGH	24	KARAN NAGAR	41	ZIND SHAH SAHIB	58	BUD DAL
8	MEHJOOR NAGAR	25	CHATTABAL	42	HASANABAD	59	LOCUT DAL
9	NATIPORA	26	SYED ALI AKBAR	43	JAMIA MASJID	60	NEW THEED
10	CHANAPORA	27	NAWAB BAZAR	44	MAKHDOOM SAHIB	61	ALASTENG
11	BAGHAT-I-BARZULLA	28	ISLAMYARBAL	45	KAWDARA	62	PALAPORA
12	RAWALPORA	29	AALI KADAL	46	ZADIBAL	63	MALOORA
13	SHEIKH DAWOOD COLONY	30	GANPATYAR	47	MADIN SAHIB	64	LAWAYPORA
14	BATAMALOO	31	BANA MOHALLA	48	NOWSHERA	65	KHUMANI CHOWK
15	ALLOCHI BAGH	32	SATHOO BARBARSHAH	49	ZOONIMAR	66	HUMHAMA
16	MAGARMAL BAGH	33	KHANKAHI-MOULLA	50	LAL BAZAR	67	PANTHA CHOWK
17	NUND RESH COLONY	34	S.R GUNJ	51	UMER COLONY	68	KHONMOH

Fig. 7.2 Study area: Srinagar Municipal Corporation

Results and Discussion

Normalized Difference Vegetation Index (NDVI)

The normalized difference vegetation index (NDVI) is a widely used remote sensing technique for assessing and monitoring vegetation health and density. In Fig. 7.3, we have analyzed the NDVI values for the Srinagar Municipal Corporation (SMC) area for the years 2001, 2011, and 2021.

The results indicate a significant change in the NDVI values over the two-decade period. The minimum NDVI values show an increasing trend, rising from −0.49 in 2001 to −0.12 in 2021. This suggests a reduction in non-vegetated or barren areas, which could be attributed to urban expansion, land-use changes, or reclamation of previously degraded lands.

On the other hand, the maximum NDVI values exhibit a fluctuating pattern, with an increase from 0.46 in 2001 to 0.71 in 2011, followed by a decrease to 0.61 in 2021. The increase in the maximum NDVI value between 2001 and 2011 may be indicative of improved vegetation health and density, possibly due to favorable climatic conditions, afforestation efforts, or better agricultural practices. However, the subsequent decline in the maximum NDVI value in 2021 could be a result of factors such as urbanization, land conversion for infrastructure development, or environmental stressors like climate change and pollution.

These findings highlight the dynamic nature of the SMC's vegetation cover over the years. The observed changes in NDVI values can have a big impact on the sustainability, ecology, and environment of the city. It is essential for policymakers and urban planners to consider these trends while formulating strategies for land-use planning, resource management, and environmental conservation in the SMC area.

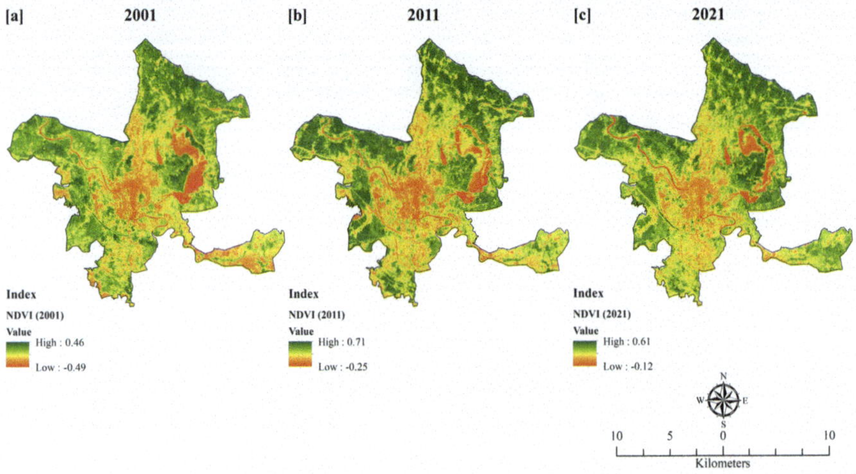

Fig. 7.3 Normalized deference vegetation index (NDVI) of Srinagar Municipal Corporation: (**a**) 2001, (**b**) 2011 and (**c**) 2021

Further research could involve a more in-depth analysis of the factors contributing to the observed NDVI changes, as well as the potential impacts on the city's ecosystem services, such as water regulation, air purification, and carbon sequestration. Future studies could explore the relationship between NDVI trends and socio-economic factors, such as population growth, urbanization, and economic development, to better understand the drivers of vegetation change in the SMC area.

Modified Normalized Difference Water Index (MNDWI)

The modified normalized difference water index (MNDWI) is a widely used remote sensing technique for mapping and monitoring water bodies (Xu, 2006). In this study, we analyze the MNDWI values for Srinagar Municipal Corporation (SMC) for the years 2001, 2011, and 2021, focusing on the changes in minimum and maximum MNDWI values over time (Fig. 7.4).

The MNDWI values for SMC in 2001, 2011, and 2021 show considerable variation. In 2001, the minimum MNDWI value was −0.34, while the maximum value was 0.61. By 2011, the minimum MNDWI value had decreased to −0.62, and the maximum value had also decreased to 0.55. In 2021, the minimum MNDWI value increased slightly to −0.39, but the maximum value decreased further to 0.26.

The decreasing trend in maximum MNDWI values from 2001 to 2021 suggests a reduction in the extent and quality of water bodies within SMC (Xu, 2006). This could be attributed to factors such as urbanization, land reclamation, and pollution, which have been known to negatively impact water bodies (Seto et al., 2012). The fluctuation in minimum MNDWI values may indicate changes in land cover types,

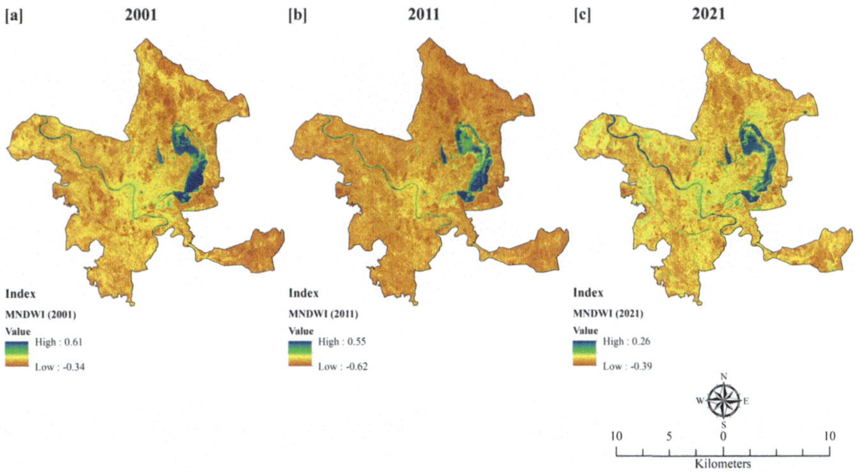

Fig. 7.4 Modified normalized deference water index (MNDWI) of Srinagar Municipal Corporation: (**a**) 2001, (**b**) 2011, and (**c**) 2021

such as an increase in impervious surfaces or vegetation, which can affect the MNDWI calculations (McFeeters, 1996).

The analysis of MNDWI values for SMC from 2001 to 2021 highlights the potential changes in water bodies over time. Further study is required to determine the root cause of these changes and to create effective management plans for safeguarding and enhancing the region's water resources.

Normalized Difference Built-Up Index (NDBI)

Figure 7.5 shows a decrease in the high NDBI values from 0.48 in 2001 to 0.22 in 2021, indicating a reduction in the extent or density of built-up areas over the 20-year period. This could be attributed to various factors such as urban planning policies, land-use changes, or economic factors affecting the growth of built-up areas.

On the other hand, the low NDBI values show a more complex trend. The value decreased from −0.31 in 2001 to −0.52 in 2011, suggesting an increase in non-built-up areas such as vegetation or water bodies. However, the value increased to −0.40 in 2021, indicating a possible decrease in non-built-up areas compared to 2011. This could be due to changes in land-use patterns, such as the transformation of agricultural land to built-up areas or the loss of vegetation due to natural or human-induced factors. Overall, the NDBI value for SMC suggests a decrease in built-up areas and a fluctuating trend in non-built-up areas over the 20-year period.

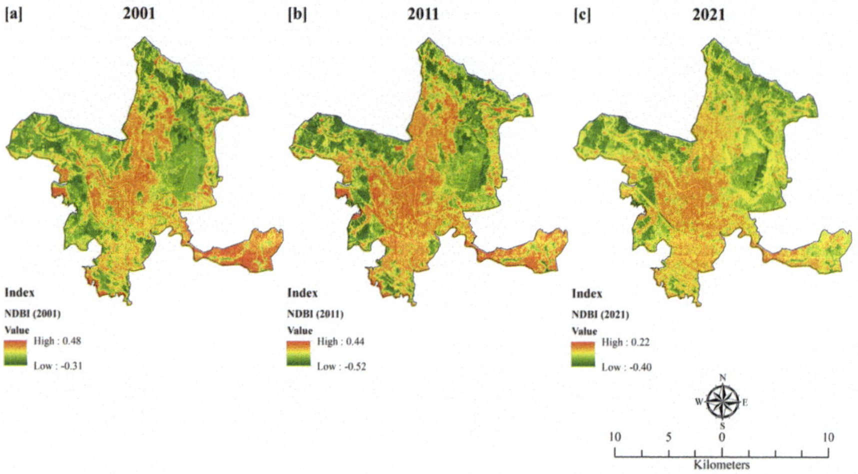

Fig. 7.5 Normalized deference built-up index (NDBI) of Srinagar Municipal Corporation: (**a**) 2001, (**b**) 2011, and (**c**) 2021

Normalized Difference Bareness Index (NDBaI)

Based on the provided NDBaI data for Srinagar MC, Fig. 7.6 shows a decrease in the high NDBaI values from 0.29 in 2001 to −0.03 in 2021, indicating a reduction in the extent or density of bare land areas over the 20-year period. This could be attributed to various factors such as land-use changes, reforestation efforts, or urban expansion.

On the other hand, the low NDBaI values show an increase from −0.81 in 2001 to −0.89 in 2011, suggesting a decrease in bare land areas and an increase in non-bare land areas such as vegetation or built-up areas. However, the value increased to −0.67 in 2021, indicating a possible increase in bare land areas compared to 2011. This could be due to changes in land-use patterns, such as the conversion of vegetation or built-up areas to bare land, or the loss of vegetation due to natural or human-induced factors. The NDBaI data for SMC suggests a decrease in bare land areas and a fluctuating trend in non-bare land areas over the 20-year period.

Land Surface Temperature (LST)

The data in Table 7.2 shows the Land Surface Temperature (LST) statistics for the Srinagar Municipal Corporation area for the years 2001, 2011, and 2021 (Fig. 7.7). The statistics include minimum temperature, temperature, mean temperature, standard deviation, and coefficient of variation.

Minimum Temperature The minimum LST has increased slightly over the years, from 17.75 °C in 2001 to 17.93 °C in 2011, and further to 18.10 °C in 2021. This

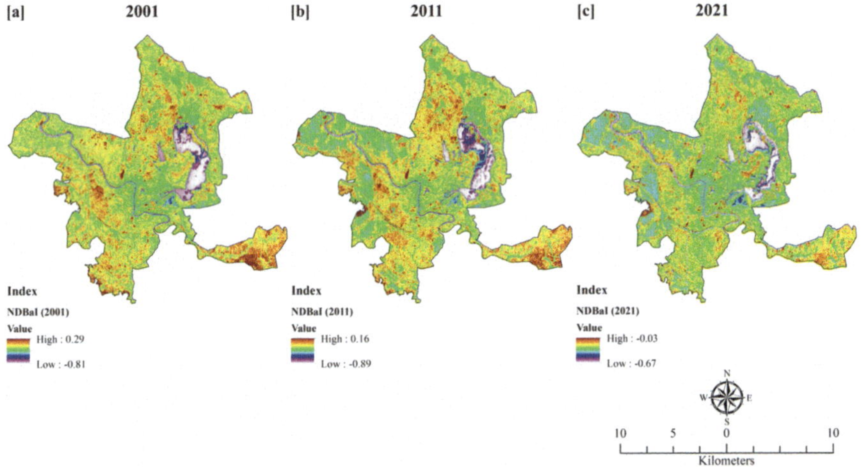

Fig. 7.6 Normalized deference bareness index (NDBaI) of Srinagar Municipal Corporation: (**a**) 2001, (**b**) 2011, and (**c**) 2021

Table 7.2 LST statistics (2001–2021)

Years	Minimum temperature (°C)	Maximum temperature (°C)	Mean temperature (°C)	Standard deviation	Coefficient of variation
2001	17.75	33.34	22.46	2.602	11.585
2011	17.93	33.72	23.02	2.571	11.169
2021	18.10	34.36	24.42	2.880	11.794

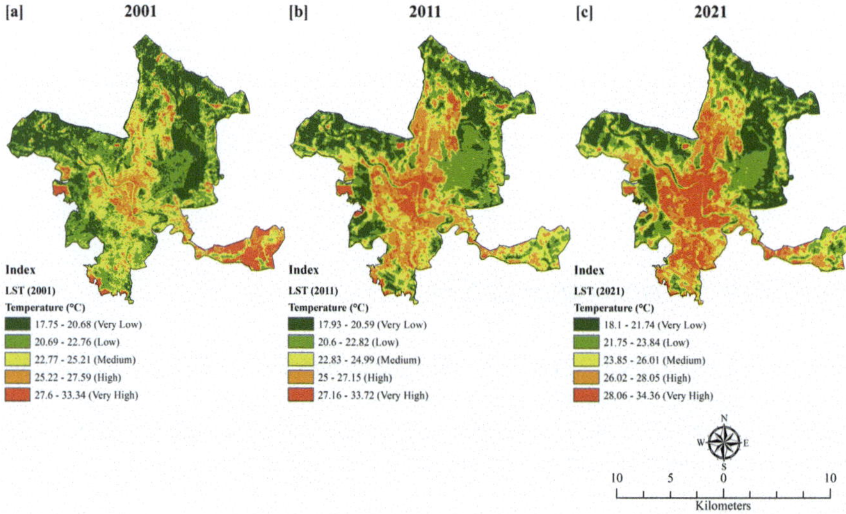

Fig. 7.7 Land surface temperature of Srinagar Municipal Corporation: (**a**) 2001, (**b**) 2011, and (**c**) 2021

indicates an increase in the lowest surface temperatures in the Srinagar Municipal Corporation area over the two decades.

Maximum Temperature The maximum LST has also increased over the years, from 33.34 °C in 2001 to 33.72 °C in 2011, and further to 34.36 °C in 2021. This suggests a consistent increase in the highest surface in the region over the 20-year period.

Mean Temperature The mean LST has increased consistently over the years, from 22.46 °C in 2001 to 23.02 °C in 2011, and further to 24.42 °C in 2021. This indicates a general warming trend in the Srinagar Municipal Corporation area over the two decades.

Standard Deviation The standard deviation, which measures the dispersion of LST values, has remained relatively stable over the years, with a slight increase from 2.602 in 2001 to 2.571 in 2011, and then a more noticeable increase to 2.880 in 2021. This suggests that the variability in surface temperatures has increased slightly over the 20-year period.

Coefficient of Variation The coefficient of variation, which measures the relative variability of LST values, has also remained relatively stable over the years, with values of 11.585 in 2001, 11.169 in 2011, and 11.794 in 2021. This indicates that the relative variability in surface temperatures has not changed significantly over the two decades.

The data reveals a general warming trend in the Srinagar Municipal Corporation area over the two decades, with an increase in minimum, maximum, and mean LST values. The variability in surface temperatures has remained relatively stable, with a slight increase in standard deviation and a relatively constant coefficient of variation. These findings have important implications for urban planning, public health, and environmental sustainability in the Srinagar Municipal Corporation area.

Area Under Different LST Categories

Figure 7.8 depicts the distribution of land surface temperature (LST) classes in the Srinagar Municipal Corporation area for the years 2001, 2011, and 2021. The corresponding data is provided in Table 7.3:

1. *Very Low LST:* The area with very low LST has decreased from 67.13 km (27.36%) in 2001 to 46.73 km (19.05%) in 2011, and then increased to

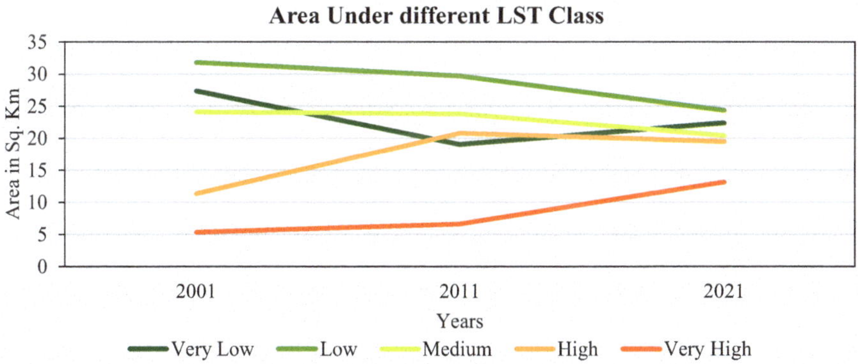

Fig. 7.8 Area under different LST categories (2001–2021)

Table 7.3 Area under different LST categories (2001–2021)

	2001		2011		2021	
LST class	Area (km)	Area (%)	Area (km)	Area (%)	Area (km)	Area (%)
Very low	67.13	27.36	46.73	19.05	54.98	22.41
Low	78.05	31.81	72.90	29.71	59.79	24.37
Medium	59.18	24.12	58.45	23.82	50.30	20.50
High	27.85	11.35	51.03	20.80	47.93	19.53
Very high	13.14	5.36	16.24	6.62	32.35	13.18

Table 7.4 Karl Pearson's coefficient of correlation between LST and LULC indices

Year	NDVI & LST	MNDWI & LST	NDBI & LST	NDBaI & LST
2001	−0.552	−0.358	0.862	0.491
2011	−0.638	−0.181	0.846	0.476
2021	−0.539	−0.283	0.783	0.407

54.98 km (22.41%) in 2021. This indicates a reduction in the extent of areas with the lowest surface temperatures over the two decades, with a slight recovery between 2011 and 2021.

2. *Low LST:* The area with low LST has also decreased from 78.05 km (31.81%) in 2001 to 72.90 km (29.71%) in 2011, and further to 59.79 km (24.37%) in 2021. This suggests a consistent decline in the extent of areas with low surface temperatures over the 20-year period.
3. *Medium LST:* The area with medium LST has remained relatively stable, with a slight decrease from 59.18 km (24.12%) in 2001 to 58.45 km (23.82%) in 2011, and then a more noticeable decrease to 50.30 km (20.50%) in 2021. This indicates a modest decline in the extent of areas with medium surface temperatures over the two decades.
4. *High LST:* The area with high LST has increased significantly from 27.85 km (11.35%) in 2001 to 51.03 km (20.80%) in 2011, and then slightly decreased to 47.93 km (19.53%) in 2021. This suggests a substantial increase in the extent of areas with high surface temperatures between 2001 and 2011, with a minor decline between 2011 and 2021.
5. *Very High LST:* The area with very high LST has increased from 13.14 km (5.36%) in 2001 to 16.24 km (6.62%) in 2011, and then more than doubled to 32.35 km (13.18%) in 2021. This indicates a consistent increase in the extent of areas with the highest surface temperatures over the 20-year period.

The data reveals a general trend of decreasing areas with very low and low LST, and increasing areas with high and very high LST in the Srinagar Municipal Corporation area over the two decades. This raises the possibility of a rise in surface temperatures in the area, which may be brought on by land cover changes, urbanization, and the urban heat island effect, among other things. These findings have important implications for urban planning, public health, and environmental sustainability in the Srinagar Municipal Corporation area.

The data provided in Table 7.4 shows the correlation between various vegetation and water indices (NDVI, MNDWI, NDBI, and NDBaI) and land surface temperature (LST) for the years 2001, 2011, and 2021. Figure 7.9 shows the simple linear regression analysis between LST and LULC indices (i.e., NDVI, MNDWI, NDBI, and NDBaI).

1. *NDVI & LST*: The negative correlation between NDVI (normalized difference vegetation index) and LST indicates that as vegetation cover increases, LST decreases. This is consistent with the cooling effect of vegetation due to shading

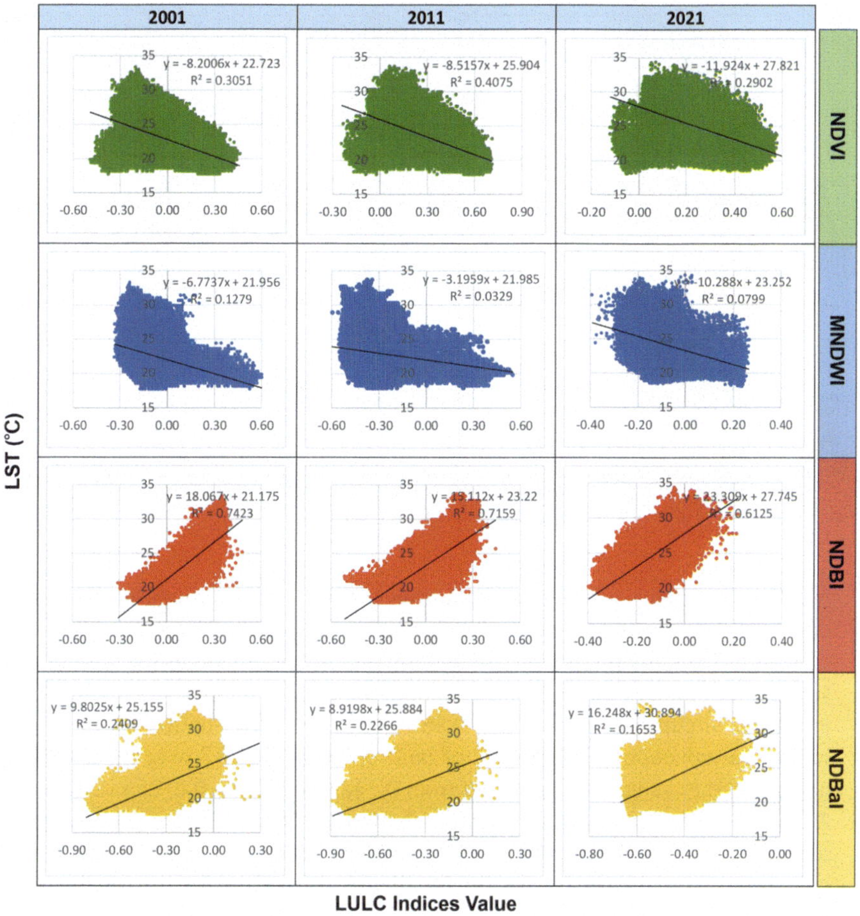

Fig. 7.9 Simple linear regression analysis between LST and LULC indices (i.e., NDVI, MNDWI, NDBI, and NDBaI)

and evapotranspiration. The correlation has become slightly less negative from −0.552 in 2001 to −0.539 in 2021, suggesting a possible decrease in the cooling effect of vegetation over time. Similarly, the regression value also shows a decreasing trend from 0.3051 to 0.2902 during 2001–2021.

2. *MNDWI & LST*: The negative correlation between MNDWI (modified normalized difference water index) and LST implies that as the presence of water bodies increases, LST decreases. This is in line with the cooling effect of water bodies due to their heat-absorbing and storing properties. The correlation has become less negative from −0.358 in 2001 to −0.283 in 2021, indicating a potential decrease in the cooling effect of water bodies over time. Also, the regression value decreased from 0.1279 to 0.0799 in 2001–2021.

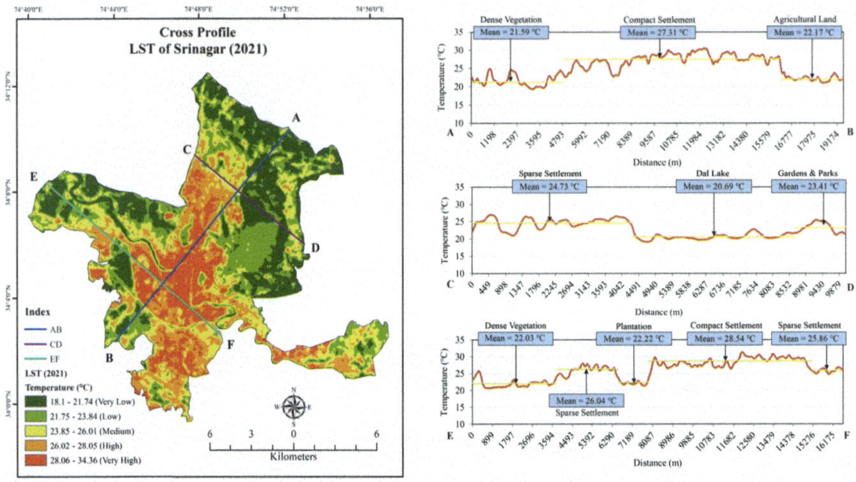

Fig. 7.10 Cross profile of land surface temperature (2021)

3. *NDBI & LST*: The positive correlation between NDBI (normalized difference built-up index) and LST suggests that as built-up areas increase, LST also increases. This is consistent with the urban heat island effect, where impervious surfaces and built-up areas contribute to higher surface temperatures. The correlation has slightly decreased from 0.862 in 2001 to 0.783 in 2021, which may indicate a reduction in the impact of built-up areas on LST over time. Consequently, the r^2 value also decreased from 0.7423 in 2001 to 0.6125 in 2021.
4. *NDBaI & LST*: The positive correlation between NDBaI (normalized difference bareness index) and LST indicates that as bare land areas increase, LST also increases. This is in line with the fact that bare land areas, with little to no vegetation or water bodies, tend to have higher surface temperatures. The correlation has decreased from 0.491 in 2001 to 0.407 in 2021, suggesting a possible reduction in the impact of bare land areas on LST over time. Hence, the linear regression value also decreased from 0.2409 to 0.1653 during the specified period.

The data shows that vegetation and water bodies have a cooling effect on LST, while built-up and bare land areas contribute to higher surface temperatures. Over the years, the correlations between these indices and LST have changed, indicating potential shifts in the impacts of land cover types on LST. Urban planners and policymakers can utilize this information to come up with plans to lessen the impact of urban heat islands and increase thermal comfort in cities.

Figure 7.10 shows the analysis of the data of three cross profiles of land surface temperature (LST) in SMC that reveal distinct patterns in LST values across different land cover types.

Cross Profile 1 (AB) shows that dense vegetation has a mean LST of 21.59 °C, which is significantly lower than the mean LST of 27.31 °C observed in compact

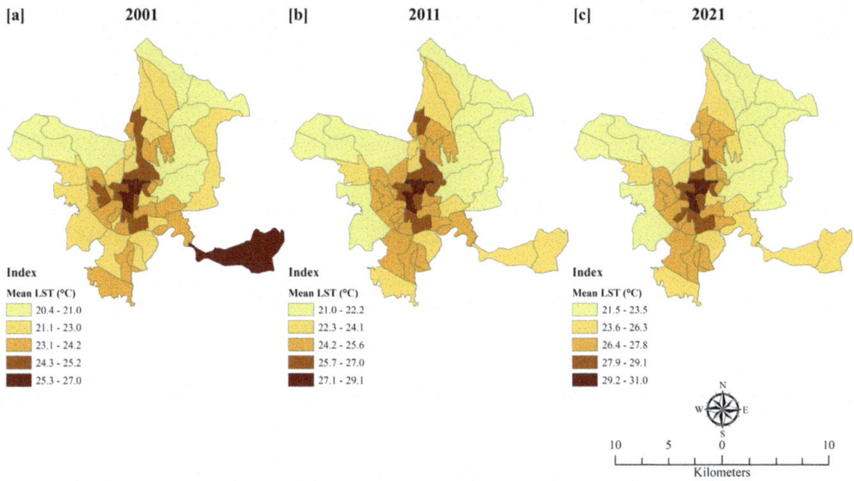

Fig. 7.11 Ward-wise LST of Srinagar Municipal Corporation: (**a**) 2001, (**b**) 2011, and (**c**) 2021

settlements. This difference can be attributed to the cooling effect of vegetation, which provides shade and promotes evapotranspiration, thereby reducing surface temperatures. Agricultural land, with a mean LST of 22.17 °C, also exhibits lower temperatures compared to compact settlements, likely due to the presence of vegetation and moisture in the soil.

Cross Profile 2 (CD) presents a different pattern, with sparse settlements having a mean LST of 24.73 °C, which is higher than the 20.69 °C observed for Dal Lake. This can be explained by the water body's ability to absorb and store heat, resulting in lower surface temperatures. Parks and gardens, with a mean LST of 23.41 °C, also exhibit lower temperatures compared to sparse settlements, highlighting the cooling effect of green spaces in urban areas.

Cross Profile 3 (EF) demonstrates a complex pattern of LST values across various land cover types. Dense vegetation has a mean LST of 22.03 °C, while sparse settlements have mean LST values of 26.04 °C and 25.86 °C. The lower temperature in dense vegetation can be attributed to the cooling effect of vegetation, as mentioned earlier. Plantations, with a mean LST of 22.22 °C, also exhibit lower temperatures, likely due to the presence of vegetation and moisture. Compact settlements, however, have the highest mean LST of 28.54 °C, indicating the strong influence of impervious surfaces and built-up areas on surface temperatures.

As a result, interpreting data of the three cross profiles of LST in SMC highlights the significant role of land cover types in influencing surface temperatures. Vegetation, water bodies, and green spaces tend to have lower LST values, while compact and sparse settlements exhibit higher temperatures. These results highlight the significance of including green areas and vegetation in urban planning to reduce the urban heat island effect and enhance thermal comfort in urban areas.

Based on the land surface temperature data of various wards in Srinagar, the following observations can be made (Fig. 7.11):

- Wards in city centers, like Khwaja Bazar, Karan Nagar, Nawab Bazar, Shaheed Gunj, Islam Yarbal, and Safa Kadal, have the highest land surface temperatures. This is likely due to higher concentration of built-up area and less vegetation in these wards.
- Wards with more open spaces, parks, and vegetation, like Hazratbal, Idgah, and Rajourikadal, have comparatively lower land surface temperatures.
- Wards on the city periphery, like Hyderpora, Bemina, and Parimpora, show land surface temperatures between 31 and 33 °C. Though less built up, these wards also have lesser vegetation.
- Wards, like Pantha Chowk and Khonmoh, lying too far from the city center, were earlier covered with bare land and had extremely high temperatures in 2001; but with the advent of time, these lands were converted for other land-use activities and thus temperature showed a slight decrease in 2011 and 2021.
- The ward-wise mean LST during the decades 2001 and 2021 showed a continuous increase from 25.3 °C to 31 °C in the areas having the maximum LST, while the mean LST increased from 20.4 °C to 23.5 °C in the areas having the minimum LST.

Recommendations

Based on the analysis, the following recommendations can be made to reduce urban heat effect:

- Increase green cover and urban forestry. Plant more trees along roads, parks, and other open spaces.
- Use cool and reflective materials for rooftops and pavements to reduce heat absorption.
- Adopt urban planning measures like increasing open spaces and water bodies to lower land surface temperatures.

Implement technology-based solutions like cool roofs, cool pavements, and urban canopies to combat urban heat.

Conclusion

The study has provided valuable insights into the spatial and temporal variations of land surface temperature (LST) in Srinagar Municipal Corporation (SMC) using remote sensing data from Landsat 7, Landsat 5, and Landsat 8 for the years 2001, 2011, and 2021, respectively. The analysis has demonstrated the significant influence of land use and land cover, vegetation cover, and urban morphology on LST, in line with previous research (Li et al., 2019; Singh et al., 2018a, b, c). It has contributed to the growing body of research on LST and urban heat island effects in rapidly urbanizing areas, especially in SMC, by highlighting the main areas and wards having the maximum and minimum LST. It has become increasingly susceptible to the negative effects of climate change and environmental degradation due to its uncontrolled urban growth and rapid population increase, which have disrupted the ecological services and hydrological procedures in the city. The study also showed that the lowest land surface temperature (LST) was in bodies of water and places with vegetation cover, while the highest LST was found in built-up areas and visible rock formations. The study also discovered a negative association between the NDVI and MNDWI indices for vegetation and water bodies and the land surface temperature. The pursuit of the Sustainable Development Goals (SDGs) by 2030 may be threatened by the continued trend of urban growth over the past several decades that has the potential to cause a catastrophic environmental imbalance. The use of remote sensing data has proven to be a valuable tool for assessing LST and its spatial and temporal variations, providing critical information for urban planning, environmental management, and public health initiatives.

The study's result has an important significance for SMC's urban planning and environmental management. By understanding the factors that contribute to higher LST and the urban heat island effect, policymakers and urban planners can develop strategies to mitigate these effects, such as increasing green spaces, promoting sustainable urban design, and implementing energy-efficient building practices. Furthermore, the results can inform public health initiatives aimed at reducing heat-related illnesses and improving overall well-being in urban areas. However, further research is needed on more detailed studies of the factors influencing LST in SMC, including the role of small-scale urban elements, such as construction supplies and street geometry. In addition, the potential effects of the urban heat island on local climate, quality of the air, and human health warrant further investigation. Long-term monitoring of LST and urban development in SMC is essential to assess the effectiveness of mitigation strategies and inform adaptive management approaches.

References

Abbas, A., He, Q., Jin, L., Li, J., Salam, A., Lu, B., & Yasheng, Y. (2021). Spatio-temporal changes of land surface temperature and the influencing factors in the Tarim Basin, Northwest China. *Remote Sensing, 13*(19). https://doi.org/10.3390/rs13193792

Ackerman, B. (1985). Temporal march of the Chicago heat Island. *Journal of Climate and Applied Meteorology, 24*, 547–554.

Bhat, R. A., & Rather, S. A. (2017). Urbanization and changing land use pattern in Srinagar city, Jammu and Kashmir, India. *Journal of Remote Sensing & GIS, 6*(2), 1–10.

Bhutiyani, M. R., Kale, V. S., & Pawar, N. J. (2007). Long-term trends in maximum, minimum and mean annual air temperatures across the Northwestern Himalaya during the twentieth century. *Climatic Change, 85*, 159–177. https://doi.org/10.1007/S10584-006-9196-1

Boschetti, M., Stroppiana, D., Brivio, P. A., & Bocchi, S. (2008). Multi-year monitoring of rice crop phenology through time series analysis of MODIS data. *International Journal of Remote Sensing, 29*(9), 2703–2718.

Census of India. (2011a). *Provisional population totals: Urban agglomerations and cities*. Office of the Registrar General and Census Commissioner, India.

Census of India. (2011b). *Srinagar District: Census 2011 data*. Retrieved from http://www.census2011.co.in/census/district/623-srinagar.html
Cohen, J. (1960). A coefficient of agreement for nominal scales. *Educational and Psychological Measurement, 20*(1), 37–46.
EPA. (2008). *Reducing Urban Heat Islands: Compendium of strategies*. Environmental Protection Agency.
Government of Jammu and Kashmir. (2000). *The Jammu and Kashmir Municipal Corporation Act, 2000*. Retrieved from http://jklaw.nic.in/pdf/MUNICIPAL%20CORPORATION%20ACT%202000.pdf
Grimmond, S. (2007). Urbanization and global environmental change: Local effects of urban warming. *The Geographical Journal, 173*(1), 83–88.
Hajat, S., O'Connor, M., & Kosatsky, T. (2010). Health effects of hot weather: From awareness of risk factors to effective health protection. *The Lancet, 375*(9717), 856–863.
Jacob, D. J., & Winner, D. A. (2009). Effect of climate change on air quality. *Atmospheric Environment, 43*(1), 51–63.
Jamal, S., & Ahmad, W. S. (2020). Assessing land use land cover dynamics of wetland ecosystems using Landsat satellite data. *SN Applied Sciences, 2*, 1891. https://doi.org/10.1007/s42452-020-03685-z
Jamal, S., Ahmad, W. S., Ajmal, U., Aaquib, M., Ashif Ali, M., Babor Ali, M., & Ahmed, S. (2022). An integrated approach for determining the anthropogenic stress responsible for degradation of a Ramsar Site–Wular Lake in Kashmir, India. *Marine Geodesy, 45*(4), 407–434.
Jamal, S., Ali, M. A., & Ahmad, W. S. (2023). Modelling transformation and its impact on water quality of the wetlands of lower Gangetic plain: A geospatial approach for sustainable resource management. *Geology, Ecology, and Landscapes*, 1–34.
Jiang, J., & Tian, G. (2010). Analysis of the impact of land use/land cover change on land surface temperature with remote sensing. *Procedia Environmental Sciences, 2*, 571–575.
Kaul, R. N. (2011). *Rediscovery of Ladakh*. Indus Publishing.
Kumar, R., Mishra, V., Buzai, G., Kumar, R., & Singh, S. (2018a). Analysis of urban heat Island (UHI) in relation to normalized difference vegetation index (NDVI): A comparative study of Delhi and Mumbai. *Environments, 5*(2), 21.
Kumar, S., Singh, R. P., & Singh, A. K. (2018b). Assessment of land surface temperature in Delhi using MODIS data. *Journal of the Indian Society of Remote Sensing, 46*(6), 947–956.
Li, X., Zhou, Y., & Li, X. (2019). Analysis of land surface temperature and its relationship with land use/land cover in urban areas. *Remote Sensing, 11*(3), 1–16.
Marazi, A., & Romshoo, S. A. (2018). Streamflow response to shrinking glaciers under changing climate in the Lidder Valley, Kashmir Himalayas. *Journal of Mountain Science, 15*, 1241–1253.
McFeeters, S. K. (1996). The use of the Normalized Difference Water Index (NDWI) in the delineation of open water features. *International Journal of Remote Sensing, 17*(7), 1425–1432.
Meraj, G., Romshoo, S.A., & Altaf, S. (2014). Inferring land surface processes from watershed characterization. Proceedings of the 16th International Association for Mathematical Geosciences - Geostatistical and Geospatial approaches for the characterization of natural resources in the environment: Challenges, processes and strategies, IAMG 2014 (pp. 430–432). https://doi.org/10.1007/978-3-319-18663-4_113/COVER
Montgomery, D. C., Peck, E. A., & Vining, G. G. (2012). *Introduction to linear regression analysis* (5th ed.). John Wiley & Sons.
Murtaza, K. O., & Romshoo, S. A. (2017). Recent glacier changes in the Kashmir Alpine Himalayas, India. *Geocarto International, 32*, 188–205. https://doi.org/10.1080/10106049.2015.1132482
Pearson, K. (1895). Notes on regression and inheritance in the case of two parents. *Proceedings of the Royal Society of London, 58*, 240–242.
Qin, Z., Karnieli, A., & Berliner, P. (2001). A mono-window algorithm for retrieving land surface temperature from Landsat TM data and its application to the Israel-Egypt border region.

International Journal of Remote Sensing, 22(18), 3719–3746. https://doi.org/10.1080/01431160010006971

Rouse, J. W., Haas, R. H., Schell, J. A., & Deering, D. W. (1974). Monitoring vegetation systems in the Great Plains with ERTS. *NASA Special Publication, 351*, 309.

Seto, K. C., Güneralp, B., & Hutyra, L. R. (2012). Global forecasts of urban expansion to 2030 and direct impacts on biodiversity and carbon pools. *Proceedings of the National Academy of Sciences, 109*(40), 16083–16088.

Shafiq, M. U., Tali, J. A., Islam, Z. U., Qadir, J., & Ahmed, P. (2022). Changing land surface temperature in response to land use changes in Kashmir valley of northwestern Himalayas. *Geocarto International, 37*, 1–20.

Singh, R. B., et al. (2018a). Urban Heat Island and land use land cover change: Study of a Metropolitan City (Delhi) in India. *Journal of Environmental Geography, 11*(1–2), 27–36.

Singh, R. B., Grover, A., & Zhan, J. (2018b). Urban heat Island and its relationship with NDVI and NDBI: A case study of Jaipur city in India. *The Egyptian Journal of Remote Sensing and Space Science, 21*(3), 269–277.

Singh, R. P., Kumar, S., & Singh, A. K. (2018c). Analysis of land surface temperature in Srinagar city using Landsat 8 data. *Journal of the Indian Society of Remote Sensing, 46*(5), 729–738.

Srinagar Municipal Corporation. (2021). *About SMC*. Retrieved from https://smcsite.org/about-smc

Weng, Q. (2009). Thermal infrared remote sensing for urban climate and environmental studies: Methods, applications, and trends. *ISPRS Journal of Photogrammetry and Remote Sensing, 64*(4), 335–344.

Weng, Q., Lu, D., & Schubring, J. (2004). Estimation of land surface temperature–vegetation abundance relationship for urban heat Island studies. *Remote Sensing of Environment, 89*(4), 467–483.

Xu, H. (2006). Modification of normalised difference water index (NDWI) to enhance open water features in remotely sensed imagery. *International Journal of Remote Sensing, 27*(14), 3025–3033.

Zha, Y., Gao, J., & Ni, S. (2003). Use of normalized difference built-up index in automatically mapping urban areas from TM imagery. *International Journal of Remote Sensing, 24*(3), 583–594.

Zhang, Y., Odeh, I. O. A., & Han, C. (2013). Bi-temporal characterization of land surface temperature in relation to impervious surface area, NDVI and NDBI, using a sub-pixel image analysis. *International Journal of Applied Earth Observation and Geoinformation, 21*, 125–138.

Part II
Adaptation to Climate Extremes: Human Response to Ecological Changes

Chapter 8
Variations in Rainfall in India: Is It Evidence of Climate Change?

Abdul Shaban and Sanjukta Sattar

Introduction

A large number of scientific research led by Inter-Governmental Panel on Climate Change (IPCC) now almost have affirmed the possibility of climate change. The tropical and subtropical countries are specifically predicted to be more affected from climate change because of their lower level of economic development (so lower capacities of adaptation and mitigation) and dependence of large share of population in these countries on primary economic activities (IPCC, 2021; UNEP, 2011; Raghunandan, 2011; Bidwai, 2011; Indian Meteorological Department, 2011). It is projected that droughts will rise throughout the twenty-first century in tropical regions (Kruger, 2006; Pal & Al-Tabbaa, 2009). There is evidence of declining rainfall in many regions, e.g., southeastern Australia (Murphy & Timbal, 2008; Manton et al., 2001). The decline in rainfall is also found in Italy (Cislaghi et al., 2005) and United Kingdom (Palmer & Raisanen, 2002). It is also predicted that the "changes in local regions can be far more dynamic than changes in global averages" (Pal & Al-Tabbaa, 2009). Fisk (1997) argues that Indian agriculture will be impacted significantly because it is largely rain-fed. The rainfall variations leading to water availability and variability may also severally impact industrial activity like power generation, having cascading effects on other sectors. Urban centers are also likely to be adversely affected due to lack of water (Shaban & Sattar, 2011). Given the differential impact which climate change may have in different regions, the degree

A. Shaban (✉)
School of Development Studies, Tata Institute of Social Sciences, Deonar, Mumbai, India
e-mail: shaban@tiss.edu

S. Sattar
Department of Geography, University of Mumbai, Mumbai, India

A. L. Singh et al. (eds.), *Climate Change, Vulnerabilities and Adaptation*,
https://doi.org/10.1007/978-3-031-49642-4_8

and nature of adaption and mitigation strategies will also be needed to be varied and context based. As rainfall is an important indicator of climate change, this research attempts to examine the annual, seasonal, and monthly rainfall behaviors in India over the 30 meteorological subdivisions during 1871 to 2008 to locate the regional and temporal trends in increase and/or decrease of rainfall, if any. This chapter is divided into six broad sections. Section "Data and Methodology" describes the data and statistical methods employed for the analysis. The trends at all-India level and in subdivisional annual rainfalls are examined in section "Trends in Annual Rainfall", while sections "Epochal Trends in Rainfall" and "Monthly and Seasonal Variations in Rainfall", respectively, analyze the epochal, and monthly and seasonal rainfall behaviors over India and at subdivisional level. The last section concludes the study.

Data and Methodology

This study uses the homogeneous region rainfall data 1871–2008 available from Indian Institute of Tropical Meteorology (IITM), Pune (IITM, 2009). This data titled 'Monthly and seasonal rainfall series for all-India homogeneous regions and meteorological sub-divisions, 1871-2008' have been compiled and constructed by Parthasarathy et al. (1994). This data series has been updated for the latter years as well, and most of the studies on rainfall in India (mentioned elsewhere in this paper) have been based on this rainfall series. Using the monthly rainfall data, the seasonal, *winter* (January–February), *pre-monsoon* (March–May), *south-west monsoon* (June–September), and *post-monsoon* (October–December), rainfalls have been computed.

This study uses various statistical methods to measure monthly, seasonal, annual, and regional rainfall trends, deviations, and variations. Some of the statistical methods used are deviation measure, coefficient of variation, trend analysis, Mann-Kendall trend test, Sen's slope and ordinary kriging.

Average annual deviations, ∂, of rainfall from grand mean, μ (rainfall during 1871–2008), for various subdivisions have been computed for different epochs to find out rainfall trend and periodic variations. The ∂ has been computed as follows:

$$\partial = \frac{\sum_{i=1}^{n}(x_i - \mu)}{n} \tag{8.1}$$

where, n is the number of years/observations in an epoch and x_i is millimeter (mm) of rainfall in ith year.

Mann-Kendall Trend Test

Mann-Kendall test statistics, *S*, is a nonparametric test to find out trend in data. It is not affected by outliers, and is computed as follows:

$$S = \sum_{k=1}^{n-1} \sum_{j=k+1}^{n} \text{sign}(x_j - x_k) \tag{8.2}$$

where, *n* is the number of observed data series and x_j and x_k are the values in the period *j* and *k* respectively; $j > k$; sign(θ) = 1 *if*$\theta > 0$; sign (θ) = 0 *if* $\theta = 0$; and sign (θ) = − 1 *if*$\theta < 0$. Mann (1945) and Kendall (1975) show that when $n \geq 10$, the *S* statistics is approximately normally distributed with *E(S)* = *0* (see Arora et al., 2005), and

$$\text{var}[S] = \left[n(n-1)(2n+5) - \sum_t t(t-1)(2t+5)\right]/18 \tag{8.3}$$

where *t* refers to the extent of any given tie and Σ_t denotes the summation over all ties. The standard normal variable *Z* is computed by (see Partal & Kahya, 2006):

$$\begin{aligned} Z &= \left\{\frac{S-1}{\text{var}(S)^{1/2}}\right\} && \text{if } S > 0 \\ Z &= 0 && \text{if } S = 0 \\ Z &= \left\{\frac{S+1}{\text{var}(S)^{1/2}}\right\} && \text{if } S < 0 \end{aligned} \tag{8.4}$$

The parameter *Z* will have positive value if there is an upward trend in the data and negative value in case of downward trend. We have computed two-tailed probability (*p*) values of the test statistics at various levels of significance to test our null hypothesis of "no trend."

To strengthen the finding and confidence about the trend in rainfall data, along with simple linear regression, We have also used Sen's slope estimator (Sen, 1968). Sen's method has an advantage over parametric simple linear regression model as it does not require the assumption of normal distribution of data (Gilbert, 1987). The Sen's nonparametric method used in this study shows change in rainfall (mm) per unit change in time (year) (Gilbert, 1987; de Lima et al. 2010; El-Nesr et al., 2010; Olofintoye & Sule, 2010: Salmi et al., 2002). In this work, Sen's slope is used in combination with Mann-Kendall trend test and is used to interpret the change per unit of time where Z value of Mann-Kendall trend test is statistically significant. In other words, when the hypothesis of no trend is rejected by Mann-Kendall test, the Sen's slope is used to quantify the trend (Patra et al., 2011).

The study has used ordinary kriging method to predict spatial value of rainfall in unknown locations. The centroids of the meteorological subdivisions have been treated as the points of incidence of rainfall and the interpolations and extrapolation of values have been made from those points (Harvard School of Public Health, 2011; Zhang, 2003; Wackernagel, 2003).

Trends in Annual Rainfall

This research, similar to that of Guhathakurta and Rajeevan (2008), shows no trend in annual rainfall in India. The simple linear regression equation yields a highly insignificant coefficient (see Table 8.1). However, there are nine meteorological subdivisions that show significant trends in annual rainfall. Out of these nine subdivisions, three (Chhattisgarh, East Madhya Pradesh, and Nagaland, Manipur, Mizoram, and Tripura) show negative trends while six subdivisions (Coastal Andhra Pradesh, Coastal Karnataka, Gangetic West Bengal, Konkan and Goa, Punjab, and Telangana) demonstrate positive trends (see Figs. 8.1, 8.2, and 8.3). Mann-Kendall test also supports the increase or decrease of annual rainfall in these subdivisions, except in Nagaland, Manipur, Mizoram, and Tripura. It also shows that there is a positive trend of rainfall in Haryana, Chandigarh, and Delhi. Sen's slope coefficient shows that rainfall in Chhattisgarh has experienced the greatest decline among all the subdivisions. The decline of rainfall over the period 1871–2008 in Chhattisgarh has been on an average, −1.440 mm per year as per Sen's slope, and 1.437 mm per year as per the simple regressions equation. As per Sen's slope, the decline in rainfall in East Madhya Pradesh has been on an average − 0.947 per year (−0.928 mm as per simple regression equation), while the decrease in the rainfall in Naga, Manipur, Mizoram, and Tripura has been at a rate of −0.758 mm (−0.870 mm) per year. However, in Nagaland, Manipur, Mizoram, and Tripura, the trend coefficient is statistically significant at 10% level of significance and the Mann-Kendall test shows the trend as insignificant.

Many meteorological subdivisions in the country show an increasing trend in annual rainfall. Most of these subdivisions, e.g., Gangetic West Bengal, Coastal Andhra Pradesh, Konkan and Goa, and Coastal Karnataka, are located along the coast. Thus, on both the eastern and western coasts (in some specific regions), the annual rainfall is showing increasing trend. Besides these coastal subdivisions, subdivisions of Punjab, Haryana, Chandigarh and Delhi, and Telangana also show increasing trends in annual rainfall. As per Sen's slope, Coastal Karnataka and Konkan and Goa are experiencing an increase in annual rainfall of above 2 mm per year, while in Gangetic West Bengal and Coastal Andhra Pradesh the increase has been 1.08 mm per year and 0.820 mm per year, respectively. In Punjab, Haryana, Chandigarh and Delhi and Telangana, the increase has been between 0.6 and 0.8 mm per year. Thus, the result shows that Central India (Chhattisgarh and Eastern Madhya Pradesh) is drying up, while some coastal areas and northwestern parts of the country are getting wetter. This result is also supported by the findings of Guhathakurta and

Table 8.1 Trends in annual rainfall in meteorological subdivisions of India, 1871–2008

S.N.	Meteorological subdivision	Division codes	Simple linear OLS regression				Mann-Kendall Z	Sen's slope
			Intercept	Trend coefficient	p-value	R-square		
1	Assam and Meghalaya	AM	2383.403	−0.474	0.389	0.005	−1.296	−0.716
2	Nagaland, Manipur, Mizoram, and Tripura	NMMT	2049.270	−0.870$	0.100	0.020	−1.443	−0.758
3	Sub Him. West Bengal and Sikkim	WS	2505.291	0.005	0.994	0.001	−0.142	−0.108
4	Gangetic West Bengal	GWB	1464.290	1.170*	0.020	0.020	2.084*	1.084
5	Orissa	ORS	1487.047	−0.038	0.930	0.000	0.013	0.008
6	Jharkhand	JKND	1348.672	−0.017	0.968	0.000	−0.077	−0.045
7	Bihar	BH	1233.052	−0.080	0.864	0.000	−0.114	−0.052
8	East Uttar Pradesh	EUP	1037.496	−0.144	0.750	0.001	−0.352	−0.175
9	West Uttar Pradesh	WUP	882.377	−0.061	0.874	0.000	−0.707	−0.296
10	Haryana, Chandigarh, and Delhi	HCD	524.183	0.501	0.105	0.019	1.947$	0.624
11	Punjab	PNJ	585.361	0.802*	0.040	0.035	2.363*	0.877
12	West Rajasthan	WR	285.300	0.130	0.579	0.002	0.370	0.086
13	East Rajasthan	ER	708.359	−0.328	0.373	0.006	−1.020	−0.398
14	West Madhya Pradesh	WMP	960.397	−−0.259	0.498	0.003	−0.913	−0.367
15	East Madhya Pradesh	EMP	1320.255	−0.928*	0.036	0.032	−2.056*	−0.947
16	Gujarat	GJ	912.198	0.003	0.995	0.000	0.052	0.037
17	Saurashtra, Kutch, and Diu	SKT	454.409	0.251	0.557	0.003	0.786	0.300
18	Konkan and Goa	KG	2405.690	1.810$	0.077	0.020	2.203*	2.120
19	Madhya Maharashtra	MM	731.149	0.113	0.718	0.001	0.190	0.056
20	Marathwara	MRT	818.220	0.201	0.669	0.001	0.342	0.165
21	Vidarbha	VDRB	1111.517	−0.379	0.388	0.005	−1.123	−0.514
22	Chhattisgarh	CHHT	1457.861	−1.437**	0.003	0.063	−2.908**	−1.440
23	Costal Andhra Pradesh	CAP	931.645	0.707$	0.084	0.022	1.960*	0.820

(continued)

Table 8.1 (continued)

S.N.	Meteorological subdivision	Division codes	Simple linear OLS regression				Mann-Kendall Z	Sen's slope
			Intercept	Trend coefficient	p-value	R-square		
24	Telangana	TLNG	841.858	0.683$	0.090	0.021	1.806$	0.761
25	Rayalaseema	RYLM	682.823	0.554	0.110	0.019	1.498	0.496
26	Tamil Nadu and Pondicherry	TNP	907.170	0.306	0.339	0.007	0.571	0.197
27	Costal Karnataka	CKT	3122.038	2.141*	0.064	0.025	2.065*	2.285
28	North Interior Karnataka	NIKT	812.863	0.265	0.405	0.005	0.757	0.275
29	South Interior Karnataka	SIKT	875.431	0.075	0.814	0.000	0.195	0.064
30	Kerala	KRL	2830.342	−0.030	0.973	0.000	−0.035	−0.025
	All-India		1089.890	−0.180	0.930	0.000	0.098	0.023

Source: Computed using data from Indian Institute of Tropical Meteorology, Pune
Note: ** significant at 1%, * significant at 5%, and $ significant at 10% level of significance

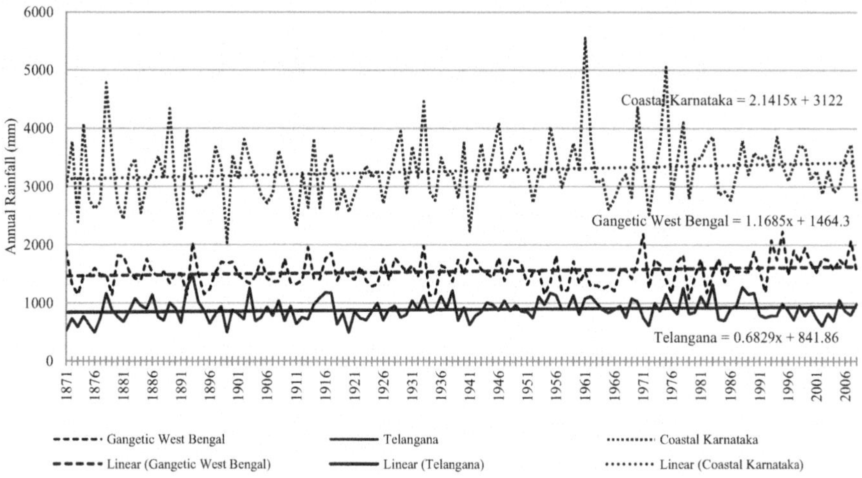

Source: Based on data from Indian Institute of Tropical Meteorology, Pune.

Fig. 8.1 Patterns and trend in annual rainfall in Gangetic West Bengal, Telangana, and Coastal Karnataka during 1871–2008. (*Source*: Based on data from Indian Institute of Tropical Meteorology, Pune)

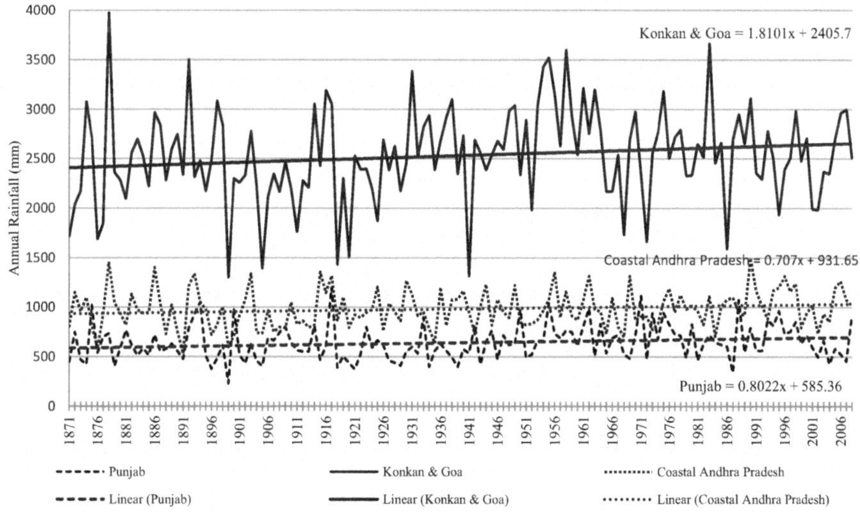

Source: Based on data from Indian Institute of Tropical Meteorology, Pune.

Fig. 8.2 Patterns and trend in annual rainfall in Punjab, Konkan and Goa, and Coastal Andhra Pradesh during 1871–2008. (*Source*: Based on data from Indian Institute of Tropical Meteorology, Pune)

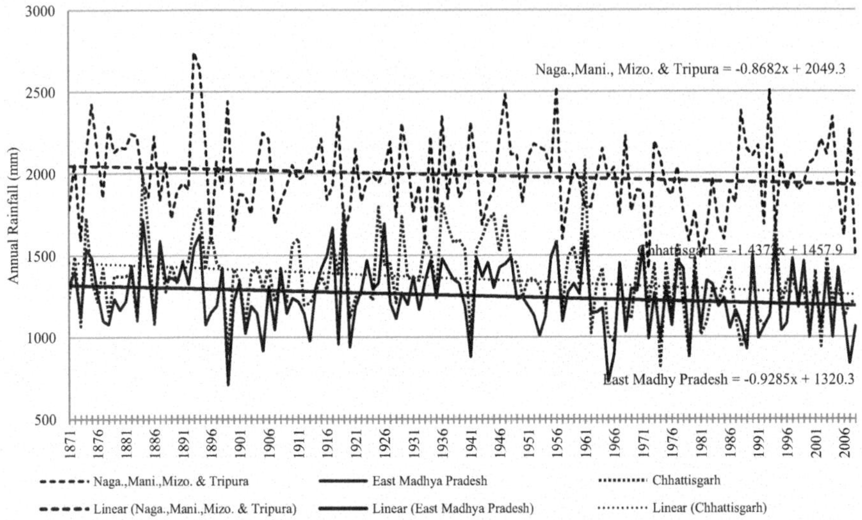

Source: Based on data from Indian Institute of Tropical Meteorology, Pune.

Fig. 8.3 Patterns and trend in annual rainfall in East Madhya Pradesh, Chhattisgarh, and Nagaland, Manipur, Mizoram, and Tripura during 1871–2008. (*Source*: Based on data from Indian Institute of Tropical Meteorology, Pune)

Rajeevan (2008), who using data for 1901–2003 showed significant and remarkable variations at regional scale in rainfall in India. Some of the studies, like that of Roy and Balling (2004), also show a decline in rainfall in northeastern states, parts of the Eastern Gangetic Plain, and Uttaranchal, while Soman et al. (1988) show a decreasing trend in rainfall in Kerala. However, these studies use different years of data and so results are not strictly comparable.

Epochal Trends in Rainfall

Annual rainfall in India shows considerable periodic variation. It is found that the annual rainfall in India largely follows 30-year cycle of alternate wet and dry period (Fig. 8.4). The cycle of 30 years is also reported by (Guhathakurta & Rajeevan, 2008). The wet periods have largely coincided with 1871–1900, 1931–1960, and 1991–2008 (the last period is likely to be till 2020), while the dry periods have largely been 1901–1930 and 1961–1990 (see Table 8.2). As can be seen from Table 8.2, at all-India level, the wet epochs have had excess in annual rainfall from the mean rainfall (except 1991–2008, this epoch may continue till 2020), while dry epochs have had deficits in rainfall. The average excess in annual rainfall at all-India level during 1871–1900 and 1931–1960 was 6.5 mm and 39.2 mm,

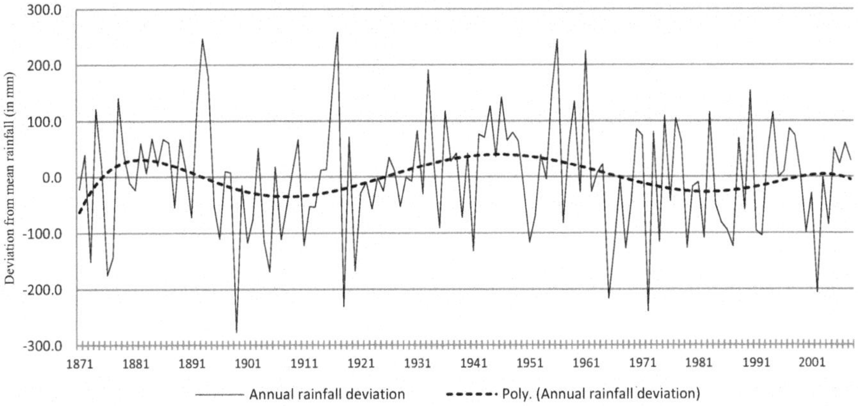

Note: the Poly.(Annual rainfall deviation) shows the polynomial regression line of 6th order to the rainfall deviation.
Source: Computed using data from Indian Institute of Tropical Meteorology, Pune.

Fig. 8.4 All-India annual rainfall deviations from mean annual rainfall during 1871–2008. *Note*: The Poly (annual rainfall deviation) shows the polynomial regression line of sixth order to the rainfall deviation. (*Source:* Computed using data from Indian Institute of Tropical Meteorology, Pune)

respectively, while an average annual deficit in rainfall during 1900–1930 and 1961–1990 has been of −25.2 mm and 16.4 mm, respectively. During 1991–2008, there was an average annual deficit in rainfall of −6.8 mm at all-India level, and as it is expected to be a wet epoch (till 2020) the deficit of rainfall may decline in later years.

During the respective wet and dry epochs, clustered positive or negative anomalies in rainfalls have had considerable incidence. However, it is not that during a wet epoch all the subdivisions have experienced excess rainfall over the grand mean of 1871–2008. Also, the excess and deficit of rainfall has taken place largely in geographically contiguous areas. During the wet epoch of 1871–1900, as many as 15 subdivisions had deficit in rainfall. The deficit of rainfall was in a contiguous geographical area in the eastern region (Assam and Meghalaya, Northern West Bengal, and Sikkim), Punjab and Rajasthan, and the area comprising four southern states of Andhra Pradesh, Karnataka, Tamil Nadu, and Kerala. During 1900–1931 dry epoch, 31 subdivisions experienced deficit in annual rainfall, but Assam and Meghalaya, Sub-Himalayan West Bengal and Sikkim, Chhattisgarh, East Madhya Pradesh, Tamil Nadu and Pondicherry, and Kerala had experienced excess of rainfall. During the wet epoch of 1931–1960, only eight subdivisions had deficit in rainfall, which included Gangetic West Bengal, East Uttar Pradesh, Haryana, Chandigarh and Delhi, Punjab, Coastal Andhra Pradesh, Rayalaseema, and Tamil Nadu and Pondicherry. In all the three epochs during 1871–1960 (that included two wet and one dry epoch), Gangetic West Bengal, Haryana, Chandigarh and Delhi, Punjab, Coastal Andhra Pradesh, and Rayalaseema had experienced deficit in rainfall, while West Rajasthan, Konkan and Goa, Coastal Karnataka, and South Interior Karnataka had experienced deficit in rainfall till 1930. In fact, the deficit in

Table 8.2 Average deviation of rainfall, coefficient of variation, and epochal trend in annual rainfall

		Average deviation of annual rainfall (mm) in different epochs						Epochal coefficient of variation (CV%)							Mann-Kendall Z, 1961–2008	Sen's slope, 1961–2008
S. No.	Subdivision	1871–1900	1900–1930	1931–1960	1961–1990	1991–2008	1961–2008	1871–1900	1901–1930	1931–1960	1961–1990	1991–2008	1961–2008	1871–2008		
1	Assam and Meghalaya	−23.60	11.20	117.80	−60.40	−75.00	−65.90	10.23	9.23	9.46	12.66	12.28	12.39	10.91	0.11	0.37
2	Nagaland, Manipur, Mizoram, and Tripura	72.50	−4.20	25.30	−98.70	8.50	−58.50	13.74	9.33	11.92	12.02	13.46	12.76	12.29	1.02	3.03
3	Sub Him. West Bengal and Sikkim	−45.80	87.40	−9.00	−36.40	6.30	−20.40	15.62	13.30	11.13	12.01	14.41	12.86	13.24	1.01	3.17
4	Gangetic West Bengal	−33.60	−30.30	−14.90	−30.50	182.10	49.20	14.89	12.18	15.00	16.45	14.55	16.82	15.19	2.78**	8.70
5	Orissa	2.20	−7.10	61.40	−90.80	57.10	−35.30	13.78	9.63	12.29	13.56	16.37	15.48	13.36	2.06*	4.76
6	Jharkhand	−6.60	15.70	22.60	−54.40	37.70	−19.90	12.01	12.30	13.91	17.52	17.27	17.57	14.49	1.56	4.61
7	Bihar	26.70	−0.90	−16.90	−8.20	−1.30	−5.60	19.11	17.43	15.05	18.70	20.08	19.02	17.78	1.05	2.67
8	East Uttar Pradesh	20.30	−17.10	29.20	−7.80	−40.90	−20.20	24.27	19.70	19.48	20.91	16.05	19.19	20.50	0.06	0.14
9	West Uttar Pradesh	35.80	−53.40	20.20	8.10	−17.90	−1.70	19.88	22.72	18.31	21.56	18.86	20.48	20.43	−0.49	−1.12
10	Haryana, Chandigarh, and Delhi	−9.00	−37.00	−4.20	30.70	32.50	31.40	25.22	27.26	27.76	25.64	21.04	23.78	25.80	−0.37	−0.53
11	Punjab	−21.60	−56.50	−0.20	58.90	32.40	49.00	31.47	29.14	24.68	27.02	23.97	25.79	27.90	−1.18	−2.57
12	West Rajasthan	−1.50	−11.90	0.10	8.20	8.60	8.30	30.88	49.76	30.31	38.30	36.07	37.09	37.18	−0.12	−0.17
13	East Rajasthan	26.90	−34.80	48.00	−7.50	−54.50	−25.10	21.75	32.46	23.65	21.71	23.42	22.36	25.07	−1.38	−2.03
14	West Madhya Pradesh	26.70	−84.40	88.70	19.50	−83.90	−19.30	14.86	20.32	16.78	18.65	19.08	19.42	18.91	−1.84$	−3.74
15	East Madhya Pradesh	46.60	0.80	50.00	−54.50	−71.40	−60.90	16.17	17.18	12.44	18.39	18.20	18.14	16.57	−0.40	−1.15

16	Gujarat	26.10	−45.10	52.70	−59.30	42.60	−21.10	28.92	34.53	28.68	30.55	27.76	29.65	30.24	1.08	3.71
17	Saurashtra, Kutch, and Diu	16.00	−29.40	1.80	−6.90	30.60	7.20	40.62	46.28	43.39	45.24	34.32	40.91	42.16	1.02	2.34
18	Konkan and Goa	−56.90	−223.80	225.40	81.00	−42.90	34.50	21.84	18.66	16.94	17.69	13.19	16.28	18.96	−0.68	−2.43
19	Madhya Maharashtra	21.70	−54.90	26.40	−14.00	34.80	4.30	22.15	20.07	18.75	16.28	19.79	17.87	19.77	0.13	0.16
20	Marathwara	28.20	−79.70	31.80	21.40	−2.70	12.40	31.27	27.87	19.57	26.46	22.82	25.00	26.28	0.10	0.24
21	Vidarbha	23.00	−43.30	72.80	−19.10	−55.70	−32.80	20.89	17.01	18.49	19.80	13.90	17.84	18.85	−0.31	−0.57
22	Chhattisgarh	48.60	12.80	116.40	−108.70	−115.10	−111.10	16.51	13.45	12.79	20.74	15.08	18.66	16.88	−0.07	−0.12
23	Costal Andhra Pradesh	−30.10	−17.90	−5.50	9.90	72.80	33.50	24.12	19.65	17.30	18.69	16.22	17.82	19.51	1.38	2.83
24	Telangana	−43.10	−44.00	51.00	73.60	−62.60	22.50	27.64	21.53	15.99	19.17	14.58	19.22	21.18	−1.68$	−3.24
25	Rayalaseema	−4.80	−30.60	−13.80	−3.00	87.00	30.80	26.89	24.40	18.16	16.26	24.45	20.81	22.51	1.44	2.43
26	Tamil Nadu and Pondicherry	−25.60	19.40	−17.40	−2.90	44.30	14.80	15.77	15.02	16.90	15.92	17.33	16.49	16.07	0.06	0.10
27	Costal Karnataka	103.90	−148.70	81.10	142.20	48.80	107.20	19.94	13.31	13.70	20.21	9.68	17.09	16.54	0.00	0.07
28	North Interior Karnataka	4.30	−53.60	29.00	24.60	−7.20	12.70	21.10	17.50	17.27	15.49	15.87	15.57	17.83	−1.54	−2.02
29	South Interior Karnataka	−2.00	−19.70	29.00	−16.70	15.70	−4.60	18.84	15.93	14.19	16.86	19.99	18.03	16.90	−0.12	−0.29
30	Kerala	−132.50	122.50	99.80	−76.60	−22.00	−56.10	16.82	12.00	13.49	15.52	10.90	13.83	14.31	0.01	0.11
	All–India	6.50	−25.20	39.20	−16.40	−6.80	−12.80	9.94	9.08	8.15	10.25	7.63	9.27	9.29	0.54	0.58

Source: Computed using data from Indian Institute of Tropical Meteorology, Pune
Note: Deviations in rainfall are computed from arithmetic mean of annual rainfall during 1871–2008

annual rainfall in Gangetic West Bengal had persisted till 1990. It is also to be noted that in many of these subdivisions like Gangetic West Bengal, Punjab, Haryana, Chandigarh and Delhi, and Rayalaseema the deficit in rainfall in earlier epochs are found due to higher rainfalls in latter epochs and because of that their mean values are affected (move upward). However, it must also be acknowledged that mean annual rainfall during 1871–2008 is a better approximation of normal rainfall than any shorter period, and that is why this mean has been used for computing the excess or the deficit in subdivisional rainfalls.

The period 1960–1990 was a dry epoch, and during this epoch 19 subdivisions had experienced deficit in annual rainfall. During this period the contiguous area comprising northeastern states, West Bengal, Orissa, Jharkhand, Bihar, and eastern Uttar Pradesh experienced deficit in rainfall. Deficit in rainfall during the period was also in contiguous area of East Madhya Pradesh, Chhattisgarh, and Vidarbha. Other subdivisions affected by the deficit in rainfall were Gujarat, Saurashtra, Kutch and Diu, East Rajasthan, Rayalaseema, Tamil Nadu and Pondicherry, South Interior Karnataka, and Kerala. The 1991–2020 epoch is considered to be a wet epoch. During 1991–2008, total 14 subdivisions had had deficit in rainfall. In fact, in many of the subdivisions, deficit in rainfall had been during both the period 1961–1990 and 1991–2008. The meteorological subdivisions with deficit in rainfall during both the periods are Assam and Meghalaya, Bihar, Eastern Uttar Pradesh, East Rajasthan, East Madhya Pradesh, Chhattisgarh, Vidarbha, and Kerala.

Another major aspect of annual rainfall in India has been that its variability (coefficient of variation, CV) has declined in majority of the meteorological subdivisions and also at all-India level over the epochs (Table 8.2). At all-India level, the CV in annual rainfall consistently decline in all the epochs, except in 1961–1990 when it rose. In majority of the subdivisions, the variability in rainfall rose during 1900–1931, but had declined thereafter. The subdivisions in which the rainfall variability has significantly increased over the period have been Assam and Meghalaya, Orissa, Jharkhand, and Tamil Nadu and Pondicherry. In all the subdivisions, the variability in annual rainfall has been higher than that at all-India level. Table 8.2 also shows the Mann-Kendall trend test and Sen's slope for the annual rainfall during 1961–2008. Only in Gangetic West Bengal (at 1% level of significance), Orissa (at 5% level), West Madhya Pradesh (at 10% level), and Telangana (at 10% level), the trends in annual rainfall during this period are statistically significant. As per the Sen's slope, in Gangetic West Bengal and Orissa, the rainfall has increased at the rate of 8.7 mm and 4.76 mm per year, respectively, while in West Madhya Pradesh and Telangana, it has declined at the rate of −3.74 mm and −3.24 mm per year during the same period.

Figure 8.5a to f are specifically useful for visualization of dynamics of rainfall in India over various epochs. The figures are based on the average annual deficit or excess of rainfall during various epochs over mean annual rainfalls of respective regions during 1871–2008. Ordinary kriging has been used to inter- and extrapolate the trend surface. The data for Jammu and Kashmir, Himachal Pradesh, Uttaranchal, and Arunachal Pradesh are not available; however, trend surface over these regions have also been computed (therefore, the trend surface forecast in these four regions

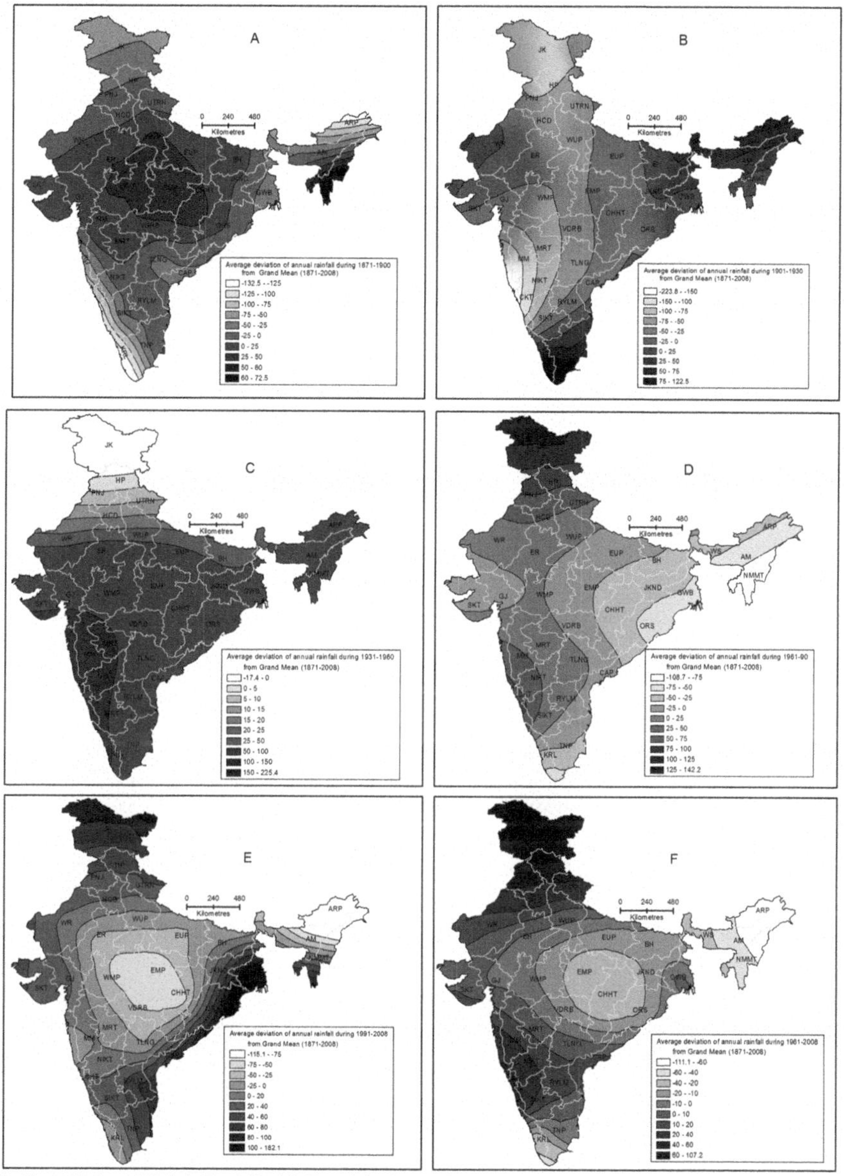

Note: The data for Jammu & Kashmir, Himachal Pradesh, Uttaranchal, and Arunachal Pradesh are not available. However trend surface for them are computed based on data from other subdivisions, and therefore prediction surface for these four states may not be very reliable. Ordinary kriging is used for prediction surface.

Source: Computed using data from Indian Institute of Tropical Meteorology, Pune.

Fig. 8.5 Trend in epochal deviations (average surplus and deficits in mm) of annual rainfall from mean of annual rainfall during 1871–2008. *Note:* The data for Jammu and Kashmir, Himachal Pradesh, Uttaranchal, and Arunachal Pradesh are not available. However, trend surface for them are computed based on data from other subdivisions, and therefore prediction surface for these four states may not be very reliable. Ordinary kriging is used for prediction surface. (*Source*: Computed using data from Indian Institute of Tropical Meteorology, Pune) may be less reliable). The figures show, as discussed above and shown in Table 8.2, excess of annual rainfall during 1871–1900 in Central India and in Nagaland, Manipur, Mizoram, and Tripura, while there was deficit in rainfall in south India and northwest India during the same period. In the next epoch, 1900–1930, the rainfall deficit was specially pronounced along the Konkan and Goa coast. Deficit in rainfall was also visible in a large part of the western flank of the country. During 1931–1960, Konkan and Goa had excess rainfalls. The eastern flank of the country experienced deficit in rainfall during 1961–1990, and during 1991–2008 the deficit in rainfall has been higher in Central India, specifically Chhattisgarh, Telangana, Madhya Pradesh and East Rajasthan, Vidarbha, and Konkan and Goa. As shown in Table 8.2, the deficit in most of these subdivisions has persisted during 1961–2008, and therefore both the Fig. 8.5e and f show rainfall hollow at Central India.

Monthly and Seasonal Variations in Rainfall

India gets a large proportion of its rainfall during summer monsoon (June to September). During 1871–2008, about 78% of the total annual rainfall in the country occurred during summer monsoon season (see Table 8.3). Post-monsoon season (October–December) during the period contributed 11.08% of the total rainfall, while the share of winter (January–February) and pre-monsoon (March–May) rainfalls was 2.18% and 8.71% respectively. It is also evident from Table 8.3 that the share of monsoon rainfall to the total annual rainfall in India is declining, while the share of other seasons is increasing. The greatest increase has been in the share of rainfall during post-monsoon season, about 1.0% between the epoch 1871–1990 and 1991–2008, while the share of summer monsoon has declined about 2% between the same epochs. Monthwise, October has experienced major increase in its share in total rainfall (about 1.0%), while the greatest decrease has been in the share of July rainfall (about 1.5%). The share of June rainfall is also showing declining trend while shares of rest of the months are either increasing or stable.

As we have seen above, some specific months have experienced increase or decrease in their share in total rainfall. However, trend analysis shows that the month-wise increase or decrease has not been homogenous across the subdivisions. The July rainfall shows statistically significant negative trend in Chhattisgarh, East Madhya Pradesh, and also at all-India level (see Table 8.4). As per the Sen's slope, the July rainfall in Chhattisgarh and East Madhya Pradesh has decreased at a rate of −0.623 mm and − 0.692 mm per year, respectively, during 1871–2008, while the decline at all-India level has been −0.127 mm per years. October rainfall has shown increasing trend in many subdivisions. However, except in Telangana (where it is significant at 5% level), in other subdivisions the trend coefficient is significant only at 10% level of significance. The cyclical nature of the summer monsoon rainfall in India is well brought out by Fig. 8.6. This chapter shows that the summer monsoon rainfall (during 1871–2008) has significant declining trends in Assam and Meghalaya, Nagaland, Manipur, Mizoram and Tripura, Chhattisgarh, and East Madhya

Table 8.3 Share of seasonal and monthly rainfalls to the total annual rainfall in India during various epochs

	Share (%) of months in total rainfall												Seasonal share (%) in total rainfall				
Epochs	Jan	Feb	Mar	Apr	May	Jun	Jul	Aug	Sep	Oct	Nov	Dec	Jan-Feb	Mar-May	Jun-Sept	Oct-Dec	Annual
1871–1900	0.89	0.98	1.38	2.25	4.90	16.21	25.25	21.59	15.99	6.87	2.55	1.14	1.87	8.53	79.04	10.56	100.00
1901–1930	1.22	1.38	1.43	2.51	4.65	14.86	25.17	22.17	15.56	6.80	3.19	1.07	2.59	8.58	77.76	11.06	100.00
1931–1960	1.09	1.20	1.36	2.39	4.92	14.30	25.44	22.30	15.73	7.63	2.78	0.86	2.29	8.67	77.77	11.27	100.00
1961–1990	0.86	1.11	1.42	2.47	4.74	14.48	24.87	23.25	15.61	6.97	2.92	1.29	1.97	8.64	78.22	11.18	100.00
1991–2008	0.86	1.32	1.45	2.69	5.25	16.29	23.75	21.77	15.10	7.76	2.73	1.03	2.18	9.39	76.91	11.51	100.00
1961–2008	0.86	1.19	1.43	2.55	4.93	15.16	24.45	22.69	15.42	7.26	2.84	1.19	2.05	8.92	77.72	11.30	100.00
1871–2008	1.00	1.18	1.40	2.44	4.86	15.13	25.00	22.25	15.64	7.16	2.84	1.08	2.18	8.71	78.03	11.08	100.00

Source: Computed using data from Indian Institute of Tropical Meteorology, Pune

Table 8.4 Man-Kendall coefficient and Sen's slope for monthly rainfall (mm) for the period 1871–2008

	January		Feb		Mar		Apr		May		Jun		Jul		Aug		Sep		Oct		Nov		Dec	
	M-K Z	Sen's slope	M-K Z	Sen's slope	M-K Z	Sen's slope	M-K Z	Sen's slope	M-K Z	Sen's slope	M-K Z	Sen's slope	M-K Z	Sen's slope	M-K Z	Sen's slope	M-K Z	Sen's slope	M-K Z	Sen's slope	M-K Z	Sen's slope	M-K Z	Sen's slope
Assam and Meghalaya	−1.813$^{\$}$	−0.038	0.845	0.038	−1.519	−0.151	0.451	0.056	−1.309	−0.227	−0.900	−0.185	0.716	0.143	−2.829**	−0.498	−1.202	−0.216	1.721$^{\$}$	0.269	0.723	0.027	0.597	0.007
Nagaland, Manipur, Mizoram, and Tripura	−1.381	−0.013	−0.567	−0.025	−1.439	−0.143	0.248	0.037	0.727	0.137	−1.132	−0.208	−0.676	−0.136	−3.000**	−0.518	−1.622	−0.250	0.324	0.045	0.410	0.016	−0.619	0.000
Sub Him. West Bengal and Sikkim	−0.431	−0.002	0.182	0.003	−0.269	−0.013	1.044	0.115	−0.002	0.000	−2.257*	−0.671	2.213*	0.798	−0.722	−0.231	−0.792	−0.244	1.697$^{\$}$	0.289	2.372*	0.017	2.12*	0.002
Gangetic West Bengal	1.059	0.008	−0.125	−0.003	0.486	0.014	1.031	0.059	−0.606	−0.073	−0.622	−0.129	0.955	0.195	0.692	0.121	2.487**	0.408	0.909	0.133	1.455	0.014	−1.005	0.000
Orissa	1.630	0.006	0.821	0.016	0.499	0.014	2.003*	0.083	−0.098	−0.008	−0.145	−0.033	−0.930	−0.185	1.335	−0.140	−0.324	−0.057	−0.747	−0.119	−0.232	−0.003	−1.846$^{\$}$	0.000
Jharkhand	0.214	0.000	−0.436	−0.012	−0.009	0.000	1.318	0.037	−0.188	−0.013	0.226	0.034	−0.806	−0.162	−1.463	−0.241	0.501	0.084	0.300	0.031	0.708	0.000	0.393	0.000
Bihar	−0.059	0.000	0.028	0.000	0.442	0.004	1.451	0.031	1.289	0.091	−0.871	−0.167	0.506	0.104	−1.208	−0.270	0.180	0.039	0.758	0.070	1.245	0.000	1.347	0.000
East Uttar Pradesh	0.177	0.004	0.622	0.009	−0.015	0.000	2.939**	0.025	0.552	0.015	0.348	0.057	−0.758	−0.175	−1.555	−0.301	1.289	0.221	0.694	0.030	1.103	0.000	1.033	0.000
West Uttar Pradesh	−1.362	−0.040	0.394	0.006	0.586	0.004	1.512	0.008	1.046	0.026	0.663	0.080	−0.884	−0.185	−0.088	−0.019	0.127	0.019	1.705$^{\$}$	0.026	0.701	0.000	−0.067	0.000
Haryana, Chandigarh, and Delhi	−0.506	−0.013	0.957	0.018	1.184	0.020	2.100*	0.013	0.790	0.020	1.149	0.093	0.436	0.065	2.014*	0.354	−0.142	−0.021	0.659	0.000	2.467**	0.000	−0.763	0.000
Punjab	−1.070	−0.044	0.970	0.039	1.996*	0.067	1.305	0.018	0.725	0.016	2.091*	0.165	0.933	0.172	1.449	0.233	0.961	0.123	1.879$^{\$}$	0.008	2.883**	0.000	−0.574	0.000
West Rajasthan	−0.268	0.000	0.633	0.000	0.983	0.000	1.979*	0.004	0.274	0.003	1.011	0.042	0.398	0.046	−0.202	−0.020	−0.777	−0.043	1.027	0.000	0.013	0.000	−0.715	0.000
East Rajasthan	−1.139	−0.007	0.393	0.000	0.478	0.000	1.844$^{\$}$	0.005	−0.166	−0.002	0.147	0.010	−0.569	−0.114	0.243	0.055	−0.692	−0.104	0.686	0.004	0.485	0.000	−1.777$^{\$}$	0.000
West Madhya Pradesh	−0.528	0.000	0.024	0.000	1.633	0.004	1.270	0.002	0.560	0.005	−1.824	−0.262	−0.924	−0.185	1.769	0.357	−1.335	−0.279	1.103	0.021	0.712	0.000	−1.081	0.000
East Madhya Pradesh	0.241	0.000	−0.006	0.000	0.652	0.005	0.963	0.008	0.742	0.011	−1.896$^{\$}$	−0.372	−2.765**	−0.692	0.364	0.085	−0.854	−0.170	0.018	0.000	−0.633	0.000	−0.481	0.000
Gujarat	−0.904	0.000	−0.711	0.000	−0.700	0.000	−0.787	0.000	−2.104*	−0.002	0.526	0.098	−1.023	−0.360	1.209	0.325	−0.580	−0.111	0.435	0.000	1.667$^{\$}$	0.000	−1.143	0.000
	0.310	0.000	−1.230	0.000	0.580	0.000	0.728	0.000	−0.659	0.000	0.849	0.094	−1.132	−0.264	1.401	0.212	0.201	0.017	0.797	0.000	0.932	0.000	−0.948	0.000

Saurashtra, Kutch, and Diu																								
Konkan and Goa	−0.374	0.000	−0.400	0.000	0.819	0.000	1.408	0.000	1.197	0.018	0.872	0.382	−0.114	−0.054	2.524**	1.117	0.425	0.148	1.047	0.123	−0.203	0.000	0.250	0.000
Madhya Maharashtra	0.944	0.000	−0.083	0.000	1.065	0.000	−0.841	−0.012	−0.114	−0.003	1.244	0.159	−1.473	−0.214	2.406*	0.287	−0.799	−0.117	0.265	0.024	−0.494	−0.007	−1.092	0.000
Marathwara	1.132	0.000	1.047	0.000	2.163*	0.003	−0.265	0.000	0.617	0.010	−0.101	−0.018	0.131	0.021	1.522	0.292	−1.905$	−0.410	1.651$	0.159	0.833	0.000	−0.713	0.000
Vidarbha	1.427	0.000	0.921	0.007	1.480	0.014	0.948	0.014	−0.355	−0.005	−0.701	−0.122	−1.878	−0.352	1.841$	0.380	−2.089*	−0.388	1.296	0.076	−0.129	0.000	−0.439	0.000
Chhattisgarh	0.490	0.000	−0.466	−0.007	0.612	0.008	1.020	0.025	0.112	0.004	−1.379	−0.337	−2.925**	−0.623	−1.410	−0.322	−0.609	−0.116	−0.180	−0.016	−0.220	0.000	−0.603	0.000
Costal Andhra Pradesh	1.092	0.003	0.980	0.000	−0.032	0.000	1.530	0.027	−0.792	−0.042	0.834	0.065	1.577	0.155	1.193	0.119	−0.199	−0.028	1.046	0.261	−0.309	−0.060	−0.295	−0.004
Telangana	2.264*	0.000	1.854	0.000	0.474	0.000	0.823	0.018	1.101	0.034	0.690	0.075	0.659	0.105	2.281*	0.358	−1.473	−0.258	2.144*	0.211	−0.458	0.000	−0.399	0.000
Rayalaseema	−0.395	0.000	0.964	0.000	1.150	0.000	1.692$	0.040	0.768	0.048	0.931	0.062	1.401	0.146	0.495	0.073	−0.946	−0.121	0.707	0.110	0.376	0.046	−0.175	0.000
Tamil Nadu and Pondicherry	−0.066	0.000	0.463	0.000	0.000	0.000	0.258	0.015	−1.081	−0.054	−1.287	−0.050	1.344	0.064	−1.066	−0.093	0.399	0.038	0.177	0.027	0.834	0.163	0.451	0.059
Costal Karnataka	0.777	0.000	−0.885	0.000	0.967	0.000	0.048	0.000	0.939	0.161	1.493	0.777	−0.473	−0.258	2.218*	0.972	−0.602	−0.166	0.711	0.168	−0.153	−0.008	−0.497	0.000
North Interior Karnataka	0.850	0.000	0.522	0.000	−0.598	0.000	0.630	0.020	1.603	0.095	0.341	0.034	−1.123	−0.158	1.901$	0.242	−0.471	−0.063	0.107	0.016	−0.922	−0.029	−0.061	0.000
South Interior Karnataka	0.486	0.000	0.348	0.000	−0.584	−0.005	1.438	0.067	−1.531	−0.114	−0.405	−0.029	−1.147	−0.126	1.088	0.115	1.097	0.137	0.193	0.025	−0.337	−0.030	−0.179	0.000
Kerala	1.387	0.008	1.554	0.032	0.832	0.042	1.563	0.173	0.090	0.026	−2.476*	−1.038	−1.296	−0.585	1.071	0.285	1.421	0.360	1.841	0.386	1.491	0.260	−0.773	−0.034
All-India	−0.412	−0.006	0.668	0.011	0.716	0.011	2.016	0.036	0.182	0.007	−0.994	−0.085	−1.887$	−0.127	1.029	0.088	−1.007	−0.082	1.139	0.070	0.359	0.012	−0.363	−0.005

Source: Computed using data from Indian Institute of Tropical Meteorology, Pune

Note: M-K *Z* is Mann-Kendall *Z*; ** significant at 1% level, * significant at 5% level, and $ significant at 10% level of significance

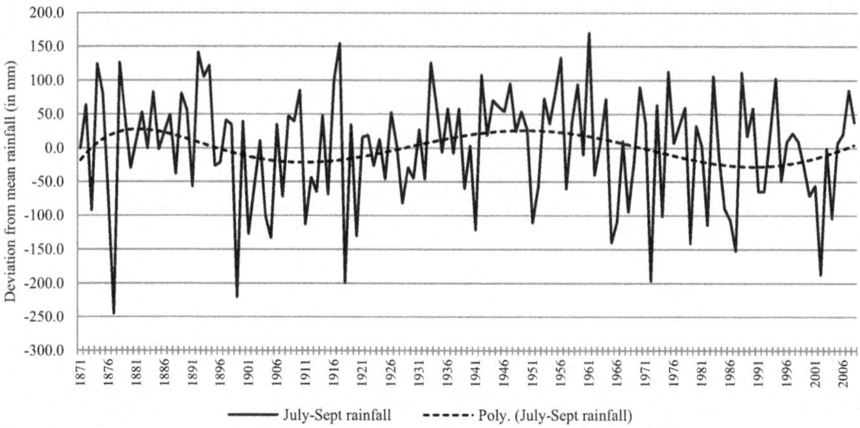

Source: Computed using data from Indian Institute of Tropical Meteorology, Pune.

Fig. 8.6 All-India summer monsoon (Jun-Sept) rainfall deviations from mean summer rainfall during 1871–2008. (*Source*: Computed using data from Indian Institute of Tropical Meteorology, Pune)

Pradesh, while it is showing increasing trends in Punjab, Gangetic West Bengal, and Coastal Andhra Pradesh (Table 8.5). Many other studies have also shown subdivisional trends in summer monsoon rainfall in India (see Kumar et al., 1992; Guhathakurta & Rajeevan, 2008).

The post-monsoon (October–December) rainfall has shown increasing trends in four subdivisions. For Sub-Himalayan West Bengal and Sikkim and Kerala, the trend coefficient is significant at 5% level, while for other two subdivisions, Assam and Meghalaya, and Marathwada, it is significant at 10% level of significance. Pre-summer monsoon (March–May) rainfall shows increasing trend in the Indo-Gangetic plain of India (except Gangetic West Bengal) and in Gujarat, Telangana, Rayalaseema, and North Interior Karnataka. Table 8.5 also shows trend in rainfall during the sub-period 1961–2008. We understand that cyclic nature of rainfall (as 1961–1990 was a dry period while 1991–2008 a wet period) cancels out the trend, still the Mann-Kendall test shows significant negative trend in summer monsoon rainfall in West Madhya Pradesh, Telangana, Tamil Nadu and Pondicherry (now called Puducherry), and positive trend in Orissa. Post-monsoon rainfall during 1961–2008 shows an increasing trend only in Kerala, but pre-monsoon rainfall is showing an increasing trend in seven meteorological sub-divisions and at the all-India level. At all-India level, as per the Sen's slope, the pre-monsoon rainfall during the period 1961–2008 increased at the rate of 0.432 mm per year.

The trend surface of average deviation of annual summer monsoon rainfall for various epochs is presented in Fig. 8.7. Given that about four-fifths of the total annual rainfall in the country takes place during summer monsoon season (June–September), the spatial pattern in monsoon rainfall strongly influences the spatial

Table 8.5 Man-Kendall coefficient and Sen's slope for seasonal rainfall (in mm) for the period 1871–2008 and 1961–2008

Sub-division	1871–2008							
	Jan-Feb		Mar-May		Jun-Sep		Oct-Dec	
	M-K Z	Sen's slope	M-K Z	Sen's slope	M-K Z	Sen's slope	M-K Z	Sen's slope
Assam and Meghalaya	−0.099	−0.005	−0.889	−0.232	−1.890$	−0.731	1.767$	0.294
Nagaland, Manipur, Mizoram, and Tripura	−0.904	−0.047	0.160	0.048	−2.923**	−1.067	0.742	0.134
Sub Him. W. Bengal and Sikkim	−0.232	−0.009	0.412	0.073	−0.834	−0.583	2.058*	0.348
Gangetic West Bengal	0.560	0.028	0.407	0.079	1.892$	0.723	1.213	−0.130
Orissa	1.502	0.075	0.876	0.092	−0.519	−0.206	−0.987	−0.201
Jharkhand	−0.514	−0.028	0.035	0.006	−0.477	−0.169	0.624	0.094
Bihar	−0.306	−0.014	1.695$	0.167	−0.615	−0.285	0.882	0.089
East Uttar Pradesh	−0.090	−0.003	1.154	0.050	−0.353	−0.151	0.589	0.042
West Uttar Pradesh	−1.093	−0.054	1.859$	0.079	−1.011	−0.390	1.495	0.073
Haryana, Chandigarh, and Delhi	−0.066	−0.003	1.879$	0.095	1.386	0.440	−0.431	−0.013
Punjab	−0.337	−0.031	1.773$	0.121	2.076*	0.764	0.460	0.017
West Rajasthan	−0.217	−0.003	1.427	0.038	−0.180	−0.042	0.458	0.004
East Rajasthan	−0.731	−0.013	0.808	0.020	−1.474	−0.529	−0.129	−0.003
West Madhya Pradesh	0.044	0.000	0.874	0.022	−1.357	−0.472	0.733	0.044
East Madhya Pradesh	0.471	0.027	0.801	0.035	−2.479**	−1.056	−0.210	−0.019
Gujarat	−0.804	0.000	−1.728$	−0.012	−0.026	−0.016	0.781	0.015
Saurashtra, Kutch, and Diu	−0.462	0.000	0.470	0.002	0.475	0.212	0.615	0.007
Konkan and Goa	−0.762	0.000	1.502	0.041	1.947$	1.692	1.318	0.192
Madhya Maharashtra	−0.091	0.000	−0.191	−0.008	0.510	0.147	−0.254	−0.034
Marathwara	1.402	0.006	0.718	0.030	−0.282	−0.108	1.660$	0.223
Vidarbha	1.200	0.036	0.633	0.030	−1.213	−0.509	0.537	0.053
Chhattisgarh	0.142	0.006	0.232	0.016	−3.138**	−1.349	−0.657	−0.077
Costal Andhra Pradesh	1.662$	0.027	0.226	0.021	1.938$	0.425	0.079	0.031
Telangana	2.481**	0.026	1.749$	0.103	0.849	0.307	1.496	0.214
Rayalaseema	−0.249	0.000	2.211*	0.168	0.550	0.154	0.434	0.100
Tamil Nadu and Pondicherry	−0.131	−0.004	−0.486	−0.045	−0.151	−0.022	0.839	0.251
Costal Karnataka	0.030	0.000	1.460	0.339	1.520	1.642	0.714	0.154
North Interior Karnataka	0.665	0.000	1.797$	0.127	−0.002	0.000	−0.191	−0.024
South Interior Karnataka	0.268	0.000	−0.668	−0.066	0.920	0.236	−0.306	−0.050
Kerala	1.300	0.050	1.027	0.312	−1.395	−1.116	2.157*	0.714
All-India	0.326	0.008	1.176	0.054	−0.990	−0.181	1.049	0.070

Sub-division	1961–2008							
	Jan-Feb		Mar-May		Jun-Sep		Oct-Dec	
	M-K Z	Sen's slope	M-K Z	Sen's slope	M-K Z	Sen's slope	M-K Z	Sen's slope
Assam and Meghalaya	2.383*	0.544	0.880	1.017	−0.924	−1.916	0.240	0.138
Nagaland, Manipur, Mizoram, and Tripura	0.596	0.178	2.053*	2.507	0.000	0.004	−0.382	−0.295
Sub Him. W. Bengal and Sikkim	1.513	0.218	1.289	1.146	0.400	1.279	0.916	0.412
Gangetic West Bengal	0.845	0.316	2.178*	1.741	1.635	3.620	0.871	0.599
Orissa	−0.249	−0.069	1.698$	0.853	1.698$	3.471	−0.604	−0.443
Jharkhand	−0.382	−0.118	1.165	0.577	0.960	2.382	−0.258	−0.170
Bihar	0.596	0.118	2.453**	1.130	1.280	2.751	−1.325	−0.743
East Uttar Pradesh	0.365	0.083	1.619	0.385	0.169	0.306	−1.307	−0.393
West Uttar Pradesh	0.463	0.091	1.832$	0.421	−0.791	−1.418	−1.041	−0.250
Haryana, Chandigarh, and Delhi	1.707$	0.400	2.881**	0.786	−0.951	−1.493	−0.997	−0.125
Punjab	1.511	0.512	0.489	0.115	−1.164	−2.669	−0.898	−0.185
West Rajasthan	1.292	0.091	1.477	0.217	−0.533	−0.525	0.598	0.031
East Rajasthan	0.552	0.032	1.940$	0.286	−1.458	−1.928	−0.303	−0.035
West Madhya Pradesh	0.098	0.000	1.050	0.125	−2.044*	−3.606	−1.423	−0.433
East Madhya Pradesh	−0.142	−0.031	0.783	0.173	−0.462	−1.026	−1.538	−0.600
Gujarat	0.102	0.000	−0.415	0.000	1.004	3.022	−0.258	−0.034
Saurashtra, Kutch, and Diu	0.348	0.000	1.351	0.043	0.951	2.333	−0.739	−0.089
Konkan and Goa	0.872	0.000	0.765	0.146	−0.915	−2.858	−1.049	−0.615
Madhya Maharashtra	0.630	0.000	−1.103	−0.227	0.382	0.646	−0.293	−0.191
Marathwara	0.568	0.000	−0.107	0.000	0.062	0.096	−0.240	−0.253
Vidarbha	1.067	0.167	−0.480	−0.103	−0.062	−0.268	−0.231	−0.136
Chhattisgarh	0.276	0.057	0.053	0.035	−0.471	−0.595	−0.427	−0.185
Costal Andhra Pradesh	1.425	0.187	1.254	0.300	0.373	0.500	0.764	1.380
Telangana	1.625	0.164	−0.222	−0.068	−2.000*	−3.068	−0.676	−0.536
Rayalaseema	1.277	0.000	1.387	0.519	0.151	0.079	1.191	0.948
Tamil Nadu and Pondicherry	0.498	0.119	0.765	0.322	−1.734$	−1.163	0.133	0.284
Costal Karnataka	1.106	0.000	0.133	0.231	−0.409	−2.252	−0.222	−0.227
North Interior Karnataka	0.355	0.000	−1.280	−0.615	−1.520	−1.589	−0.667	−0.433
South Interior Karnataka	0.299	0.000	−0.302	−0.135	0.027	0.134	−0.107	−0.183
Kerala	−0.169	−0.027	−0.053	−0.077	−1.298	−4.977	2.204*	3.380
All-India	1.541	0.146	2.063*	0.432	−0.231	−0.123	−0.098	−0.027

Source: Computed using data from Indian Institute of Tropical Meteorology, Pune
Note: M-K *Z* is Mann-Kendall *Z*; ** significant at 1% level, * significant at 5% level, and $ significant at 10% level of significance

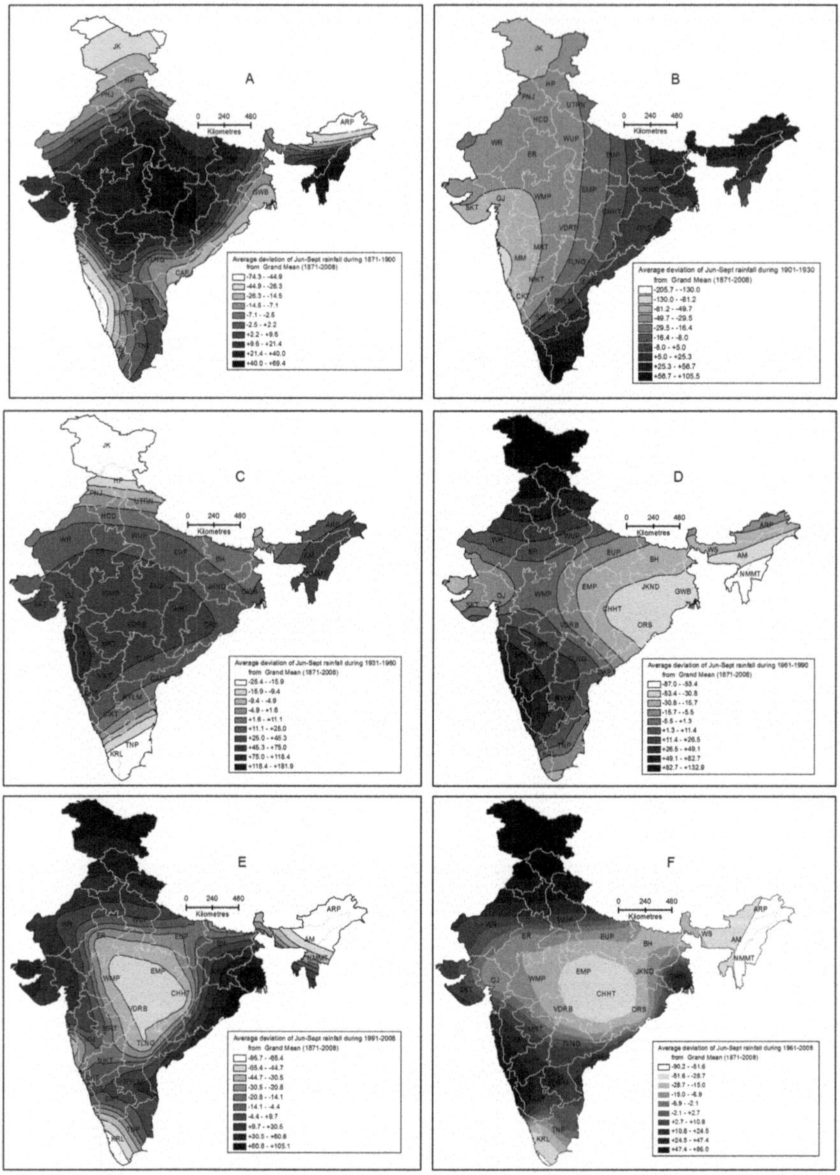

Note: The data for Jammu & Kashmir, Himachal Pradesh, Uttaranchal, and Arunachal Pradesh are not available. However, prediction surface for them are computed based on data from other subdivisions, and therefore prediction surface for these four states may not be very reliable. Ordinary kriging is used for trend surface.
Source: Computed using data from Indian Institute of Tropical Meteorology, Pune.

Fig. 8.7 Trend in epochal deviations (average surplus and deficits in mm) in summer monsoon rainfall from mean of summer monsoon rainfall during 1871–2008. *Note*: The data for Jammu and Kashmir, Himachal Pradesh, Uttaranchal, and Arunachal Pradesh are not available. However, prediction surface for them are computed based on data from other subdivisions, and therefore prediction surface for these four states may not be very reliable. Ordinary kriging is used for trend surface. (*Source*: Computed using data from Indian Institute of Tropical Meteorology, Pune) pattern in total annual rainfall. This is why the spatial pattern in average deviation in summer monsoon rainfall is largely similar to that of annual rainfall presented in Fig. 8.5. Figure 8.6 shows the temporal cycle in the monsoon rainfall in India, while Fig. 8.7 depicts the spatial cycle of the rainfall. The Central India had excess summer monsoon rainfall during 1871–1900, then again excess during 1931–1960, followed by deficit during 1961–1990 (see also Tables 8.6a and 8.6b). The spatial cycle has largely followed the wet and dry epochal cycle. However, the worrying aspect is that in more than half of the wet cycle of 1991–2008, the deficit in rainfall in Central India has persisted. However, deficit in rainfall over two cycles (one wet and another dry, for over 60 years) is not unusual. Both Figs. 8.5 and 8.7 show deficit in rainfall in Konan and Goa subdivision during 1981–1930 (two cycles). However, it received excess rainfall (both annual and monsoonal) during 1931–1990 (another two cycle of 30 years each, one wet and another dry). During 1991–2008, Konkan and Goa has been deficient in rainfall (in both annual and monsoonal). Kerala had had deficit in annual rainfall during 1871–1900, but had excess of rainfall during 1901–1960, then again has had deficit during 1961–2008. This is an indication that some of the meteorological subdivisions in the country may be going through 60-year cycle (rather than 30-year cycle visible at the all-India level) in annual and monsoonal rainfalls.

Conclusion

This study shows some very interesting aspects of rainfall in India. The rainfall at all-India level continues without any trend. However, there are nine meteorological subdivisions which have shown significant trends in annual rainfall. Out of these nine subdivisions, three (Chhattisgarh, East Madhya Pradesh, and Nagaland, Manipur, Mizoram and Tripura) show negative trend while six subdivisions (Coastal Andhra Pradesh, Coastal Karnataka, Gangetic West Bengal, Konkan and Goa, Punjab, and Telangana) demonstrate positive trends. Thus, the result shows that Central India (Chhattisgarh and Eastern Madhya Pradesh) is drying up, while some coastal areas and northwestern part of the country are getting wetter. The annual and summer monsoon rainfall at all-India level show 30-year cycle of alternate wet and dry epochs. The wet epochs have been 1871–1900, 1931–1960 and 1991–2008 (the last wet epoch is expected to continue till 2020), while the dry epochs have been 1900–1930 and 1961–1990. It is not that during a wet epoch all the subdivisions experienced excess rainfall. There is also strong evidence that the excess and deficit of rainfall have taken place largely in geographically contiguous areas. Summer monsoon rainfall contributes about 78% of the total rainfall in the country. However, there is strong evidence that the share of summer monsoon rainfall is decreasing mainly due to decline in rainfall in July, while the share of post-monsoon (October–December) rainfall has considerably increased. The study also shows that the summer monsoon rainfall (during 1871–2008) has significant declining trend in subdivisions of Assam and Meghalaya, Nagaland, Manipur, Mizoram and Tripura,

Table 8.6a Average deviation of epochal January–February and March–May rainfalls (mm)

Divisions	Division codes	January–February						March–May					
		1871–1900	1900–1930	1931–1960	1961–1990	1991–2008	1961–2008	1871–1900	1900–1930	1931–1960	1961–1990	1991–2008	1961–2008
Assam and Meghalaya	AM	0.4	1.5	−0.6	−9.0	12.7	−0.9	10.6	−15.2	44.1	−41.4	3.0	−24.7
Nagaland, Manipur, Mizoram, and Tripura	NMMT	6.8	−1.4	−0.3	−9.5	7.6	−3.1	14.4	−6.0	−17.5	−19.3	47.3	5.7
Sub Him. West Bengal and Sikkim	WS	0.5	1.5	2.8	−5.1	0.6	−3.0	−5.0	3.5	13.6	−13.6	2.6	−7.5
Gangetic West Bengal	GWB	−3.1	4.5	1.6	−6.6	6.1	−1.9	4.3	−1.7	−16.3	−7.5	35.4	8.6
Orissa	ORS	−10.8	5.9	3.0	4.6	−4.4	1.2	5.5	−7.2	−14.6	−2.8	32.0	10.2
Jharkhand	JKND	−7.7	11.8	9.1	−7.4	−9.6	−8.2	7.9	−3.2	−5.6	−2.0	4.7	0.5
Bihar	BH	0.2	−1.1	6.9	−5.0	−1.8	−3.8	2.4	−6.5	−12.8	4.3	20.9	10.6
East Uttar Pradesh	EUP	−0.8	1.3	5.7	−5.3	−1.4	−3.9	−1.6	−0.2	−2.7	1.1	5.8	2.8
West Uttar Pradesh	WUP	1.5	3.8	3.0	−5.6	−4.3	−5.1	−3.5	0.7	−6.3	4.4	7.9	5.7
Haryana, Chandigarh, and Delhi	HCD	0.6	−0.4	4.1	−5.6	2.3	−2.6	−3.9	−2.5	−8.1	3.6	18.2	9.1
Punjab	PNJ	5.0	−1.2	3.5	−12.7	8.9	−4.6	−8.9	3.5	−10.1	11.9	5.9	9.6
West Rajasthan	WR	0.4	−1.1	1.2	−1.5	1.7	−0.3	−2.0	−0.2	−7.1	6.0	5.6	5.9
East Rajasthan	ER	−0.5	1.3	1.9	−1.6	−1.7	−1.6	0.1	1.5	−6.3	1.5	5.2	2.9
West Madhya Pradesh	WMP	−1.3	1.4	1.5	−0.6	−1.6	−1.0	−2.4	0.6	1.1	−0.8	2.5	0.4
East Madhya Pradesh	EMP	−6.8	0.0	9.0	3.9	−10.3	−1.4	−5.3	5.2	1.3	−2.9	2.9	−0.7
Gujarat	GJ	0.2	0.8	−0.3	−2.3	2.7	−0.4	−1.8	3.0	−1.2	0.6	−1.0	0.0
	SKT	−0.1	1.2	0.4	−1.6	0.1	−1.0	−2.9	−0.4	3.8	−3.6	5.3	−0.3

Saurashtra, Kutch, and Diu													
Konkan and Goa	KG	6.0	−1.0	−2.4	−3.4	1.3	−1.6	−8.3	−2.1	0.5	2.3	12.7	6.2
Madhya Maharashtra	MM	1.6	2.2	−1.6	−2.1	−0.2	−1.4	−0.2	−3.4	6.1	1.7	−7.1	−1.6
Marathwara	MRT	−4.7	4.8	−0.8	1.0	−0.5	0.4	−2.3	−1.9	2.5	3.3	−2.9	1.0
Vidarbha	VDRB	−6.7	5.2	1.9	0.0	−0.6	−0.2	−0.3	−2.8	3.7	0.5	−2.0	−0.4
Chhattisgarh	CHHT	−8.9	10.4	6.9	−5.4	−5.1	−5.3	−4.8	4.0	10.7	−6.1	−6.2	−6.1
Costal Andhra Pradesh	CAP	−9.6	7.5	−4.1	0.3	9.8	3.9	−5.1	−4.5	2.0	5.8	3.0	4.7
Telangana	TLNG	−8.4	5.0	1.7	−2.0	6.1	1.0	−10.9	−2.3	11.7	2.8	−2.1	1.0
Rayalaseema	RYLM	−4.9	12.9	−2.6	−3.6	−2.8	−3.3	−5.7	−13.9	10.6	2.2	11.3	5.6
Tamil Nadu and Pondicherry	TNP	−14.2	18.1	0.7	−3.7	−1.5	−2.9	0.8	−13.0	18.8	−13.7	11.8	−4.1
Costal Karnataka	CKT	2.0	0.6	−0.1	−1.9	−0.9	−1.5	−17.7	−42.5	37.1	12.1	18.3	14.4
North Interior Karnataka	NIKT	−2.1	3.9	−1.1	−0.5	−0.4	−0.5	−6.1	−12.3	5.9	11.6	1.6	7.8
South Interior Karnataka	SIKT	−1.7	4.7	0.0	−2.6	−0.6	−1.9	3.2	−6.5	12.2	−7.5	−2.2	−5.5
Kerala	KRL	−9.0	5.9	1.8	0.7	1.1	0.8	−29.6	−32.2	87.1	−16.8	−14.2	−15.8
All-India		−3.2	3.8	2.1	−2.6	−0.1	−1.7	−1.4	−3.5	3.0	−2.2	6.8	1.2

Source: Computed using data from Indian Institute of Tropical Meteorology, Pune
Note: Deviations in rainfall are computed from arithmetic means of respective seasons during 1871–2008

Table 8.6b Average deviation of epochal June–September and October–December rainfalls (mm)

Subdivision	June–September						October–December					
	1871–1900	1900–1930	1931–1960	1961–1990	1991–2008	1961–2008	1871–1900	1900–1930	1931–1960	1961–1990	1991–2008	1961–2008
Assam and Meghalaya	−3.3	24.0	47.7	−19.3	−81.7	−42.7	−31.4	1.0	26.6	9.3	−9.1	2.4
Nagaland, Manipur, Mizoram, and Tripura	69.4	−0.1	36.5	−82.5	−38.8	−66.1	−18.1	3.3	6.6	12.7	−7.5	5.1
Sub Him. West Bengal and Sikkim	−7.7	66.4	−25.4	−27.9	−9.1	−20.9	−33.6	15.9	0.1	10.3	12.2	11.0
Gangetic West Bengal	−21.5	−19.5	−17.0	−5.1	105.1	36.2	−13.4	−13.6	16.9	−11.3	35.5	6.3
Orissa	−4.0	4.7	51.5	−72.4	33.5	−32.7	11.4	−10.4	21.5	−20.1	−4.0	−14.1
Jharkhand	4.6	5.6	8.9	−44.1	41.6	−11.9	−11.4	1.5	10.2	−0.9	1.0	−0.2
Bihar	26.4	14.5	−10.3	−23.4	−12.1	−19.2	−2.4	−7.7	−0.7	15.9	−8.4	6.8
East Uttar Pradesh	20.3	−21.4	23.5	−5.8	−27.6	−14.0	2.4	3.3	2.6	2.2	−17.6	−5.2
West Uttar Pradesh	46.9	−61.9	11.6	8.9	−9.2	2.1	−9.1	4.0	12.0	0.4	−12.3	−4.3
Haryana, Chandigarh, and Delhi	−3.3	−37.8	−3.4	34.3	16.9	27.8	−2.4	3.7	3.2	−1.6	−4.8	−2.8
Punjab	−11.4	−59.9	−5.2	59.9	27.6	47.8	−6.3	1.0	11.6	−0.3	−10.0	−3.9
West Rajasthan	1.0	−11.0	7.6	3.4	−1.8	1.5	−0.9	0.4	−1.7	0.3	3.1	1.3
East Rajasthan	30.7	−39.9	52.2	−10.6	−53.9	−26.8	−3.3	2.4	0.2	3.2	−4.1	0.4
West Madhya Pradesh	36.3	−84.3	78.3	14.6	−74.8	−18.9	−6.0	−2.2	7.8	6.3	−10.0	0.2
East Madhya Pradesh	60.4	−8.6	33.6	−55.0	−50.8	−53.4	−1.7	4.1	6.0	−0.6	−13.2	−5.3
Gujarat	30.6	−45.5	49.4	−62.4	46.4	−21.6	−2.9	−3.5	4.8	4.8	−5.4	1.0
Saurashtra, Kutch, and Diu	20.7	−25.7	−1.8	−9.7	27.4	4.2	−1.7	−4.4	−0.6	8.0	−2.2	4.2
Konkan and Goa	−32.4	−205.7	181.9	89.9	−56.2	35.1	−22.0	−15.0	45.3	−7.8	−0.8	−5.2
Madhya Maharashtra	13.3	−35.4	6.6	−3.1	30.8	9.6	7.0	−18.4	15.2	−10.5	11.2	−2.4
Marathwara	39.8	−61.7	23.1	7.6	−14.8	−0.8	−4.6	−21.0	6.9	9.5	15.5	11.7
Vidarbha	30.9	−36.8	63.3	−24.3	−55.2	−35.9	−0.9	−8.9	3.8	4.7	2.1	3.7
Chhattisgarh	62.4	−5.6	87.6	−87.0	−95.7	−90.2	−0.1	3.9	11.2	−10.2	−8.1	−9.4
Costal Andhra Pradesh	−22.8	−7.6	−1.6	18.4	22.7	20.0	7.3	−13.2	−1.8	−14.6	37.3	4.9
Telangana	−15.2	−34.5	37.5	57.9	−76.2	7.6	−8.7	−12.1	0.1	14.9	9.7	12.9

Rayalaseema	6.3	−23.1	−7.7	6.5	29.9	15.3	−0.5	−6.5	−14.1	−8.0	48.6	13.2
Tamil Nadu and Pondicherry	0.6	8.9	−19.3	14.3	−7.4	6.2	−12.8	5.4	−17.6	0.2	41.4	15.6
Costal Karnataka	−74.3	−91.9	28.6	132.9	7.8	86.0	−13.8	−14.8	15.4	−0.9	23.6	8.3
North Interior Karnataka	2.9	−27.0	20.6	17.2	−22.8	2.2	9.7	−18.2	3.5	−3.6	14.4	3.1
South Interior Karnataka	−0.7	−13.4	1.7	9.6	4.7	7.7	−2.8	−4.4	15.1	−16.2	13.8	−4.9
Kerala	−7.4	105.5	−20.8	−29.3	−80.0	−48.3	−86.5	43.2	31.7	−31.1	71.1	7.2
All−India	16.1	−22.5	27.6	−10.8	−17.4	−13.3	−5.0	−3.0	6.5	−0.8	3.9	1.0

Source: Computed using data from Indian Institute of Tropical Meteorology, Pune

Note: Deviations in rainfall are computed from arithmetic means of respective seasons during 1871–2008

Chhattisgarh, and East Madhya Pradesh, while it is showing increasing trend in Punjab, Gangetic West Bengal, and Coastal Andhra Pradesh. There is also strong evidence that instead of 30-year cycle (which is evident at all-India level), some of the meteorological subdivisions in India follow 60-year cycle of wet and dry spells in annual and summer monsoon rainfalls, which any studies, including those of Guhathakurta and Rajeevan (2008), have not reported. However, one is not sure about how the changing climatic system due to changing atmospheric chemistry and circulation will impact various meteorological subdivisions. It will be a worrying aspect if the annual and monsoonal rainfall deficit in some of the subdivisions, like those in Central India (East Madhya Pradesh, Chhattisgarh, and Vidarbha), Eastern Uttar Pradesh, Bihar, and Kerala persists beyond 2020, the two cycle period of 60 years.

The climate change and specially rainfall variations may have significant impact on the livelihoods of people. The farm sector in India is already in crisis because of variability in rainfall and frequent drought-like situations in many regions of the country. The semi-arid areas of the country are most impacted in this regard. Punjab, Haryana, Western Uttar Pradesh, Rajasthan, Madhya Pradesh, Chhattisgarh, Interior Odisha, Vidarbha, Marathwada, Interior Karnataka, Interior Andhra Pradesh, Telangana, and Central Tamil Nadu seem to be much impacted from rainfall variability. This has led to increase in the farmers' suicide in the semi-arid belt of India, besides leading to climate induced migration from these regions to urban areas or other states in search for livelihoods. There is a need that the impact of climate change be mitigated either through agricultural practices which are water-efficient, crops which can withstand the rainfall variability or through developing non-farm sectors in these areas. Further, a larger number of industrial and commercial towns are located in the semi-arid belt of the country. The decline in rainfall may lead to water availability and adequacy issue in many towns in this region. Further, the groundwater has already reached a critical level of no perennial reserve. This requires that the surface water sources be rapidly developed and protected, and effective use of water be promoted for sustainable development.

References

Arora, M., Goel, N. K., & Singh, P. (2005). Evaluation of temperature trends over India. *Hydrological Sciences Journal, 50*, 81–93.

Bidwai, P. (2011). Durban: Road to Nowhere. *Economic & Political Weekly xlvI*, 10–12.

Cislaghi, M., De Michele, C., Ghezzi, A., & Rosso, R. (2005). Statistical assessment of trends and oscillations in rainfall dynamics: analysis of long daily Italian series. *Atmospheric Research, 77*, 188–202.

de Lima, M. I. P., Carvalho, S. C. P., de Lima, J. L. M. P., & Coelho, M. F. E. S. (2010). Trends in precipitation: analysis of long annual and monthly time series from mainland Portugal. *Advances in Geosciences, 25*, 155–160.

El-Nesr, M., Alazba, A., & Abu-Zreig, M. (2010). Analysis of evapotranspiration variability and trends in the Arabian Peninsula. *American Journal of Environmental Sciences, 6*, 535–547.

Fisk, D. (1997). *Climate change and its impacts: A global perspective*. Department of the Environment, Transport and the Regions, the Met Office Hadley Centre Brochure, UK.
Gilbert, R. O. (1987). *Statistical methods for environmental pollution monitoring*. John Wiley & Sons.
Guhathakurta, P., & Rajeevan, M. (2008). Trends in the rainfall patterns over India. *International Journal of Climatology, 28*, 1453–1469.
Harvard School of Public Health. (2011). *Geographic Information System (GIS) in Public Health Research*. Available at: http://www.hsph.harvard.edu/research/gis/arcgis-tips/kriging/index.html (Accessed 24 Dec 2011).
IITM. (2009). Monthly and seasonal rainfall series for All-India Homogeneous Regions and Meteorological Subdivisions, 1871–2008. Available at: http://www.tropmet.res.in/static_page.php?page_id=53 (Accessed 15 May 2009).
Indian Meteorological Department. (2011). *Annual climate summary, 2010*. National Climate Centre.
IPCC. (2021). *AR6 climate change 2021: The physical science basis*. https://www.ipcc.ch/report/ar6/wg1/ (Accessed 13 Feb 2022).
Kendall, M. G. (1975). *Rank Correlation Methods* (4th ed.). Charles Griffin.
Kruger, A. C. (2006). Observed trends in daily precipitation indices in South Africa: 1910–2004. *International Journal of Climatology, 26*, 2275–2285.
Kumar, R., Pant, G. B., Parthasarathy, B., & Sontakke, N. A. (1992). Spatial and sub-seasonal patterns of the long term trends of Indian summer monsoon rainfall. *International Journal of Climatology, 12*, 257–268.
Mann, H. B. (1945). Non-parametric tests against trend. *Econometrica, 13*, 245–259.
Manton, M. J., Della-Marta, P. M., Haylock, M. R., Hennessy, K. J., Nicholls, N., Chambers, D. A., Collins, D. A., Daw, G., et al. (2001). Trends in extreme daily rainfall and temperature in Southeast Asia and the south Pacific, 1961-1998. *International Journal of Climatology, 21*, 269–284.
Murphy, B. F., & Timbal, B. (2008). A review of recent climate variability and climate change in south eastern Australia. *International Journal of Climatology, 28*, 859–879.
Olofintoye, O. O., & Sule, B. F. (2010). Impact of global warming on the rainfall and temperature in The Niger Delta of Nigeria. *Journal of Research Information in Civil Engineering, 7*, 33–48.
Pal, I., & Al-Tabbaa, A. (2009). Regional changes in extreme monsoon rainfall deficit and excess in India. *Dynamics of Atmosphere and Oceans, 49*, 206–214.
Palmer, T. N., & Raisanen, J. (2002). Quantifying the risk of extreme seasonal precipitation events in a changing climate. *Nature, 415*, 512–514.
Partal, T., & Kahya, E. (2006). Trend Analysis in Turkish Precipitation Data. *Hydrological Processes, 20*, 2011–2026.
Parthasarathy, B., Munot, A., & Kothawale, D. R. (1994). All-India Monthly and Seasonal Rainfall Series 1887-1993. *Theoretical and Applied Climatology, 49*, 217–224.
Patra, J. P., Mishra, A., Singh, R., & Raghuwanshi, N. S. (2011). Detecting rainfall trends in twentieth century (1871–2006) over Orissa State, India. *Climatic Change, 111*, 1–17. https://doi.org/10.1007/s10584-011-0215-5
Raghunandan, D. (2011). Durban Platform: Kyoto Negotiations Redux. *Economic & Political Weekly, xlvI*, 13–15.
Roy, S. S., & Balling, J. R. (2004). Trends in extreme daily precipitation indices in India. *International Journal of Climatology, 24*, 457–466.
Salmi, T., Määttä, A., Anttila, P., Ruoho-Airola, T., & Amnell, T. (2002). Detecting trends of annual values of atmospheric pollutants by the Mann-Kendall Test and Sen's Slope Estimates -The Excel Template Application Makesens, Finnish Meteorological Institute Publications on Air Quality No. 31, Helsinki, Finland. Available at: http://www.ilmanlaatu.fi/ilmansaasteet/julkaisu/pdf/MAKESENS-Manual_2002.pdf (Accessed 25 Dec 2011).
Sen, P. K. (1968). Estimates of the regression coefficient based on Kendall's Tau. *Journal of American Statistical Association, 63*, 1379–1389.

Shaban, A., & Sattar, S. (2011). Water security and sustainability in urban India. *International Journal of Global Environmental Issues, 11*, 31–54.
Soman, M. K., Kumar, K. K., & Singh, N. (1988). Decreasing trend in the rainfall of Kerala. *Current Science, 57*, 7–12.
UNEP. (2011). *Bridging the Emissions Gap, United Nations Environment Programme (UNEP)*. Available at: http://www.unep.org/pdf/UNEP_bridging_gap.pdf (Accessed 28 Dec 2011).
Wackernagel, H. (2003). *Multivariate geostatistics: An introduction with application*. Springer.
Zhang, H. (2003). Optimal interpolation and the appropriateness of cross-validating variogram in spatial generalized linear mixed models. *The Journal of Computational and Graphical Statistics, 12*, 698–713.

Chapter 9
Potential Effects of Climate Change in Saline Shallow Lakes in the North of Chile (Salar de Atacama, 23°S, Chile) and South Lipez of Bolivia (Khalina Lake, 22.61°S)

Patricio R. De los Rios-Escalante, Carlos Esse, Francisco Correa-Araneda, Lien Rodríguez, Carla E. Fernandez, and Pablo E. Prado

Introduction

The Andes mountains of the south of Peru, north of Chile, northwest Argentina, and southwest Bolivia are characterized by the presence of many shallow saline lakes associated with saline deposits of volcanic origin. These lakes are poorly studied, mainly due to access difficulties, and many of these studies are based on the results of expeditions that described mainly biological components (Hurlbert et al., 1986; Williams et al., 1995; De los Ríos-Escalante & Woelfl, in press). From an ecological viewpoint, these sites are feeding and nesting areas of aquatic birds such as the Chilean flamingo (*Phoenicopterus chilensis;* Molina, 1782), the Andean flamingo (*Phoenicoparrus andinus;* Philippi, 1854), or the James flamingo (*Phoenicoparrus jamesi*; Sclater, 1866; all considered endangered species; Jaramillo, 2003).

P. R. De los Rios-Escalante (✉)
Departamento de Ciencias Biológicas y Químicas, Universidad Católica de Temuco, Facultad de Recursos Naturales, Temuco, Chile

Universidad Católica de Temuco, Núcleo de Estudios Ambientales, Temuco, Chile
e-mail: prios@uct.cl

C. Esse · F. Correa-Araneda
Instituto Iberoamericano de Desarrollo Sostenible (IIDS), Unidad de Cambio Climático y Medio Ambiente (UCCMA), Facultad de Arquitectura, Construcción y Medio Ambiente, Universidad Autónoma de Chile, Temuco, Chile

L. Rodríguez
Universidad San Sebastián, Concepción, Chile

C. E. Fernandez · P. E. Prado
Universidad Mayor San Simón, Unidad de Limnología y Recursos Acuáticos, Cochabamba, Bolivia

A. L. Singh et al. (eds.), *Climate Change, Vulnerabilities and Adaptation*,
https://doi.org/10.1007/978-3-031-49642-4_9

In this sense, due to the low quantity of detailed limnological studies that involved descriptions of seasonal or monthly variations in biotic and abiotic parameters, the use of remote sensing techniques is suggested, considering that the optical properties of lakes can be associated with biological parameters such as plankton structure (De los Ríos-Escalante et al., 2022). The basis of the use of remote sensing in limnological studies consists of collecting available satellite images and performing computational-specific algorithms to extract information such as surface temperature and chlorophyll-a concentrations. Accordingly, it is possible to get long-term temporal and spatial scale variations in limnological parameters of water bodies, such as mountain lakes (Rodriguez-Lopez et al., 2020). Furthermore, it is possible to use satellite images to obtain photosynthetic pigment concentration in order to prepare sampling programs for studying the water quality in lakes exposed to pollution problems (Topp et al., 2020).

The aim of this study is to first characterize the monthly variations of chlorophyll-a concentration, surface temperature, wind speed, relative humidity, and atmospheric pressure in two lagoons in north of Chile (Los Flamencos, Salar de Atacama, 23°S, Chile) and South Lipez of Bolivia (Khalina lake, 22.61°S) using satellite images (Landsat 8), as a potential tool for limnological studies in lakes with access difficulties.

Material and Methods

Site Descriptions

The Los Flamencos Lake, located in Los Flamencos National Reserve at 4322 m a.s.l. in the Salar de Atacama of northern Chile, is a hypersaline lake and has the highest content of sodium, sulfate, magnesium, bicarbonate, chlorine, nitrate, and arsenic ions when compared with other lakes, ponds, and springs found in Salar de Atacama (Dorador et al., 2009). This site has important populations of brine shrimp *Artemia franciscana* that would graze probably on halophilic bacteria and microalgae (De los Rios-Escalante & Woelfl, in press).

Lake Bush or Khalina is an hyperhaline lake (conductivity ≈ 89,200 $\mu S\ cm^{-1}$, data from 2011) situated at an altitude of 4542 m a.s.l. in the southwestern Andean range in Bolivia. The volcanic origin of the bedrock and the endorheic situation of the basin determine the chemistry of the water. Bicarbonates, carbonates, chlorides, sulfates, calcium, magnesium, and sodium are the dominant ionic species of the water. Algal composition of the lake is very poor with the exception of diatoms (Bacillariophyta) which can reach more than a hundred species. The biomass of phytoplankton generally is low (40.4 $\mu g\ l^{-1}$ of chlorophyll-a in 2011), but some genera are particularly characteristic and restricted to this area of saline lakes in Bolivia, such as *Dunaliella*, *Chloromonas,* and *Planochloris*. Zooplankton is composed of around 40 species with a clear dominance of rotifers and copepods, principally *Boeckella poopoensis*. Benthic organisms are dominated by insects (principally Chironomidae) and crustaceans (Ostracoda and Hyalellidae; Maldonado et al., 2012). Only two species of fishes were described in this area, *Orestias agassi* and *Trichomycterus* sp.

Climatic Variables

The required time series of climatological variables were acquired from the NASA Prediction of Worldwide Energy Resources (2022). The parameters used are based on solar radiation obtained from satellite observation and meteorological data from assimilation models. The parameters are available in climatological, monthly, annual, and daily time series formats. Seventeen variables were selected for each month for the period 2021 (Table 9.1); the final data set consisted of 12 months, including the 17 climatological variables for each georeferenced sample point.

Download and Satellite Image Preprocessing

Fifty-six multispectral images, using LANDSAT 8 corresponding to level 1 of collection number 2, were orthorectified and corrected in function of land characteristics. These were downloaded for free from the US Geological Service (https://landlook.usgs.gov/) on February 3, 2023. The studied sites were covered by two satellite scenes each, for Flamencos Natural Reserve path 232 and row 076, and path 233 and row 076, respectively, whereas for Khalina images, the scenes were path 232 and row 076 and path 233 and row 075, respectively. The images were selected in quality function, specifically for cloudiness percentage (lower than 20%) and covering one year (2022), including all seasons. All images were corrected to the atmosphere bottom (BOA) with software ACOLITE (version 20211124.0) (https://github.com/acolite, accessed February 4, 2023). The atmospheric correction by defect used the Dark Spectrum Fitting method (Algorithm DSF), that was selected in the processor ACOLITE (Vanhellemont, 2020). The obtained ACOLITE products corresponded to reflectance at surface level for LANDSAT 8 images.

Algorithm for Parameters Estimation

ACOLITE includes several algorithms for recovering derived parameters from reflectance, temperature, chlorophyll-a, and turbidity. The products and algorithms for the satellite mission used were the following: "tur_dogliotti2015," "chl_oc2," "chl_oc3," and "bt10." Chlorophyll-a concentration (mg/m^3) was estimated using the algorithm between bands blue and green. The algorithms "oc2" and "oc3" utilized second and third bands respectively (Franz et al., 2015). Turbidity

Table 9.1 List of the 17 climatic variables included in this work (annual average)

Climatological variables	Description	Chaxa average (± sd)	Khalina average (± sd)	Units
PS	Surface Pressure	73.99 ± 0.09	57.71 ± 0.04	kPa
TS	Earth Skin Temperature	16.27 ± 4.70	2.95 ± 4.27	°C
T2M	Temperature at 2 Meters	14.73 ± 2.98	1.09 ± 2.80	°C
QV2M	Specific Humidity at 2 Meters	2.56 ± 1.24	2.32 ± 1.24	gKg^{-1}
RH2M	Relative Humidity at 2 Meters	19.69 ± 6.32	35.38 ± 12.33	%
WS2M	Wind Speed at 2 Meters	2.52 ± 0.34	3.36 ± 0.50	ms^{-1}
GWETTOP	Root Zone Soil Wetness	0.10 ± 0.03	0.25 ± 0.09	0–1
T2M_MAX	Temperature at 2 Meters Maximum	31.33 ± 3.08	18.12 ± 3.29	°C
T2M_MIN	Temperature at 2 Meters Minimum	−1.32 ± 3.06	−14.88 ± 2.87	°C
GWETPROF	Profile Soil Moisture	0.32 ± 0.00	0.35 ± 0.02	0–1
GWETROOT	Root Zone Soil Wetness	0.32 ± 0.00	0.35 ± 0.02	0–1
PRECTOTCORR	Precipitation Corrected	0.02 ± 0.03	0.38 ± 0.56	mm day^{-1}
ALLSKY_SFC_UVA	All Sky Surface UVA Irradiance	17.43 ± 4.61	17.16 ± 4.00	Wm^{-2}
ALLSKY_SFC_UVB	All Sky Surface UVB Irradiance	0.50 ± 0.19	0.52 ± 0.18	Wm^{-2}
ALLSKY_SRF_ALB	ALLSKY_SRF_ALB	0.20 ± 0.01	0.20 ± 0.01	dimensionless
PRECTOTCORR_SUM	Precipitation Corrected Sum	8.19 ± 1.08	139.43 ± 16.71	mm

(Formazin Nephelometric Units [FNU]) was estimated using the algorithm of Dogliotti et al. (2015). The algorithm of change utilized the red band for "red&lt" and the NIR (Near Infrared) band for "red&0.07" with an initial ponderation intermediate. The luminosity of thermal infrared sensor (TIRS) in Kelvin was used to determine temperature, which is converted to Celsius degrees. Image processing extracted images of medium pixels (50 points for each site) with a 3 × 3 window, with SNAP software version 8.0 (https://step.esa.int/main/download/snap-download/, accessed February 5, 2023). All algorithms were used to represent and include the spatiotemporal distribution of parameters such as temperature, turbidity, and chlorophyll-a during entire year season.

Data Analysis

The obtained data was processed using machine learning tools available in the Python software (Van Rossum & Drake, 1995), with the libraries "numpy" (Harris et al., 2020), "pandas" (McKinney, 2010), "matplotlib" (Hunter, 2007), "seaborn" (Waskom, 2021), and "statsmodel" (Seabold & Perktold, 2010). K-means analysis was applied to determine the potential groups between available data (VanderPlas, 2017).

Results

The temperature results for Los Flamencos Lake revealed fluctuations for all sites that varied mainly between −11.5 and 15 °C, with the exception of a few stations with maximum temperatures higher than 20 °C. Chlorophyll-a values varied mainly between 7.28 and 127 μg l^{-1}, with half sites having fluctuations between 250 and 322 μg l^{-1}; for chlorophyll-3, values varied mainly between 7.55 and 56 μg l^{-1}, with half sites having fluctuations between 80 and 100 μg l^{-1}. Finally, turbidity varied mainly between 76 and 288 FNU; five sites presented low values higher than 10 FNU (Fig. 9.1).

For this site, the lower values were between −7.59 and 3 °C in winter season, whereas the maximum values between 5 and 25 °C were reported for spring. For chlorophyll-a, the maximum values were in autumn, between 16.5 and 322.12 μg l^{-1}, whereas the low values were in summer varying between 7.28 and 30 μg l^{-1}; for chlorophyll-3, the maximum values were in autumn and fluctuated between 15 and 97.4 μg l^{-1}, and the lower values were for winter and spring varying between 7.73 and 15.72 μg l^{-1}. Finally, turbidity has high values in spring and summer varying between 275.83 and 287.84 FNU, with spring having a low value of 12.03 FNU (Figs. 9.2 and 9.3).

The temperature results for the Khalina lake revealed fluctuations for all sites that varied mainly between −8.5 and 19.77 °C; the chlorophyll-a values varied mainly

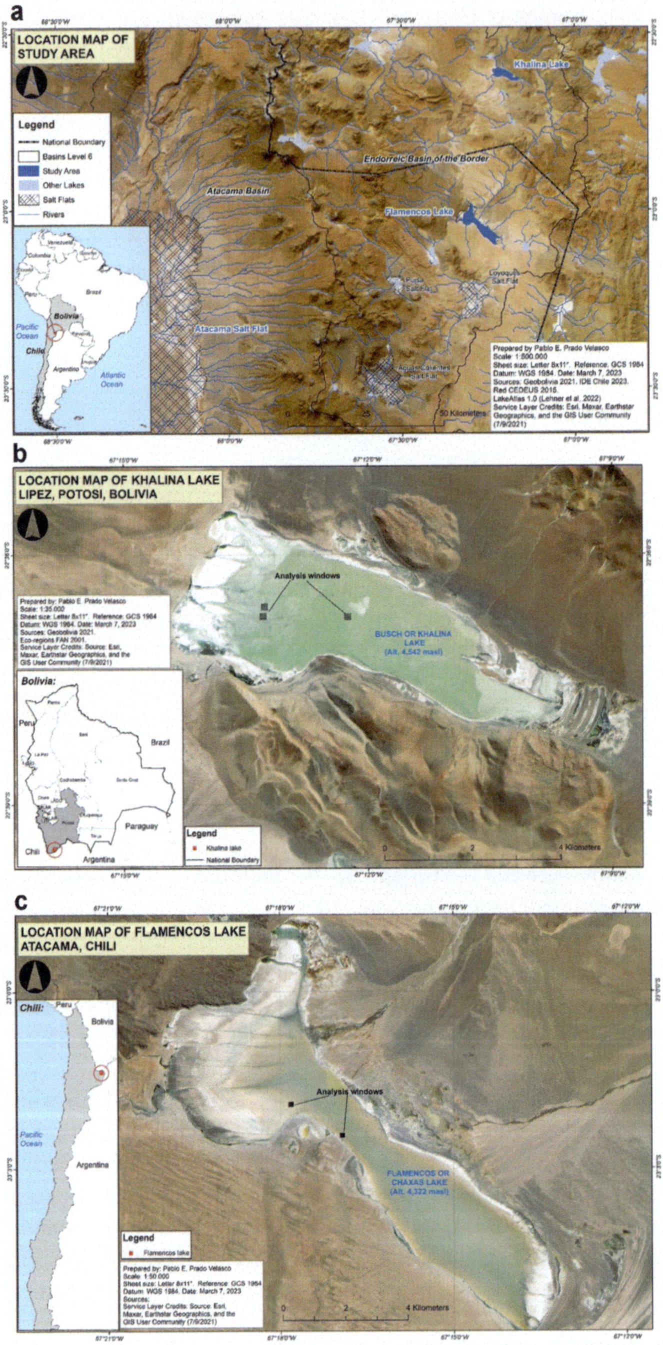

Fig. 9.1 Map of studied sites: (**a**) location map of study area; (**b**) location map of Khalina Lake, Lipez, Potosi, Bolivia; and (**c**) location map of Flamencos Lake, Atacama, Chili

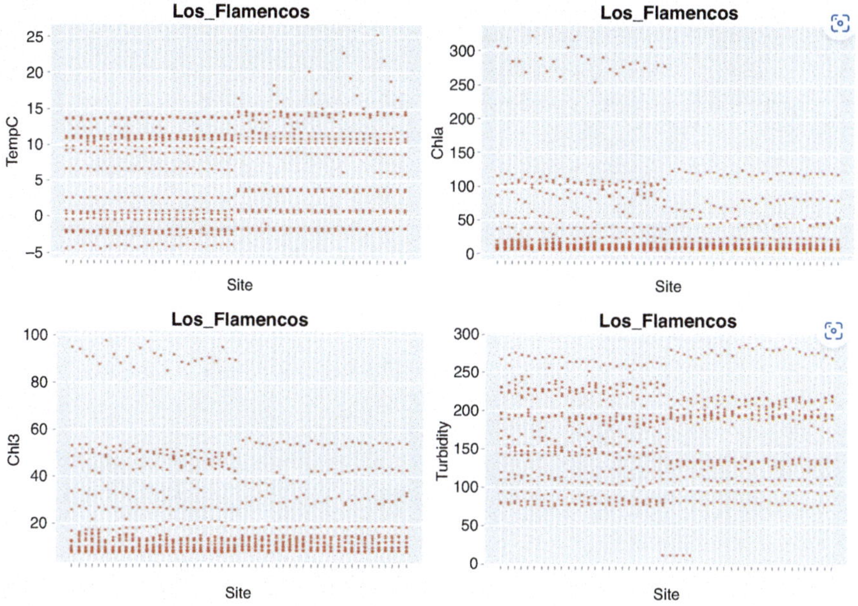

Fig. 9.2 Temperature (°C), chlorophyll-a (μg l^{-1}), chlorophyll-3 (μg l^{-1}), and turbidity for sites in Los Flamencos lagoon in Salar de Atacama

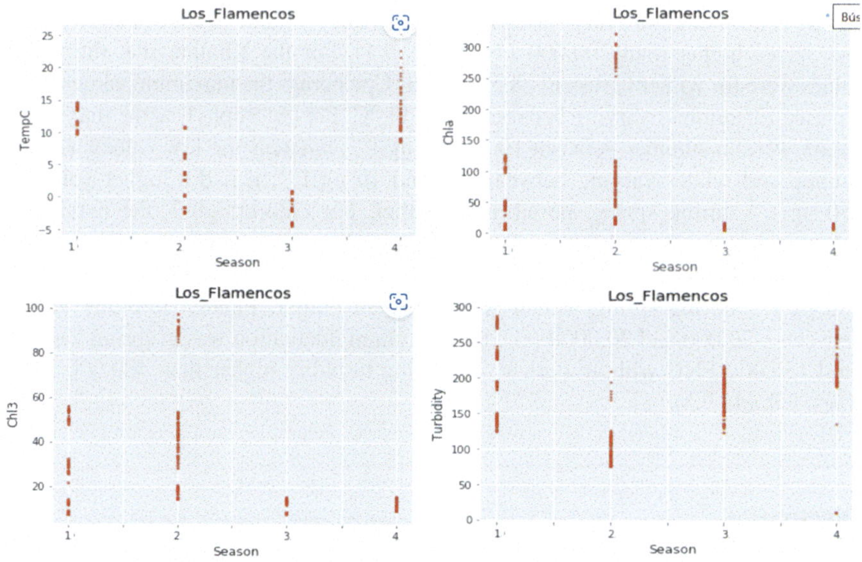

Fig. 9.3 Temperature (°C), chlorophyll-a (μg l^{-1}), chlorophyll-3 (μg l^{-1}), and turbidity for sampling seasons (1: summer; 2: autumn; 3: winter; 4: spring) at Los Flamencos lagoon in Salar de Atacama

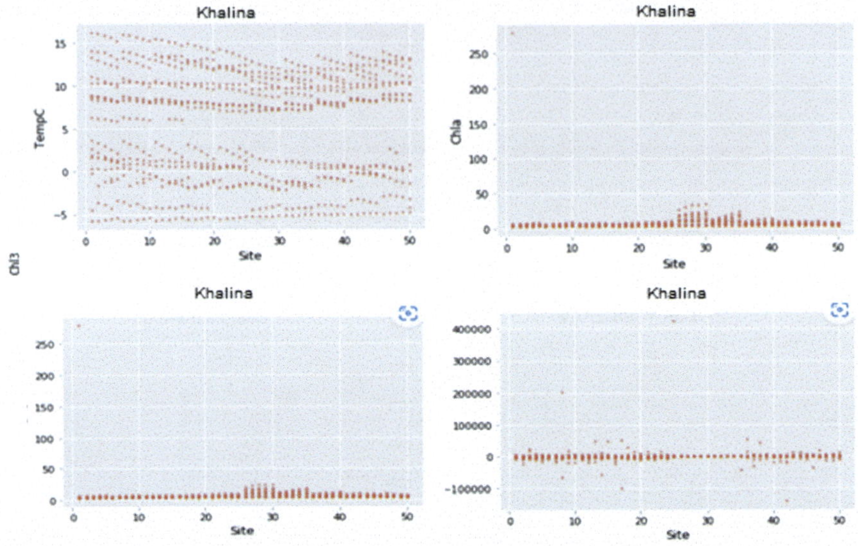

Fig. 9.4 Temperature (°C), chlorophyll-a (μg l^{-1}), chlorophyll-3 (μg l^{-1}), and turbidity for sites in Khalina lagoon

between 2.83 and 35 μg l^{-1}, with few sites around 50 μg l^{-1}; for chlorophyll-3, the values varied mainly between 3.23 and 33.8 μg l^{-1}, with one site having higher than 250 μg l^{-1}. Finally, turbidity varied mainly between −10 and 10 FNU, and two sites had values higher than 200,000 FNU (Fig. 9.1). For the Khalina site, the lower values were for winter between −8.5 to −1.6 °C, whereas the maximum values were reported for summer varying between 7 and 17 °C. For chlorophyll-a, the maximum values were in summer between 10 and 57 μg l^{-1}, whereas the low values were in autumn and winter varying between 2.8 and 10 μg l^{-1}; it had a higher value of 250 μg l^{-1} during spring, possibly an outlier. For chlorophyll-3, the maximum values were in summer between 10 and 33.8 μg l^{-1}, whereas the low values were in autumn and winter varying between 3.23 and 9.74 μg l^{-1}; it had a higher value of 250 μg l^{-1} during spring, which is possibly an outlier. Finally, turbidity varied between −20,000 and 20,000 FNU; the maximum fluctuation was in spring varying until 150,000 FNU with an outlier site having turbidity higher than 400,000 FNU (Figs. 9.4 and 9.5).

Discussion

The exposed results revealed marked fluctuations in temperature at seasonal level, that indicate that temperature does not affect turbidity. The scarce literature about seasonal studies in saline lakes revealed marked fluctuations in temperature and salinity level mainly due to precipitations during January and February, the so-called

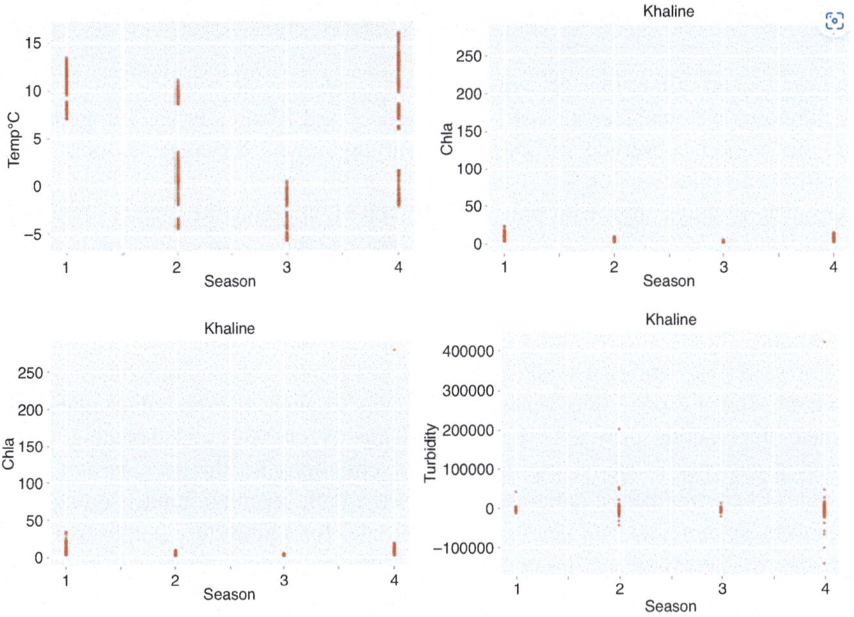

Fig. 9.5 Temperature (°C), chlorophyll-a (μg l^{-1}), chlorophyll-3 (μg l^{-1}), and turbidity for sampling seasons (1: summer; 2: autumn; 3: winter; 4: spring) in Khalina lake

"Bolivian winter" (Zúñiga et al., 1994). These precipitations generate a decrease in salinity and increase in phytoplankton biomass, whereas in middle year (July–August), due to the absence of precipitation, there is an increase in salinity and low phytoplankton activities (Zúñiga et al., 1991, 1994). Also, the marked temperature fluctuations, on daily and seasonal scale, are similar to the descriptions for saline lakes at north of Salar de Atacama (Zúñiga et al., 1994).

The limnological fluctuations in saline lakes, can be studied by remote sensing techniques, based in studies of evaporation rate and surface fluctuations, where it was possible marked associations between these two variables (Darehshouri et al., 2023). Other similar studies on saline lakes using remote sensing denoted the detection of chlorophyll-a fluctuations associated with environmental parameters with consequent effects on *Artemia* genus populations used as live resource for aquaculture (Tian et al., 2023). Regarding former studies for saline lakes in the north of Salar de Atacama (Zúñiga et al., 1994) and south of Torres del Paine National Park (Campos et al., 1996), it would be interesting visit these sites at monthly or seasonal scale, and in different sampling stations. These data can be compared with spectral properties (Darehshouri et al., 2023; Tian et al., 2023). On this basis, the studies based on remote sensing techniques and their potential associations with field works, can be as important useful techniques for long-term studies of potential effects of global climate changes on limnological characteristics of standing waters.

Conclusion

The results and discussions presented in this chapter provide valuable insights into the limnological characteristics of Los Flamencos and Khalina lakes. The temperature fluctuations observed at both sites, varying across different seasons, have significant implications on the overall water quality and dynamics of these saline lakes. The findings align with previous research highlighting the impact of precipitation patterns on temperature and salinity levels in these ecosystems, particularly during the Bolivian winter. These fluctuations, in turn, influence the turbidity and phytoplankton biomass, creating a complex interplay between various environmental factors.

The scarcity of seasonal studies in saline lakes emphasizes the importance of this research. The use of remote sensing techniques to analyze evaporation rates and surface fluctuations proves to be a promising avenue for understanding these dynamic systems. Notably, the detection of chlorophyll-a fluctuations and their linkages to environmental parameters underscore their potential impacts on aquatic organisms such as *Artemia* genus, which are vital for aquaculture purposes. Comparisons with previous studies conducted in different regions, such as the north of Salar de Atacama and south of Torres del Paine National Park, could offer valuable insights into the broader context of limnological changes across distinct geographical areas.

As the global climate continues to evolve, the integration of remote sensing techniques with field studies emerges as a critical approach for comprehending the long-term effects of climate change on the limnological characteristics of standing waters. The findings presented in this chapter shed light on the intricate relationships between temperature, salinity, phytoplankton biomass, and other environmental variables, thereby contributing to a deeper understanding of the complex dynamics within these unique saline lake ecosystems. Further research endeavors that build upon these findings and employ interdisciplinary approaches will be essential for addressing the challenges posed by climate change and ensuring the sustainable management of such vital aquatic systems.

Acknowledgments The study was financed by project MECESUP UCT 0804, and the collaboration of the PT09BD001 project supported by the Swedish International Development Cooperation Agency (SIDA). The main author expresses his gratitude to M.I. and S.M.A., for their valuable comments in improving the chapter.

References

Campos, H., Soto, D., Parra, O., Steffen, W., & Agüero, G. (1996). Limnological studies of Amarga lagoon, Chile: A saline lake in Patagonia, South America. *International Journal of Salt Lake Research, 4*, 301–314. https://doi.org/10.1007/BF01999114

Darehshouri, S., Michelsen, N., Schüth, C., Tajrishy, M., & Schulz, S. (2023). Evaporation from the dried-up lake bed of Lake Urmia, Iran. *Science of The Total Environment, 858*, 159960. https://doi.org/10.1016/j.scitotenv.2022.159960

De los Rios-Escalante, P., Contreras, A., Lara, G., Latsague, M., & Esse, C. (2022). Associations between spectral properties, bacteriological characteristics, chlorophyll and zooplankton communities in two north Patagonian lakes. *Animal Biology, 73*(1), 65–84. https://doi.org/10.1163/15707563-bja10097

De los Rios-Escalante, P., & Woelfl, S. (In press). *A review of zooplankton research in Chile.* Limnologica.

Dogliotti, A. I., Ruddick, K., Nechad, B., Doxaran, D., & Knaeps, E. (2015). A single algorithm to retrieve turbidity from remotely-sensed data in all coastal and estuarine waters. *Remote Sensing of Environment, 156*, 157–168. https://doi.org/10.1016/j.rse.2014.09.020

Dorador, C., Meneses, D., Urtuvia, V., Demergasso, C., Vila, I., Witzel, K.-P., & Imhoff, J. F. (2009). Diversity of Bacteroidetes in high-altitude saline evaporitic basins in northern Chile. *Journal of Geophysical Research, 114*, G00D05. https://doi.org/10.1029/2008JG000837

Franz, B. A., Bailey, S. W., Kuring, N., & Werdell, P. J. (2015). Ocean color measurements with the Operational Land Imager on Landsat-8: Implementation and evaluation in SeaDAS. *Journal of Applied Remote Sensing, 9*(1), 096070. https://doi.org/10.1117/1.JRS.9.096070

Harris, C. R., Millman, K. J., van der Walt, S., Gommers, R., Virtalen, P., Cournapeau, D., Wieser, E., Taylor, J., Berg, S., Smith, N. J., Kern, R., Picus, M., Hoyer, S., van Kerkwijk, M. H., Brett, M., Haldane, A., Fernandez del Río, J., Wiebe, M., Peterson, P., Gérard-Marchant, P., Sheppars, K., Reddy, T., Weckesser, W., Abbasi, H., Gohlke, C., & Oliphant, T. E. (2020). Array programming with NumPy. *Nature, 585*, 357–362. https://doi.org/10.1038/s41586-020-2649-2

Hunter, J. D. (2007). Matplotlib: A 2D graphics environment. *Computer Science Engineering, 9*, 90–95. https://doi.org/10.1109/MCSE.2007.55

Hurlbert, S. H., Loayza, W., & Moreno, T. (1986). Fish-flamingo-plankton interactions in the Peruvian Andes. *Limnology & Oceanography, 31*(3), 457–468. https://doi.org/10.4319/lo.1986.31.3.0457

Jaramillo. (2003). *Aves de Chile*. Lynx ediciones.

Maldonado, M., Navarro, G., Acosta, F., Aguilera, X., De la Barra, N., Cadima, M., & Goitia, E. (2012). *Humedales y cambio climático en los Altos Andes de Bolivia*. Unidad de Limnología y Recursos Acuáticos. Universidad Mayor de San Simón.

McKinney, W. (2010). Data structures for statistical computing in python. *Proceedings 9th Python in Sciences Conferences, 445*, 51–56. https://doi.org/10.25080/Majora-92bf1922-00a

Rodríguez-López, L., Duran-Llacer, I., González-Rodríguez, L., Abarca-del-Rio, R., Cárdenas, R., Parra, O., Martínez-Retureta, R., & Urrutia, R. (2020). Spectral analysis using LANDSAT images to monitor the chlorophyll-a concentration in Lake Laja in Chile. *Ecological Informatics, 60*, 101183. https://doi.org/10.1016/j.ecoinf.2020.101183

Seabold, S., & Perktold, J. (2010). *Statsmodels: Econometric and statistical modeling with python.* In 9th Python in Science Conference. https://doi.org/10.25080/Majora-92bf1922-011

Tian, L., Tian, J., Wang, J., Wang, X., & Li, W. (2023). A novel remote sensing index for brine shrimp (Artemia) slick detection in salt lakes. *Remote Sensing of Environment, 286*, 113428. https://doi.org/10.1016/j.rse.2022.113428

Topp, S. N., Pavelsky, T. M., Jensem, D., Simard, M., & Ross, M. R. V. (2020). Research trends in the use of remote sensing for inland water quality science: moving towards multidisciplinary applications. *Water, 12*, 169. https://doi.org/10.3390/w12010169

Vander Plas, J. (2017). *Python data science handbook* (p. 529). O'Reilly Media, Inc., 1005 Gravenstein Highway North, Sebastopol, CA 95472.

Vanhellemont, Q. (2020). Sensitivity analysis of the Dark Spectrum Fitting atmospheric correction for metre- and decametre-scale satellite imagery using autonomous hyperspectral radiometry. *Optics Express, 27*(20), A1372–A1399.

Van Rossum, G., & Drake, F. L., Jr. (1995). *Python reference manual*. Centrum voor Wiskunde en Informatica Amsterdam.

Waskom, M. L. (2021). Seaborn: statistical data visualization. *Journal of Open Software, 6*, 3021. https://doi.org/10.21105/joss.0302

Zuñiga, L., Campos, V., Pinochet, H., & Prado, B. (1991). A limnological reconnaissance of Lake Tebenquiche, Salar de Atacama, Chile. *Hydrobiologia, 210*, 19–25. https://doi.org/10.1007/BF00014320

Zuñiga, O., Wilson, R., Ramos, R., Retamales, E., & Tapia, L. (1994). Ecología de *Artemia franciscana* en la laguna Cejas, Salar de Atacama, Chile. *Estudios Oceanologicos, 13*, 71–84.

Chapter 10
Spatiotemporal Analysis of the Karakoram Dynamics: A Case Study of the Ghulkin Glacier, Gilgit Baltistan, Pakistan

Muhammad Amin and Aqil Tariq

Introduction

A glacier is a big, slow-moving mass of thick ice that slides downhill. A glacier's upper half is known as the accumulation (input) zone, and it is here that snow and ice are added to the glacier by compression and snowfall. Ablation (output) zone describes the bottom part of the glacier where ice is lost due to melting and evaporation (Majeed et al., 2023). An equilibrium line of height separates the ablation and accumulation zone, where the annual balance between accumulation and ablation occurs (Esmaeili et al., 2024). Research on mass balance is critical for understanding glacial responses to recent climatic changes (Jalayer et al., 2023; Sharifi et al., 2022a), particularly how an increase in temperature and humidity causes more melting of the glacier and the formation of new glacier lakes, which results in a decrease in a glacier's mass (Basharat et al., 2022; Tariq et al., 2023). Simulation models imitate complex meteorological, glaciological, periglacial, and hydrological interactions (Basharat et al., 2022; Bokhari et al., 2023; Tariq et al., 2023). Simulating glacier dynamics at a regional scale, however, requires the adoption of "direct" approaches (e.g., in glaciological and budget methods) to meet model calibration requirements over large distances and on a longer timescale (Imran et al., 2022; Sharifi et al., 2022b; Tariq et al., 2023b; Tariq & Qin, 2023;

M. Amin
Institute of Geo-Information & Earth Observation, PMAS Arid Agriculture University Rawalpindi, Rawalpindi, Pakistan

A. Tariq (✉)
Department of Wildlife, Fisheries and Aquaculture, College of Forest Resources, Mississippi State University, Mississippi State, MS, United States

State Key Laboratory of Information Engineering in Surveying, Mapping and Remote Sensing, Wuhan University, Wuhan, China
e-mail: at2139@msstate.edu; aqiltariq@whu.edu.cn

A. L. Singh et al. (eds.), *Climate Change, Vulnerabilities and Adaptation*,
https://doi.org/10.1007/978-3-031-49642-4_10

Zamani et al., 2022). Such labor-intensive approaches have not been considered practical for poorly surveyed rugged areas of glaciated ranges such as Himalaya Karakoram and Hindu Kush (HKH) (Baqa et al., 2022; Majeed et al., 2022b; Moazzam et al., 2022), which are melting at a fast rate due to local and global warming (Hussain et al., 2022a; Khan et al., 2022; Shah et al., 2022).

Among the HKH glaciers, the Karakorum glaciers in Pakistan are greatly affected by high incoming radiations, humid conditions, and heatwaves (Ghaderizadeh et al., 2022; Tariq et al., 2022b, c; Wahla et al., 2022). This is due to the increase in glacial lake formation trends during recent years, which include a high frequency of avalanches, landslides, and outburst floods from the new glacial lakes, termed glacial lake outburst floods (GLOFs). All these pose continuous hazards to the lowland communities (Fu et al., 2022; Haq et al., 2022; Jalayer et al., 2022; Majeed et al., 2022a). Understanding the current shifts in glaciers' mass balance and the resulting risks is hampered by a lack of yearly and seasonal data on glaciological observations (such as land cover changes, debris, glacial lakes, and climate factors) (Bera et al., 2022; Hussain et al., 2022b; Khalil et al., 2022b; Tariq et al., 2022a; Ullah et al., 2022).

For broad areas where field-based glaciological measurements are scarce, remote sensing (RS) has become increasingly popular due to its high spatial and temporal resolution, which allows for the estimation of glaciological mass balance and hydrometeorological changes (da Silva Monteiro et al., 2022; Islam et al., 2022; Khalil et al., 2022a; Tariq et al., 2022d). Mainly, it has been effective for real-time glacier monitoring concerning the regional and local climate. To estimate changes in the glacier surface, i.e., retreat, advance, and glacier lakes, several remote sensing techniques have been used, such as the Normalized Differentiate Water Index (NDWI) to detect the glacier lake and the normalized difference snow index (NDSI) for snow cover changes (Baloch et al., 2021; Khan et al., 2022; Majeed et al., 2021; Shah et al., 2022). In addition to that, these indicators assist in the evaluation of the hydrometeorological limits in order to estimate dangers such as GLOFs (Felegari et al., 2021; Mousa et al., 2020; Sharifi et al., 2021; Tariq, 2023).

The study aims to analyze spatiotemporal changes in the Ghulkin glacier concerning climate change that has been taking place since 1990. To do so, (i) extensive field surveys have been conducted in the Ghulkin region to estimate ice mass/volume losses at the glacier ablation zone, (ii) RS data has been used to estimate changes in the glacial land cover due to climate change, and (iii) the glacial mass losses due to climate change have been projected. The findings of this study show that from 1990 to 2015, there has been an increase in surface temperature and glacial lake formation at high altitudes and a decrease of ice mass from the side moraine at the ablation zone Ghulkin glacier.

Materials and Methods

Study Area

The research area encompasses the entirety of the Ghulkin glacier and the Hussaini settlement in the upper Hunza valley (Fig. 10.1). The latitude of the research region is 5.8026°N, and the longitude is 74.9832° E. The glacier has a total area of 28.5 km^2 in its entirety. The Ghulkin glacier is about 17 km long and approximately 70 m wide. The glacier ablation zone is about 6 km long, but varies with time. The glacier snout is advanced due to its steep slope, while the tendency of glacier lake formation has increased, and the ice mass from side moraines has decreased. The GLOF events are occurring more frequently in HKH glaciers, and the emergence of glacial ablation zones is the evidence (Felegari et al., 2023; Hu et al., 2021; Siddiqui et al., 2020). The Gulkin glacier is approximately 17 km in length and 70 m in width. The settlement of Hussaini can be found on the glacier's northern flank, while the village of Gulkin can be found on its southern flank. The Karakoram Highway and the Upper Hunza River can be found to the east, and the glacier's source peak can be found to the west.

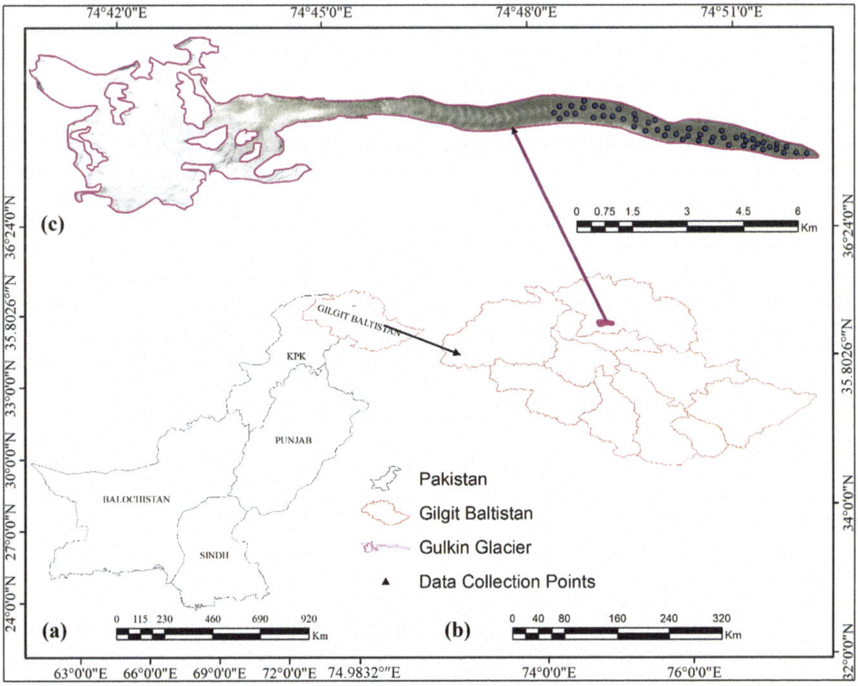

Fig. 10.1 (**a**) Boundary of Pakistan; (**b**) Gilgit Baltistan; and (**c**) description of the study area and data collection points at the ablation zone of Ghulkin glacier

Table 10.1 Historical GLOF events from Ghulkin Glacier

Year	Date of event	Glacier (GLOF) location	River
2007	05-April	Gulkin	Hunza
2009	03-May		
2010	04-April		
2011	28-April		
2012	14 May		
2014	27-May		

The glacier's melting zone extends for approximately 6 km in length. It has been noted that the glacier is melting at an alarming rate; this can be deduced from the appearance of supraglacial lakes that have formed in the ablation zone. It is observed that five lakes have been included in the glaciers. The construction of lakes on the glacier is so that two lakes, namely Ashore Baig and Ghowj, are on the northern side toward Hussaini village at an elevation of 2734 m and 2876 m, respectively. Ashore Baig is 200 m long and 20 m wide; its depth reaches up to 15 to 20 m, while other lakes, namely Roud 1 and 2 are located at an elevation of 2856 m in the southern side toward Gulkin village (Asif et al., 2023; Shah et al., 2021; Tariq & Shu, 2020). The average length of these two lakes is 40 m and 30 m, respectively, while the width is 25 and 20 m, respectively. These lakes show the formation of supraglacial lakes on glacier relatively oval shape, and the depth reaches up to 10 m when filled. The critical point is that all these lakes are formed near the lateral moraines of the glacier's left and right flanks. The lakes have been developed in the ablation zone, where melting is rapidly increasing (Basharat et al., 2022; Tariq et al., 2023, 2023a). The debris layer covers glaciers also washed out; this can also activate the melting rate. In the terminus point of the glacier, medium to large craves have been developed. These crevasses squeeze in winter. It reopens when temperature increases (freezing and thawing) and gives way to the flow of melting water. On the Hussaini side near the Ashore Baig lake, the crevasse is traversed in North-East (NE), and the melting flows into Ashore Baig Lake. In the same line, another lake, Ghowj, is located at a distance of 1 km upstream. From this lake, a large crevasse is observed to pass toward the Gulkin side in the direction of North to South East (N-SE) (CERT monitoring team). A parallel crack in the order of NS is also evident toward Gulkin near the terminus (see Fig. 10.4). Lateral moraines on either side of the glacier also spark evidence of the glacier's melting. It is observed that the moraines are exposed to 10 to 14 m high, while at the head of the Gulkin village, the moraines are superimposed by the glacier as the moraines are subsiding/sliding down due to the pressure of the glacier, as a result of that the moraines are diminishing and seepage from the glacier.

The villages of Sosot and Khalti in Tehsil Gupis were hit by two GLOFs in 1999. The Shimshal valley was the site of GLOF incidents in 2000. Six GLOF episodes, all originating from the Ghulkin glacier between 2007 and 2009 (Table 10.1), caused extensive damage to the Karakoram highway and local property. In 2009, the Passu glacier was responsible for one GLOF occurrence. Hundreds of trees were destroyed in 2012 due to a GLOF event that originated from the Ghulkin glacier. As a result, the number of GLOF incidents in the region has increased since 2007, especially those caused by HKH glaciers like the Ghulkin glacier (Mihalcea et al., 2008). Therefore, the spatiotemporal effects of climate change on the Ghulkin glacier's mass balance and the accompanying hazards must be studied.

Data Sets

Remote Sensing Data

Satellite imagery of Landsat were obtained from the NASA-EROS for the years 1990, 1995, 2000, 2010, and 2015 (Table 10.2). Moreover, four spectral bands with a 30-m spatial resolution of Landsat-8 were used to cover the 1990–2015 time periods (Quincey et al., 2009). The geometric and radiometric errors were removed through pre-processing techniques before utilizing the data for image analysis.

Elevation Data

This research used the digital elevation model (DEM) data of Alospalsar, which was obtained from the NASA (Round et al., 2017). Recent research have made extensive use of the data provided by the Alospalsar DEM instrument in order to investigate spatiotemporal differences in glaciers in relation to variations in the climate of the Himalayan ranges (Pimentel & Flowers, 2011). High-resolution DEM data, in comparison to low-resolution ASTER DEM, was used to perform the following glaciological measurements on the Ghulkin glacier: calculating the total mass-loss volume at the glacier ablation zone and creating the slope and aspect surfaces for predicting the glacier mass-loss areas due to climate change.

Table 10.2 Image data specifications of Landsat-5 thematic mapper (TM), Landsat-7 enhanced thematic mapper plus (ETM+), and Landsat-8 operational land imager

Data acquisition	Satellite	Sensor	SR	TR	Path	Row	Source
22/7/1995	Landsat-4 and -5	TM	30 m	16 days	150	35	USGS (EROS)
2/7/2000	Landsat-4 and -5	TM	30 m	16 days	150	35	
17/7/2005	Landsat-7	ETM+	30 m	16 days	150	35	
14/7/2010	Landsat-7	ETM+	30 m	16 days	150	35	
14/7/2015	Landsat-8	OLI/TIRS	30 m	16 days	150	35	
11/11/2010	DEM	Aster	30 m				Earth Data

DEM digital elevation model, *ETM+* enhanced thematic mapper plus (ETM+), *OLI* operational land imager, *SR* spatial resolution, *TM* thematic mapper, *TR* temporal resolution

Mass Balance Data

With the help of the ICIMOD, its geologists, and the community watch group (CWG), accurate mass-balance data for the Ghulkin glacier's ablation zone was gathered through the use of GPS and Rangefinder. FOCUS Pakistan established the CWG group in 2008 to monitor glacier changes in Gilgit-Baltistan and Chitral under the project co-funded by ICIMOD. Data collected for 36 heights of side moraines on the left and right of the glacier is shown in Fig. 10.1.

Methods

Glacier Volume Calculation

Using direct measurements (obtained through GPS and Rangefinder), the total volume/mass loss was estimated at the glacier ablation zone. First, the inverse distance weighted (IDW) interpolation method was used to generate the mass-loss surface from the glacier's 36 mass-loss measurements. The mass-loss interpolated surface was converted into the triangular irregular network (TIN) using heights in the GPS-based field data to calculate volume loss at the glacier ablation zone. The TIN data structure allows for the inclusion of information about each pixel's x, y, and z values, which are required to calculate two- and three-dimensional areas and three-dimensional volumes. Finally, the derived TIN data was used to calculate glacier mass loss from the moraines' sides. These glacier mass losses from moraine sides were validated using field photographs collected by FOCUS Pakistan geologists and CWG members. Furthermore, the TIN surface was used to calculate the total mass-loss volume at the glacier ablation zones through the functional surface tool of three-dimensional extensions in the ArcGIS environment.

Estimating Land Cover Change of the Ghulkin Glacier

Remote sensing images were analyzed for the delineation of water bodies at the ablation zone of the glacier and the land cover classification for estimating spatio-temporal changes in the debris and water bodies from 1990 to 2015. At the same time, extractions of surface temperatures were also recorded.

Delineation of Water Bodies at the Ablation Zone of the Glacier

Landsat 4, 5, 7, and 8 images were stacked, and the NDWI was calculated as NDWI = (Green − NIR / Green + NIR), where NIR is the near-infrared band of Landsat images of 4, 5, 7, and 8. The threshold value is accurately segmented as 0.4 as this threshold of NDWI values represents water bodies; therefore, they were used to delineate temporal glacial lakes between 1990 and 2015. In addition, the segregation of the resulting images for glacial lakes was performed by correlating them with field pictures from 2007 to 2016 which were obtained from the FOCUS Humanitarian Assistance, Pakistan.

Land Cover Classification of the Ghulkin Glacier

Land cover maps of 1990 and 2015 were prepared using image classification techniques. To perform supervised classification, training data was collected from field surveys by collaborating with geologists. In total, 36 ground control points were observed from the glacier ablation zone between 2008 and 2015. Three land cover classes were mapped, including snow cover, debris, and lake waters. These classes were manually delineated with Google imagery to calculate an accurate area of each land cover class. Finally, NDWI maps were used to delineate recently formed glacial lakes. Statistics obtained from land cover maps were used to compare area changes in the glacier snow, debris, and glacier lakes between 1990 and 2015. The results of the NDWI maps were validated through high-resolution Google imagery.

Extraction of Surface Temperature on the Ghulkin Glacier

For the same period 1990–2015, thermal bands of Landsat 4–8 were used to extract temporal variations of the surface temperatures on the Ghulkin glacier. Initially, the top-of-atmosphere (TOA) radiance was extracted from the reflectance values. Equation (10.1) was rendered as follows.

$$T = K_2/(Ln\ (1\ K_1 + (\text{radiance})) - 273.15 \tag{10.1}$$

where T is at-satellite brightness temperature (Celsius), K_2denotes the specific constant band for thermal conversion from the metadata (K_2_CONSTANT_BAND_x, where x is the band number), and K_1 is the band-specific constant for thermal conversion from the metadata (K_1_CONSTANT_BAND_x, where x is the band number). The spectral radiance is calculated as in Eq. (10.2):

$$(((L\max\ - L\min)/254) * (DN - 1)) + Lm \tag{10.2}$$

where the values are provided in the Landsat metadata file.

Table 10.3 Weights assigned to various inputs in the weighted overlay analysis for identifying the extensive mass-loss (snow melting) areas of the Ghulkin glacier due to climate change

Inputs	Input values	Weights
	0–20	9
Slope	20–30	6
	Above 30	1
	South East	9
Aspect	East	8
	Another slope aspect	1
	Above 15 feet	9
Predicted mass-loss	10 feet	6
	Below 5	3
	0.3 above	9
NDWI	0 to 0.3	6
	Below 0	0

Weighted Overlay Analysis

Weighted overlay analysis was performed to identify the potential site for newly formed glacier lakes. We have used four datasets in raster format, namely slope, aspect, NDWI, and interpolation surface of mass loss from the side moraine of the Ghulkin glacier. These raster surfaces were reclassified and the weights were assigned in the ArcGIS environment. The consequences (0–9) for the four surfaces are shown in Table 10.3.

Results

Volume/Mass Loss at the Ablation Zone of the Ghulkin Glacier

Figure 10.2a shows the mass-loss surface of the Ghulkin glacier interpolated from the glacier's 36 mass-loss measurements. Comparing the mass-loss sites with field photographs of the ablation zone during the years 2009, 2011, 2012, and 2013 in Fig. 10.2b indicates that the ablation zone of the glacier has lost significant glacier mass (through snow melting), which developed into new water bodies on the glacier surface. The ablation zone of the glacier had losses of 163,853 cubic meters of glacier ice mass from the side moraine of the glacier. This can be viewed from landscape changes in the aerial imagery from 2009, 2011, 2012, and 2013 (see Fig. 10.2b).

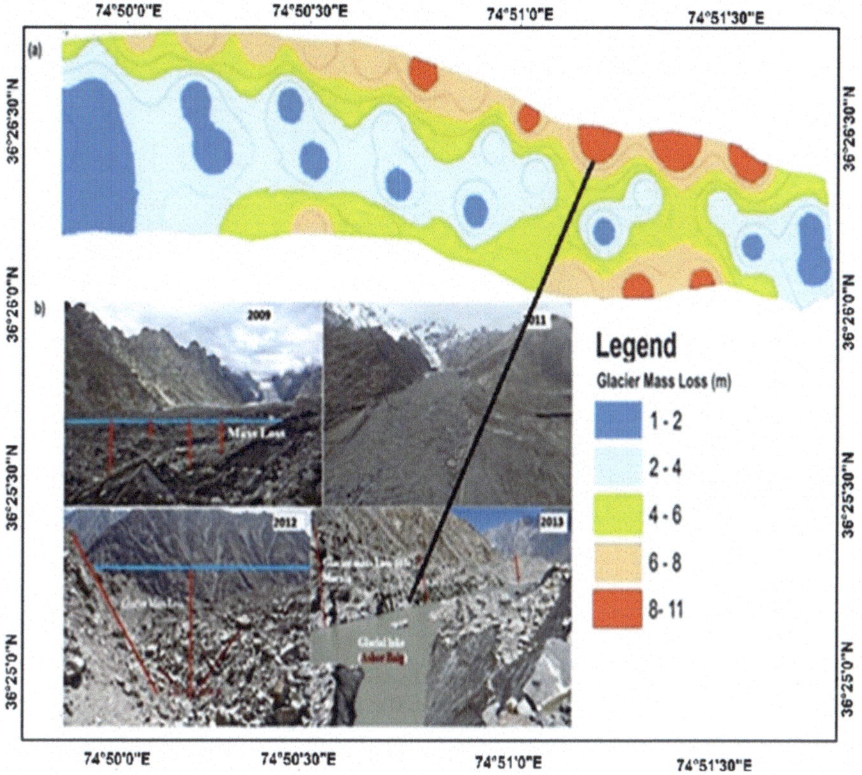

Fig. 10.2 (**a**) Map showing mass loss (meters) at the ablation zone of the Ghulkin glacier and (**b**) the field photographs of the ablation zone in 2009, 2011, 2012, and 2013

Estimating Land Cover Change of the Ghulkin Glacier due to Climate Change Through Remote Sensing

Figure 10.3 shows the temporal maps of NDWI for the period 1990 and 2015, derived from Landsat 4, 5, 7, and 8 data. The glacier lake area/size results were also validated through high-resolution Google imagery. Figure 10.4 shows spatiotemporal changes in the glacier lakes of the Ghulkin glacier in the years 1990, 1995, 2000, 2010, and 2015. This indicates that the total area covered by Ghulkin lakes has increased by 18,990 km^2 from 1990 to 2015. The surface temperature of the glacier has also increased between 1990 and 2015. This is undeniable evidence of the effects of global warming on glacier mass balance, the formation of high-altitude glacier lakes, and the progression of the snow line into the accumulation zone.

Figure 10.5a, b show the land cover change of the Ghulkin glacier from 1990 to 2015. This indicates that the fresh snow covered 87.17% of the glacier in 1995, and this was reduced to 82.68% in 2015 (i.e., a snow loss of 4.49% in the last 25 years).

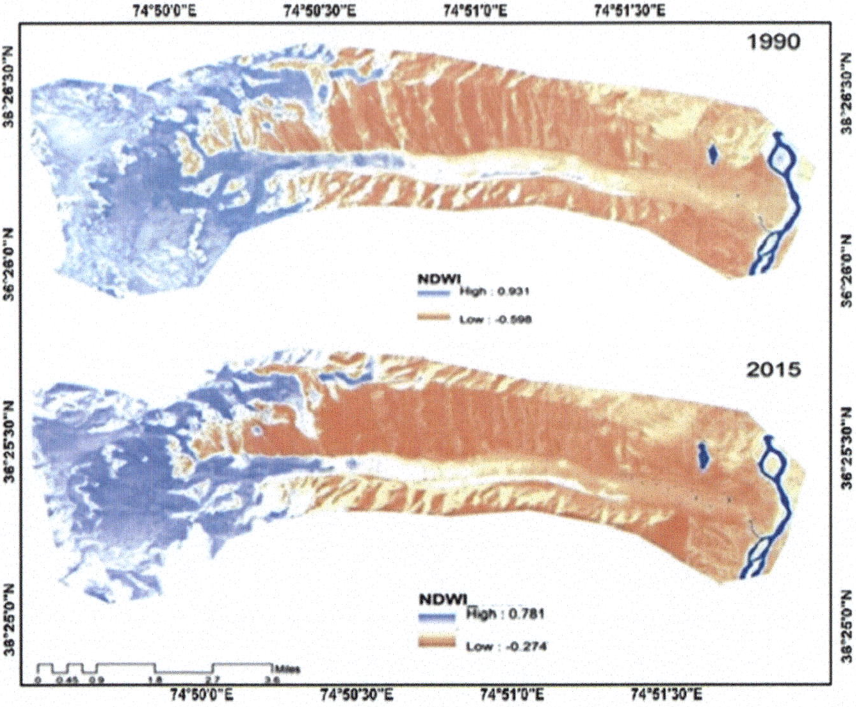

Fig. 10.3 Temporal maps of the NDWI for the period 1990–2015 derived from Landsat 4, 5, and 8 data

Consequently, the total area of the Ghulkin glacier (i.e., glacier and snow cover) decreased from 24.84 to 23.56 km^2 from 1990 to 2015 (see Fig. 10.5b). The snow line uplifting at a higher altitude was due to the increase in temperature. This can be seen in an increase in the debris cover on the glacier from 12.82% to 17.26% (i.e., a rise of 4.44%). The 0.005 km^2 area of glacier lakes in 1990 increased to 0.018 km^2 by 2015 (Fig. 10.5b).

Figure 10.6 shows the temporal variation of surface temperatures on the Ghulkin glacier extracted from the thermal bands of Landsat-4 and -8 for the same period (1990–2015). Clearly, this increase in temperature explains the increase in the formation of new glacier lakes during this period. A temperature increase of about 1 to 2 °C is observed locally, particularly at the ablation zone of the glacier from 1990 to 2015, compared to a global temperature increase of 0.8 °C for the same period (Lutz et al., 2014). Figure 10.7 compares locally varying surface temperatures of the Ghulkin glacier for the years 1990 and 2015. The blue line indicates 1990 data, while the red line shows the data for 2015.

This graph shows the comparison of area coverage of temperature between 1990 and 2015. In 1990, the temperature range −19 to −1 °C covered 2365.3 acres of the glacier, 0 °C covered 1501 acres, and + 1 to +27 °C covered 2621.1 acres, while in

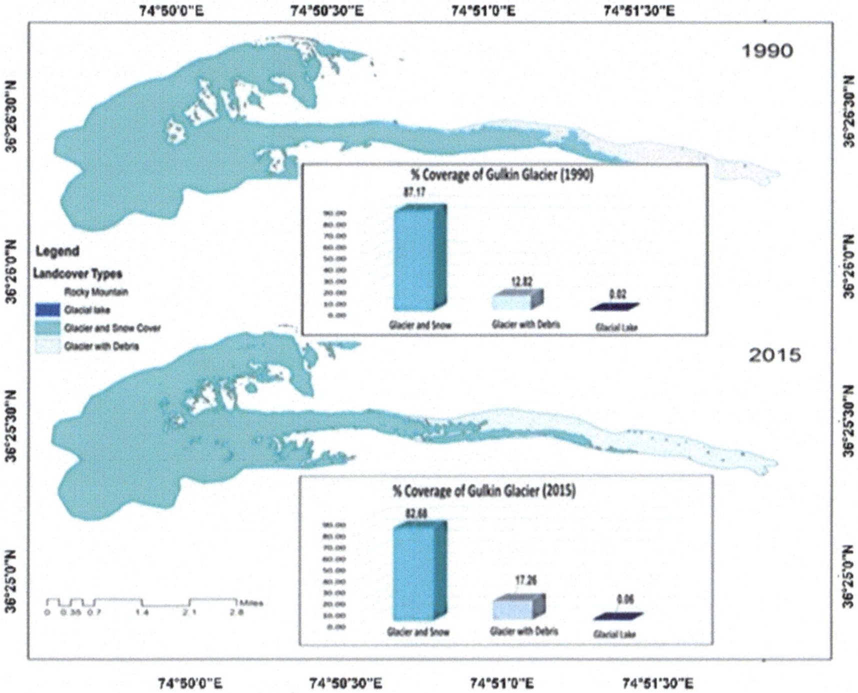

Fig. 10.4 Spatiotemporal changes in the areas of the Ghulkin glacial lakes (in sq. m) between 1990 and 2015

2015, the temperature range −19 to −1 °C covered 1645.3 acres of glacier, 0 °C covered 1623.2 acres, and + 1 to +27 °C covered 3218.9 acres. The results show a significant temperature change with respective area coverage between 1990 and 2015.

Topographic Analysis of the Ghulkin Glacier

Figure 10.8 shows the results of a weighted overlay analysis of the Ghulkin glacier mass loss. The areas within the brown to red zones (5–7) have a higher probability of developing new glacier lakes due to extensive snow-cover loss in the ablation zone. The yellow zones with a value of 4 have a moderate likelihood of such change. The greener zones (within a range of 0–3) have a very low probability of changing the glacier ablation zone into glacier lakes.

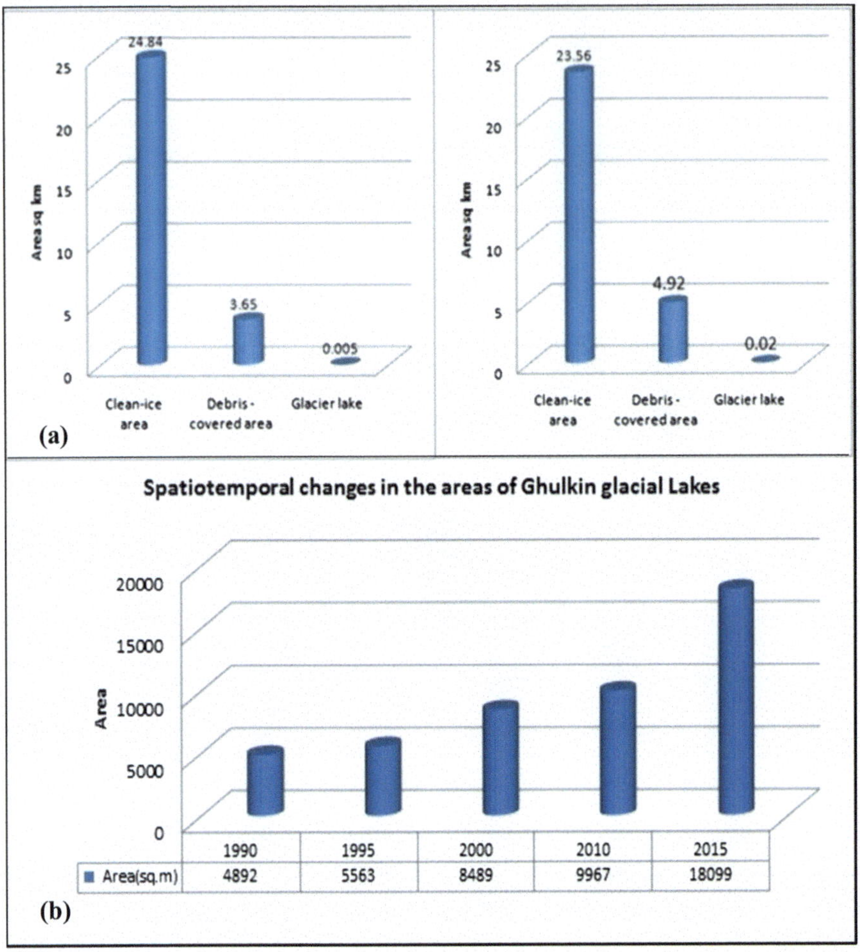

Fig. 10.5 (**a**) Land cover classification of percentage coverage of glacier snow, debris, and glacial lakes in 1995 and 2015 and (**b**) comparison of statistics for clean ice area, debris cover area, and glacial lakes in 1995 and 2015

The slope map in Fig. 10.9a shows that the ablation zone of the Ghulkin glacier existed within the range of 0–26 degrees. This is a moderate slope which is favorable for the formation of glacier lakes. The aspect map in Fig. 10.9b shows the slope directions of the glacier. The central part of the ablation zone has an east, southeast, and south aspect. These aspects have a higher tendency of solar insolation in comparison to a north slope aspect. These topographic factors clearly show that there is a greater chance of glacier lake formations.

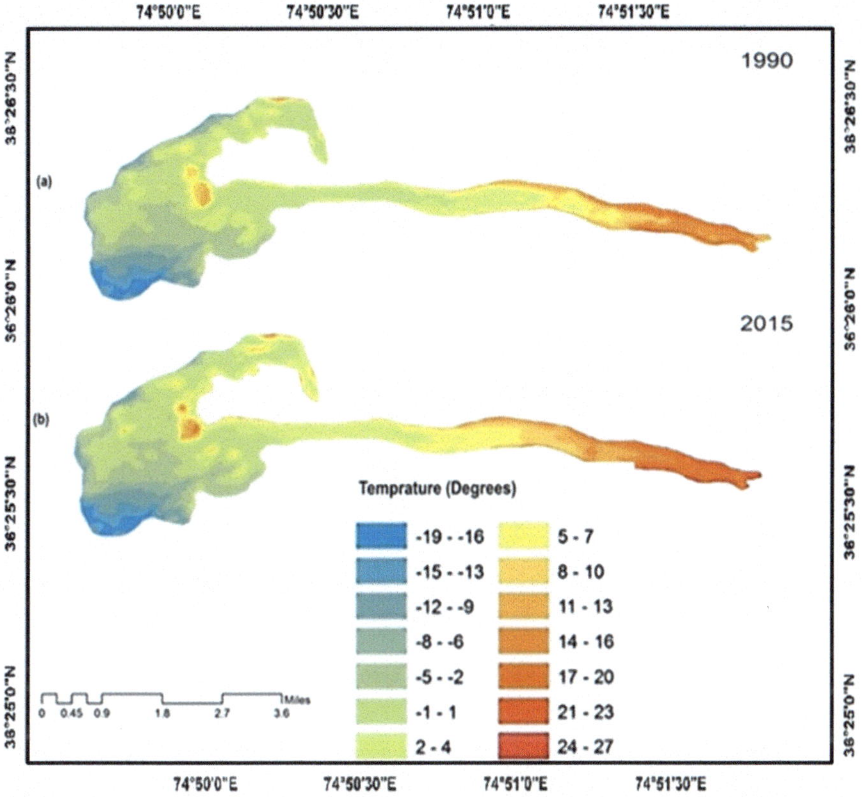

Fig. 10.6 Spatiotemporal distribution of the land surface temperatures in 1999 and 2015 derived from Landsat-4 and 8 thermal bands (TIRS)

Discussion

This study assessed the impact of climate change on the Ghulkin glacier's mass balance by using remote sensing techniques coupled with field measurements. Such mass-balance estimations are often carried out using (1) volume-area scaling techniques that infer ice volume from the glacier area through scaling relationships (Quincey et al., 2011), (2) RS geodetic methods, which estimate changes in elevation using DEMs on a pixel-by-pixel basis; and (3) accumulation area ratio (AAR) or equilibrium line altitude (ELA) methods, which infer yearly balances of mass from two parameters associated to a glacier's mass balance; AAR and ELA were calculated from field observations or satellite imagery. To estimate the glacier's mass balance on the scale of the region, the ELA technique is used with field data to determine the glacier's annual mass balance variation.

To delineate debris cover of the Ghulkin glacier, ASTER-derived DEM and field photographs were used to generate training data for the land cover classification. This is because the presence of debris cover poses problems in the image

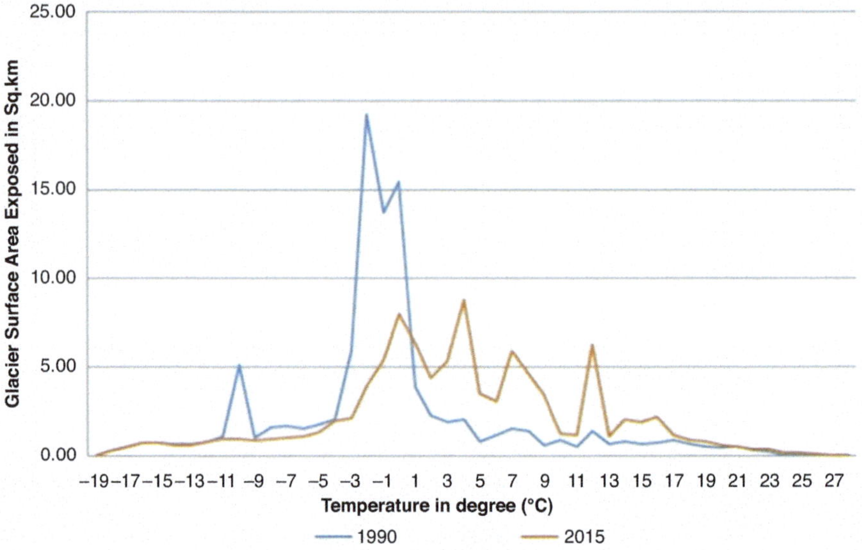

Fig. 10.7 Comparison of temperature statistics derived from Landsat images for the years 1999 and 2015

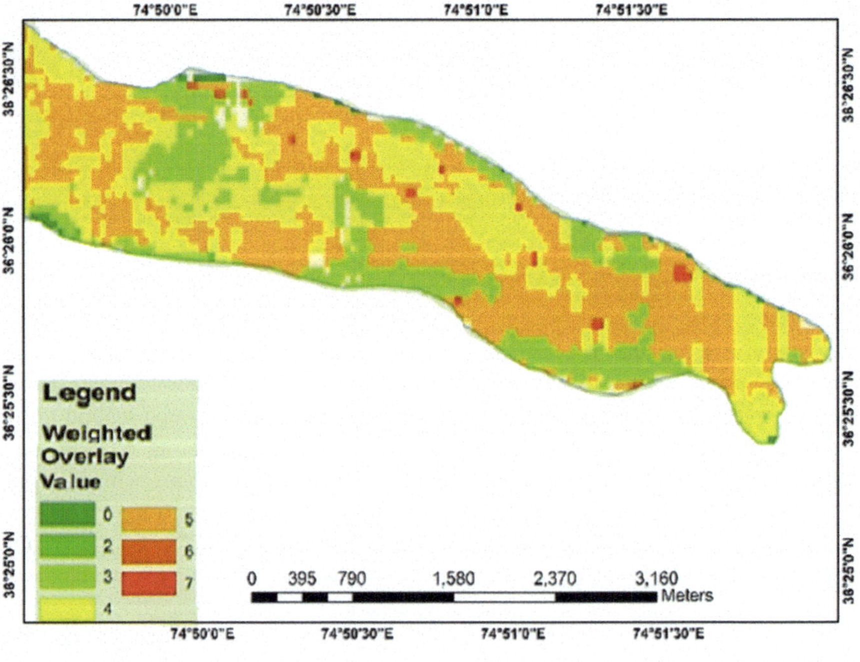

Fig. 10.8 Projected mass-loss of the Ghulkin glacier calculated using weighted overlay analyses of the slope, aspect, the NDWI, and mass-loss interpolation surfaces

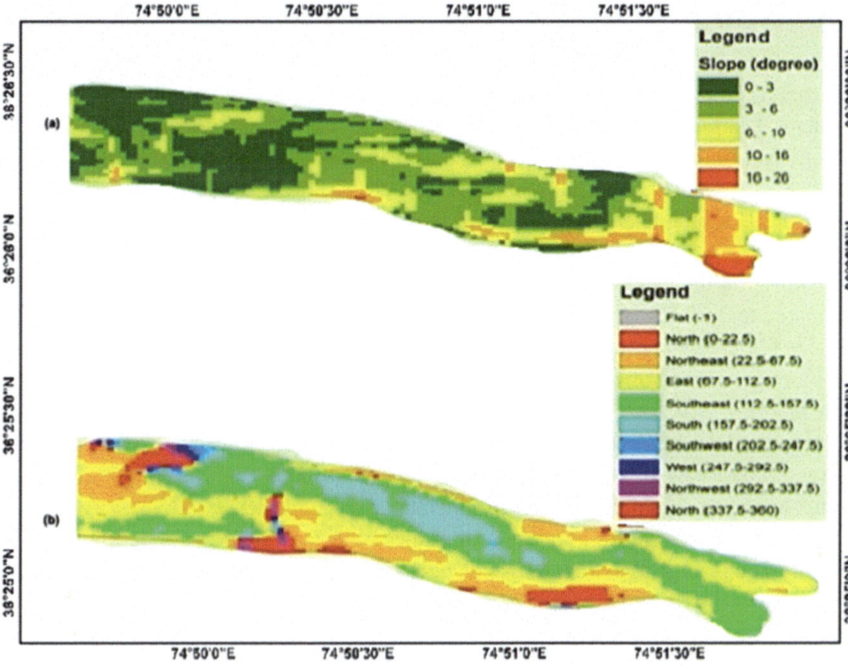

Fig. 10.9 Maps of slope (**a**) and aspect (**b**) of the ablation zone of the Ghulkin glacier

classification, and therefore in the automatic mapping of the glaciers. Moreover, widespread debris cover on HKH glacial ranges reduces their retreat rates, which are inappropriate as indicators of recent climate change (Quincey & Luckman, 2014). Nevertheless, the volume-area scaling techniques are not able to take into account such debris cover effects.

In addition to the link to changes in glacier measurements, further experimentation with this approach is needed to assess the impact of climatic variables on the mass balance of the Ghulkin glacier. This should include the following: (1) evaluating soil types to investigate further the impact of changes in debris cover on the glacier movement, (2) using other climatic parameters such as precipitation along with utilizing runoff data which could be helpful for better estimation of glacier lakes and GLOFs, and (3) coupling field data with high-resolution imagery which may be helpful to include the effect of localized seasonal variations over low-lying areas.

An important motivation for this study was to project GLOF sites over the Karakoram ranges in northern Pakistan. Using the multitemporal earth observation data, the area of the Ghulkin glacier lakes was found to have increased from 0.0065 km^2 to 0.02 km^2 between 1990 and 2015. Consequently, the devastating GLOF trend that started in 1999 continues under the present climate conditions. Our future work will extend this approach to model GLOF sites that will affect human settlements shortly.

Conclusion

The spatiotemporal analysis reveals that there has been a decrease in the overall glacier mass. During the last 25 years, the total clean-ice area decreased by 4.51%. This glacier mass loss caused an increase of 4.44% in the debris-covered area and an increase of 0.04% in the glacial lakes. The glacier snow areas have expanded from 1990 to 2015, as evidenced by the overlaying maps of glacier surface temperature and land cover. Specifically, the glacier snow areas measured 9.56 km^2 compared to 7.85 km^2 for the colder temperature range of −19 °C to −1 °C, and 6.07 km^2 compared to 2.90 km^2 for 0 °C. In contrast, for the warmer temperature range (+1 °C to +27 °C), the glacier areas increased from 10.61 km^2 to 15.49 km^2. Conclusively, the more generous temperature range contained more glacier surface areas in 2015 than in 1990. The total area of the Ghulkin glacier (i.e., glacier and snow cover) underwent a marked retreat of about 1.28 km^2 between 1990 and 2015, with an average retreat rate of 0.05 km^2 per year. This is obvious from the frequent GLOF events and corresponding changes in the landscape since 1999.

The devastating GLOF trend recorded since 1999 will continue under the present climatic conditions. This research will help the local government and other stakeholders with timely and efficient monitoring of glacier lakes and associated hazards. Thus, lowland communities can be ready to adapt to the effects of climate change on the glaciers in a more effective manner.

Acknowledgments This research received funding from the USAID Partnerships for Enhanced Engagement in Research (PEER) cycle 5, under project "Satellite Enhanced Snow-Melt Flood and Drought Predictions for the Kabul River Basin with Surface and Ground Water." The authors acknowledge the support of geologists and members of Community Watch groups (CWG) during field data collection and field analysis related to mass glacier retreats.

Credit Authorship Contribution Statement Muhammad Amin and Aqil Tariq did all the analysis and contributed equally to this chapter. All authors have read and agreed to the published version of the chapter.

Availability of Data and Material The authors would like to pay special and heart-whelming thanks to NASA Earth data (https://lpdaacsvc.cr.usgs.gov/appeears/) for providing the Grace satellite data. The datasets generated and/or analyzed during this study are not publicly available but are available from the corresponding author on reasonable request.

Competing Interests The authors declare that there is no conflict of interest in this chapter's publication. Moreover, the writers have thoroughly addressed ethical issues, including plagiarism, informed consent, fraud, data manufacturing and/or falsification, dual publication and/or submission, and redundancy.

Data Availability Statement The authors would like to pay special and heart-whelming thanks to NASA USGS (https://earthexplorer.usgs.gov/) for providing the Landsat data.

References

Asif, M., Kazmi, J. H., & Tariq, A. (2023). Traditional ecological knowledge based indicators for monitoring rangeland conditions in Thal and Cholistan Desert, Pakistan. *Environmental Challenges, 13*, 100754. https://doi.org/10.1016/j.envc.2023.100754

Baloch, M. Y. J., Zhang, W., Chai, J., Li, S., Alqurashi, M., Rehman, G., Tariq, A., Talpur, S. A., Iqbal, J., Munir, M., & Hussein, E. E. (2021). Shallow groundwater quality assessment and its suitability analysis for drinking and irrigation purposes. *Water (Switzerland), 13*, 1–25. https://doi.org/10.3390/w13233361

Baqa, M. F., Lu, L., Chen, F., Nawaz-ul-Huda, S., Pan, L., Tariq, A., Qureshi, S., Li, B., & Li, Q. (2022). Characterizing spatiotemporal variations in the urban thermal environment related to land cover changes in Karachi, Pakistan, from 2000 to 2020. *Remote Sensing, 14*, 2164. https://doi.org/10.3390/rs14092164

Basharat, M., Khan, J. A., Khalil, U., Tariq, A., Aslam, B., & Li, Q. (2022). Ensuring earthquake-proof development in a swiftly developing region through neural network modeling of earthquakes using nonlinear spatial variables. *Buildings, 12*, 1713. https://doi.org/10.3390/buildings12101713

Bera, D., Das Chatterjee, N., Mumtaz, F., Dinda, S., Ghosh, S., Zhao, N., Bera, S., & Tariq, A. (2022). Integrated influencing mechanism of potential drivers on seasonal variability of LST in Kolkata Municipal Corporation, India. *Landscape, 11*, 1461. https://doi.org/10.3390/land11091461

Bokhari, R., Shu, H., Tariq, A., Al-Ansari, N., Guluzade, R., Chen, T., Jamil, A., & Aslam, M. (2023). Land subsidence analysis using synthetic aperture radar data. *Heliyon, 9*, e14690. https://doi.org/10.1016/j.heliyon.2023.e14690

da Silva Monteiro, L., de Oliveira-Júnior, J. F., Ghaffar, B., Tariq, A., Qin, S., Mumtaz, F., Correia Filho, W. L. F., Shah, M., da Rosa Ferraz Jardim, A. M., da Silva, M. V., de Barros Santiago, D., Barros, H. G., Mendes, D., Abreu, M. C., de Souza, A., Pimentel, L. C. G., da Silva, J. L. B., Aslam, M., & Kuriqi, A. (2022). Rainfall in the urban area and its impact on climatology and population growth. *Atmosphere (Basel)., 13*, 1610. https://doi.org/10.3390/atmos13101610

Esmaeili, M., Abbasi-Moghadam, D., Sharifi, A., Tariq, A., & Li, Q. (2024). ResMorCNN Model: Hyperspectral Images Classification Using Residual-Injection Morphological Features and 3DCNN Layers. *IEEE Journal of Selected Topics in Applied Earth Observations and Remote Sensing, 17*, 219–243. https://doi.org/10.1109/JSTARS.2023.3328389

Felegari, S., Sharifi, A., Moravej, K., Amin, M., Golchin, A., Muzirafuti, A., Tariq, A., & Zhao, N. (2021). Integration of Sentinel 1 and Sentinel 2 Satellite Images for Crop Mapping. *Applied Sciences, 11*, 10104. https://doi.org/10.3390/app112110104

Felegari, S., Sharifi, A., Khosravi, M., Sabanov, S., Tariq, A., & Karuppannan, S. (2023). Using Sentinel-2 data to estimate the concentration of heavy metals caused by industrial activities in Ust-Kamenogorsk, Northeastern Kazakhstan. *Heliyon. 9*(11):e21908. https://doi.org/10.1016/j.heliyon.2023.e21908

Fu, C., Cheng, L., Qin, S., Tariq, A., Liu, P., Zou, K., & Chang, L. (2022). Timely plastic-mulched cropland extraction method from complex mixed surfaces in arid regions. *Remote Sensing, 14*, 4051. https://doi.org/10.3390/rs14164051

Ghaderizadeh, S., Abbasi-Moghadam, D., Sharifi, A., Tariq, A., & Qin, S. (2022). Multiscale dual-branch residual spectral-spatial network with attention for hyperspectral image classification. *IEEE Journal of Selected Topics in Applied Earth Observations and Remote Sensing*, 1–14. https://doi.org/10.1109/JSTARS.2022.3188732

Haq, S. M., Tariq, A., Li, Q., Yaqoob, U., Majeed, M., Hassan, M., Fatima, S., Kumar, M., Bussmann, R. W., Moazzam, M. F. U., & Aslam, M. (2022). Influence of edaphic properties in determining forest community patterns of the Zabarwan Mountain Range in the Kashmir Himalayas. *Forests, 13*, 1214. https://doi.org/10.3390/f13081214

Hu, P., Sharifi, A., Tahir, M. N., Tariq, A., Zhang, L., Mumtaz, F., & Shah, S. H. I. A. (2021). Evaluation of vegetation indices and phenological metrics using time-series MODIS data for monitoring vegetation change in Punjab, Pakistan. *Water, 13*, 2550. https://doi.org/10.3390/w13182550

Hussain, S., Lu, L., Mubeen, M., Nasim, W., Karuppannan, S., Fahad, S., Tariq, A., Mousa, B. G., Mumtaz, F., & Aslam, M. (2022a). Spatiotemporal variation in land use land cover in the response to local climate change using multispectral remote sensing data. *Landscape, 11*, 595. https://doi.org/10.3390/land11050595

Hussain, S., Qin, S., Nasim, W., Bukhari, M. A., Mubeen, M., Fahad, S., Raza, A., Abdo, H. G., Tariq, A., Mousa, B. G., Mumtaz, F., & Aslam, M. (2022b). Monitoring the dynamic changes in vegetation cover using spatio-temporal remote sensing data from 1984 to 2020. *Atmosphere (Basel)., 13*, 1609. https://doi.org/10.3390/atmos13101609

Imran, M., Ahmad, S., Sattar, A., & Tariq, A. (2022). Mapping sequences and mineral deposits in poorly exposed lithologies of inaccessible regions in Azad Jammu and Kashmir using SVM with ASTER satellite data. *Arabian Journal of Geosciences, 15*, 538. https://doi.org/10.1007/s12517-022-09806-9

Islam, F., Riaz, S., Ghaffar, B., Tariq, A., Shah, S. U., Nawaz, M., Hussain, M. L., Amin, N. U., Li, Q., Lu, L., Shah, M., & Aslam, M. (2022). Landslide susceptibility mapping (LSM) of Swat District, Hindu Kush Himalayan region of Pakistan, using GIS-based bivariate modeling. *Frontiers in Environmental Science, 10*, 1–18. https://doi.org/10.3389/fenvs.2022.1027423

Jalayer, S., Sharifi, A., Abbasi-Moghadam, D., Tariq, A., & Qin, S. (2022). Modeling and predicting land use land cover spatiotemporal changes: A case study in Chalus Watershed, Iran. *IEEE Journal of Selected Topics in Applied Earth Observations and Remote Sensing, 15*, 5496–5513. https://doi.org/10.1109/JSTARS.2022.3189528

Jalayer, S., Sharifi, A., Abbasi-Moghadam, D., Tariq, A., & Qin, S. (2023). Assessment of spatiotemporal characteristic of droughts using in situ and remote sensing-based drought indices. *IEEE Journal of Selected Topics in Applied Earth Observations and Remote Sensing, 16*, 1483–1502. https://doi.org/10.1109/JSTARS.2023.3237380

Khalil, U., Azam, U., Aslam, B., Ullah, I., Tariq, A., Li, Q., & Lu, L. (2022a). Developing a spatiotemporal model to forecast land surface temperature: A way forward for better town planning. *Sustainability, 14*, 11873. https://doi.org/10.3390/su141911873

Khalil, U., Imtiaz, I., Aslam, B., Ullah, I., Tariq, A., & Qin, S. (2022b). Comparative analysis of machine learning and multi-criteria decision making techniques for landslide susceptibility mapping of Muzaffarabad district. *Frontiers in Environmental Science, 10*, 1–19. https://doi.org/10.3389/fenvs.2022.1028373

Khan, A. M., Li, Q., Saqib, Z., Khan, N., Habib, T., Khalid, N., Majeed, M., & Tariq, A. (2022). MaxEnt modelling and impact of climate change on habitat suitability variations of economically important Chilgoza Pine (Pinus gerardiana Wall.) in South Asia. *Forests, 13*, 715. https://doi.org/10.3390/f13050715

Lutz, A. F., Immerzeel, W. W., Shrestha, A. B., & Bierkens, M. F. P. (2014). Consistent increase in High Asia's runoff due to increasing glacier melt and precipitation. *Nature Climate Change, 4*, 587–592. https://doi.org/10.1038/nclimate2237

Majeed, M., Tariq, A., Anwar, M. M., Khan, A. M., Arshad, F., Mumtaz, F., Farhan, M., Zhang, L., Zafar, A., Aziz, M., Abbasi, S., Rahman, G., Hussain, S., Waheed, M., Fatima, K., & Shaukat, S. (2021). Monitoring of land use–Land cover change and potential causal factors of climate change in Jhelum district, Punjab, Pakistan, through GIS and multi-temporal satellite data. *Landscape, 10*. https://doi.org/10.3390/land10101026

Majeed, M., Lu, L., Haq, S. M., Waheed, M., Sahito, H. A., Fatima, S., Aziz, R., Bussmann, R. W., Tariq, A., Ullah, I., & Aslam, M. (2022a). Spatiotemporal distribution patterns of climbers along an abiotic gradient in Jhelum District, Punjab, Pakistan. *Forests, 13*, 1244. https://doi.org/10.3390/f13081244

Majeed, M., Tariq, A., Haq, S. M., Waheed, M., Anwar, M. M., Li, Q., Aslam, M., Abbasi, S., Mousa, B. G., & Jamil, A. (2022b). A detailed ecological exploration of the distribution patterns of wild Poaceae from the Jhelum District (Punjab), Pakistan. *Sustainability, 14*, 3786. https://doi.org/10.3390/su14073786

Majeed, M., Lu, L., Anwar, M. M., Tariq, A., Qin, S., El-Hefnawy, M. E., El-Sharnouby, M., Li, Q., & Alasmari, A. (2023). Prediction of flash flood susceptibility using integrating analytic hierarchy process (AHP) and frequency ratio (FR) algorithms. *Frontiers in Environmental Science, 10*, 1–14. https://doi.org/10.3389/fenvs.2022.1037547

Mihalcea, C., Mayer, C., Diolaiuti, G., D'Agata, C., Smiraglia, C., Lambrecht, A., Vuillermoz, E., & Tartari, G. (2008). Spatial distribution of debris thickness and melting from remote-sensing and meteorological data, at debris-covered Baltoro glacier, Karakoram, Pakistan. In *Annals of glaciology* (pp. 49–57). https://doi.org/10.3189/172756408784700680

Moazzam, M. F. U., Rahman, G., Munawar, S., Tariq, A., Safdar, Q., & Lee, B. (2022). Trends of rainfall variability and drought monitoring using standardized precipitation index in a scarcely gauged basin of Northern Pakistan. *Water, 14*, 1132. https://doi.org/10.3390/w14071132

Mousa, B. G., Shu, H., Freeshah, M., & Tariq, A. (2020). A novel scheme for merging active and passive satellite soil moisture retrievals based on maximizing the signal to noise ratio. *Remote Sensing, 12*, 1–23. https://doi.org/10.3390/rs12223804

Pimentel, S., & Flowers, G. E. (2011). A numerical study of hydrologically driven glacier dynamics and subglacial flooding. In *Proceedings of the Royal Society A: Mathematical, Physical and Engineering Sciences* (pp. 537–558). The Royal Society Publishing. https://doi.org/10.1098/rspa.2010.0211

Quincey, D. J., & Luckman, A. (2014). Brief communication: On the magnitude and frequency of Khurdopin glacier surge events. *The Cryosphere, 8*, 571–574. https://doi.org/10.5194/tc-8-571-2014

Quincey, D. J., Copland, L., Mayer, C., Bishop, M., Luckman, A., & Belò, M. (2009). Ice velocity and climate variations for Baltoro Glacier, Pakistan. *Journal of Glaciology, 55*, 1061–1071. https://doi.org/10.3189/002214309790794913

Quincey, D. J., Braun, M., Glasser, N. F., Bishop, M. P., Hewitt, K., & Luckman, A. (2011). Karakoram glacier surge dynamics. *Geophysical Research Letters, 38*, 2–7. https://doi.org/10.1029/2011GL049004

Round, V., Leinss, S., Huss, M., Haemmig, C., & Hajnsek, I. (2017). Surge dynamics and lake outbursts of Kyagar Glacier, Karakoram. *The Cryosphere, 11*, 723–739. https://doi.org/10.5194/tc-11-723-2017

Shah, S. H. I. A., Yan, J., Ullah, I., Aslam, B., Tariq, A., Zhang, L., & Mumtaz, F. (2021). Classification of aquifer vulnerability by using the DRASTIC index and geo-electrical techniques. *Water, 13*, 2144. https://doi.org/10.3390/w13162144

Shah, S. H. I. A., Jianguo, Y., Jahangir, Z., Tariq, A., & Aslam, B. (2022). Integrated geophysical technique for groundwater salinity delineation, an approach to agriculture sustainability for Nankana Sahib Area, Pakistan. *Geomatics, Natural Hazards And Risk, 13*, 1043–1064. https://doi.org/10.1080/19475705.2022.2063077

Sharifi, A., Felegari, S., Tariq, A., & Siddiqui, S. (2021). Forest cover change detection across recent three decades in Persian oak forests using convolutional neural network. In *Climate impacts on sustainable natural resource management* (pp. 57–73). Wiley. https://doi.org/10.1002/9781119793403.ch4

Sharifi, A., Felegari, S., & Tariq, A. (2022a). Mangrove forests mapping using Sentinel-1 and Sentinel-2 satellite images. *Arabian Journal of Geosciences, 15*, 1593. https://doi.org/10.1007/s12517-022-10867-z

Sharifi, A., Mahdipour, H., Moradi, E., & Tariq, A. (2022b). Agricultural field extraction with deep learning algorithm and satellite imagery. *Journal of the Indian Society of Remote Sensing, 50*, 417–423. https://doi.org/10.1007/s12524-021-01475-7

Siddiqui, S., Ali Safi, M. W., Rehman, N. U., & Tariq, A. (2020). Impact of climate change on land use/land cover of Chakwal District. *International Journal of Economic and Environmental Geology, 11*, 65–68. https://doi.org/10.46660/ijeeg.vol11.iss2.2020.449

Tariq, A (2023). Quantitative comparison of geostatistical analysis of interpolation techniques and semiveriogram spatial dependency parameters for soil atrazine contamination attribute. In: Stathopoulos N, Tsatsaris A, Kalogeropoulos KBT-G for G (eds) *Geoinformatics for Geosciences*. Elsevier, pp 261–279

Tariq, A., & Qin, S. (2023). Spatio-temporal variation in surface water in Punjab, Pakistan from 1985 to 2020 using machine-learning methods with time-series remote sensing data and driving factors. *Agricultural Water Management, 280*, 108228. https://doi.org/10.1016/j.agwat.2023.108228

Tariq, A., & Shu, H. (2020). CA-Markov chain analysis of seasonal land surface temperature and land use landcover change using optical multi-temporal satellite data of Faisalabad, Pakistan. *Remote Sensing, 12*, 1–23. https://doi.org/10.3390/rs12203402

Tariq, A., Mumtaz, F., Zeng, X., Baloch, M. Y. J., & Moazzam, M. F. U. (2022a). Spatio-temporal variation of seasonal heat islands mapping of Pakistan during 2000–2019, using day-time and night-time land surface temperatures MODIS and meteorological stations data. *Remote Sensing Applications: Society and Environment, 27*, 100779. https://doi.org/10.1016/j.rsase.2022.100779

Tariq, A., Siddiqui, S., Sharifi, A., & Shah, S. H. I. A. (2022b). Impact of spatio-temporal land surface temperature on cropping pattern and land use and land cover changes using satellite imagery, Hafizabad District, Punjab, Province of Pakistan. *Arabian Journal of Geosciences, 15*, 1045. https://doi.org/10.1007/s12517-022-10238-8

Tariq, A., Yan, J., Gagnon, A. S., Riaz Khan, M., & Mumtaz, F. (2022c). Mapping of cropland, cropping patterns and crop types by combining optical remote sensing images with decision tree classifier and random forest. *Geo-spatial Information Science, 00*, 1–19. https://doi.org/10.1080/10095020.2022.2100287

Tariq, A., Yan, J., Ghaffar, B., Qin, S., Mousa, B. G., Sharifi, A., Huq, M. E., & Aslam, M. (2022d). Flash flood susceptibility assessment and zonation by integrating analytic hierarchy process and frequency ratio model with diverse spatial data. *Water, 14*, 3069. https://doi.org/10.3390/w14193069

Tariq, A., Hashemi Beni, L., Ali, S., et al. (2023). An effective geospatial-based flash flood susceptibility assessment with hydrogeomorphic responses on groundwater recharge. *Groundwater for Sustainable Development*, *5*, 100998. https://doi.org/10.1016/j.gsd.2023.100998

Tariq, A., Jiango, Y., Li, Q., Gao, J., Lu, L., Soufan, W., Almutairi, K. F., & Habib-ur-Rahman, M. (2023a). Modelling, mapping and monitoring of forest cover changes, using support vector machine, kernel logistic regression and naive Bayes tree models with optical remote sensing data. *Heliyon, 9*, e13212. https://doi.org/10.1016/j.heliyon.2023.e13212

Tariq, A., Jiango, Y., Lu, L., Jamil, A., Al-ashkar, I., Kamran, M., & El Sabagh, A. (2023b). Integrated use of Sentinel-1 and Sentinel-2 data and open-source machine learning algorithms for burnt and unburnt scars. *Geomatics, Natural Hazards and Risk, 14*, 28. https://doi.org/10.1080/19475705.2023.2190856

Ullah, I., Aslam, B., Shah, S. H. I. A., Tariq, A., Qin, S., Majeed, M., & Havenith, H.-B. (2022). An integrated approach of machine learning, remote sensing, and GIS data for the landslide susceptibility mapping. *Landscape, 11*, 1265. https://doi.org/10.3390/land11081265

Wahla, S. S., Kazmi, J. H., Sharifi, A., Shirazi, S. A., Tariq, A., & Joyell Smith, H. (2022). Assessing spatio-temporal mapping and monitoring of climatic variability using SPEI and RF machine learning models. *Geocarto International, 0*, 1–20. https://doi.org/10.1080/10106049.2022.2093411

Zamani, A., Sharifi, A., Felegari, S., Tariq, A., & Zhao, N. (2022). Agro climatic zoning of saffron culture in Miyaneh City by using WLC method and remote sensing data. *Agriculture, 12*, 1–15. https://doi.org/10.3390/agriculture12010118

Chapter 11
Monitoring Spatio-Temporal Pattern of Meteorological Drought Stress Using Standardized Precipitation Index (SPI) over Bundelkhand Region of Uttar Pradesh, India

Afia Aslam and Adeeba Parveen

Introduction

Water is the most crucial element to maintain the life process on Earth. It is the surviving source for the biotic environment and all economic and non-economic activities of human beings. There is not a single substitute of water available on the face of Earth. Ensuring an adequate supply of water in the soil is a fundamental requirement for plant growth and agricultural sustainability. Conversely, water scarcity gives rise to a catastrophic phenomenon known as drought. On contrary of all human advancement in twenty-first century, whole world is suffering from water scarcity due to unchecked water pollution and scanty precipitation. Rainfall and glaciers are the only sources of freshwater, but alteration in long-term climate has severely impacted both the sources (Machiwal & Jha, 2012). Climate change can cause significant variation in hydro-meteorological parameters such as rainfall, temperature and humidity, but among all the parameters, rainfall is valued most relevant due to its role in the water cycle. Therefore, disasters such as flood and drought are highly associated with rainfall (Wang et al., 2017). Hydro-meteorological hazards caused due to variability in rainfall have severe effects on the environment, community, economy, health and agriculture. Millions of lives are impacted due to loss in biodiversity, economic loss, crop failure and mental disorder caused by these hazards (Miah et al., 2017). It is a well-established fact that the climate change has adversely affected the pattern of precipitation in most of the regions, which has worsened the situation of food security. Excess of anthropogenic interference in the natural environment has increased the frequency of extreme

A. Aslam (✉) · A. Parveen
Department of Geography, Aligarh Muslim University, Aligarh, UP, India

A. L. Singh et al. (eds.), *Climate Change, Vulnerabilities and Adaptation*,
https://doi.org/10.1007/978-3-031-49642-4_11

climatic events (floods and drought), which has depressed the crop yield and productivity (Fischer et al., 2005).

Drought is a natural hazard initiated due to lack of precipitation that leads to water scarcity, affecting the associated population at large. It is a natural reoccurring phenomenon that takes place due to long dry spells and inadequate rainfall. This event happens when evaporation and transpiration exceed the amount of rainfall for a long period of time at a specific location (Rahman & Lateh, 2016). On the basis of the affected areal extent, drought is ranked the second most severe natural disaster in the world (Nagarajan, 2009). In India, Approximately 68% of India's territory is plagued by recurring yearly droughts due to unpredictable monsoon patterns (Patel & Yadav, 2015). Drought is a complex disaster, which alters the whole system of a region wherever it occurs. Living organisms and the environment are more affected by drought than any other natural hazard. Drought has different categories based on the method of its measurement, i.e., meteorological drought, agricultural drought, hydrological drought and socio-economic drought. Meteorological drought can be defined as the intensity and duration of drought when precipitation falls below the normal level. It is usually considered when the amount of rainfall decreases by twenty-five percent. Then comes hydrological drought that shows the consequences of low precipitation in the form of a shortage in surface water and groundwater. Agriculture drought is also a consequence of meteorological and hydrological drought which can be seen in the form of low soil moisture to sustain the agriculture activity. Socio-economic drought is based on a shortage of goods due to a deficit of water supply and insufficient agricultural commodities (Ebi & Bowen, 2016).

Drought impact assessment is based on various parameters such as its frequency, time period, severity and area coverage. Drought assessment is a complex process, as every type of drought has its own identifying variables. To monitor the meteorological drought, various indices are used such as Standardized Precipitation Index (SPI) (McKee et al., 1993), Deciles Index (DI), Drought Area Index (DAI), Percent of Normal (PN), Palmer Drought Severity Index (PDSI) (Palmer, 1965), China-Z index (CZI), Effective Drought Index (EDI), Reconnaissance Drought Index (RDI) (Tsakiris et al., 2007) and, the most recently developed, Standardized Precipitation Evapotranspiration Index (SPEI). The calculation of these indices is performed using historical time series data of rainfall and temperature (Danandeh Mehr & Vaheddoost, 2020). Standardized Precipitation Index (SPI) is the most used and recognized drought index in the world. World Meteorological Organization (WMO) has widely accepted and suggested this method of drought monitoring due to its easy calculation and accuracy (WMO, 2012). In recent times, SPI (standardized precipitation index) is commonly employed by researchers in drought assessment due to its efficiency and simpleness. It has a specific feature in its evaluation that it can be computed at any time scale for any location and required only one parameter, which is rainfall (Komuscu 1999; Anctil et al., 2002; Lana et al., 2001; Min et al., 2003; Domonkos, 2003; Bonaccorso et al., 2003). This method has another peculiar advantage of standardization that it helps in identifying the frequency and intensity of extreme drought events at different time scales and locations. It also helps in studying the history of drought events if the values of SPI are plotted against time series (Rahman & Lateh, 2016).

Bundelkhand is one of the most drought-prone regions in the country having a background of reoccurring drought events. The region featured semi-arid and sub-humid climate, undulating topography and very few patches of forest. The major constraints of agriculture practice in this region are the soil with poor water holding capacity and lack of irrigation facility. Along with a long history of recurrent drought events and depleting water levels, this has created a threat to communities involved in agriculture (Kundu et al., 2021). The frequency of drought events has increased from one to three during the period of 1968–1992 (Singh et al., 2003). Recurrent events of drought have been recorded between 2004 and 2007 due to scanty rainfall, which has disturbed the crop planning in this region (Patel & Yadav, 2015).

The main objective of this study is to screen meteorological drought in the Bundelkhand region of Uttar Pradesh during the period of 1980 to 2020. To study the framed objective, a well-established meteorological drought index, i.e. SPI (Standardised Precipitation Index) has been employed. The monthly rainfall data from all seven districts of the study area have been used at different time scales of 3, 6 and 12 months. The values of SPI at different time scales have been categorized into four categories, i.e. normal, moderate, severe and extreme drought. Meteorological drought is basically insufficient rainfall over an area for a certain time period. Among all the indices used to detect meteorological drought, SPI remains the most sustainable due to its easy-to-calculate property. It simply uses monthly rainfall data to detect meteorological drought, and it also helps in exploring the spatial and temporal extent (Hayes et al., 1999; Heim, 2000; Ibrahim et al., 2009; Wilhite et al., 2000). Forty-six years of precipitation data from seven stations of Ankara Province, Turkey, was used to calculate SPI drought index at 3-, 6- and 12-month time scales. Drought events were determined in the province with respect to each time scale (Danandeh Mehr & Vaheddoost, 2020).

Study Area

Bundelkhand region of Uttar Pradesh is situated between the Indo-Gangetic plain in the north and Vindhyan Mountains to the north-west and south. It is spread over 29,400 sq. km of land in seven districts—Jhansi, Jalaun, Lalitpur, Hamirpur, Banda and Chitrakoot—accommodating around 97 lakh people (Census, 2011). This region has been characterized by a hot and semi-humid climate with hot and dry summers and equally cool winters. The annual precipitation varies from 60 cm to 100 cm, the majority of which is received from the south-west monsoons between the months of July and August (Planning Department, Govt. of U.P., 2020). The rainfall received is unpredictable and highly variable; as a result, the region experiences recurring situations of famine, droughts and limited agricultural production.

Material and Methods

Data Used

The monthly rainfall (mm) data of all seven districts has been used in this study. The required data has been downloaded from Climate Research Unit (CRU) at 0.5° × 0.5° spatial resolution from 1980 to 2020. For this study, time series gridded data available on the version CRU TS 4.04 (https://crudata.uea.ac.uk/), grid-box was used. Various research works have established close agreement between Indian Meteorology Department (IMD) and Climate Research Unit (CRU). CRU datasets have been used over IMD due to their finer resolution (Verma & Ghosh, 2019).

Method

Standardized Precipitation Index (SPI)

Over the time, climatologists and meteorologists had used different drought indices to monitor the drought intensity in a region. Many drought indices such as Palmer Drought Severity Index, precipitation percentiles, percentage of normal precipitation etc. were used for screening the drought severity. But meteorologists needed a simpler method that was convenient to calculate and statistically relevant and significant. It led American scientists McKee, Doesken and Kleist to create the Standardized Precipitation Index (SPI) in 1993 (World Meteorological Organization, 2012).

Standardized Precipitation Index (SPI) is a commonly used drought index based on normalization of precipitation probabilities. It studies the precipitation deficit and performs drought monitoring at different time scales. SPI is largely computed for monthly data, but it can also be used to evaluate daily and weekly precipitation data. Due to its simple handling, the World Meteorological Department has recommended the compulsory use of this index in every country monitoring meteorological drought (Danandeh Mehr and Vaheddoost, 2020).

McKee and others (1993) have developed the classification system for the values derived by SPI calculation. They defined that drought event occurs when SPI values remain continuously negative and reach an intensity of −0.1 or less, and it ends when values become positive. Therefore, each drought event has a duration given by the starting and ending points, and intensity for every month. Table 11.1 shows drought severity.

The major advantage of SPI is its statistical consistency and ability to detect both short- and long-term drought impacts with different time scales. Its intrinsic probability nature makes SPI suitable for drought risk analysis (Guttman, 1999). Hayes et al. (1999) in their paper found that SPI can be used for detecting the starting point of drought and its spatio-temporal progression. SPI is based on an equi-probability

Table 11.1 SPI values

2.0+	Extremely wet
1.5 to 1.99	Very wet
1.0 to 1.49	Moderately wet
−0.99 to 0.99	Near normal
−1.0 to −1.49	Moderately dry
−1.5 to −1.99	Severely dry
−2 and less	Extremely dry

transformation of aggregated monthly precipitation into a standard normal variable. It is assumed that aggregated rainfall to be gamma distributed and maximum likelihood method is used to calculate the parameters of distribution (Thomas et al., 2015b). According to McKee et al. (1993), SPI is computed by fitting the gamma distribution function. It is then transformed into a standard normal distribution where the mean value is zero and the variance is one.

The gamma distribution is defined by the probability density function (p.d.f.) as follows:

$$g(x) = [1/\beta^{\alpha} \times \Gamma(\alpha)] \times x^{\alpha-1} \times e^{-x/\beta} \tag{11.1}$$

where $\alpha > 0$, α is sharp parameter; $\beta > 0$, β is scale parameter; $x > 0$, x is precipitation amount

$$(\alpha) = \int_0^{\infty} y^{\alpha-1} e^{-y} \mathrm{d}y, \ (\alpha) \text{ is a gamma function.}$$

The cumulative probability distribution function G(x) is obtained from integrating p.d.f. as follows:

$$G(x) = \int_0^x g(x)dx = \frac{1}{\widehat{\beta}^{\alpha}\Gamma(\widehat{\alpha})}\int_0^x x^{\widehat{\alpha}-1} e^{-x/\widehat{\beta}} dx \tag{11.2}$$

where

$$\widehat{\alpha} = \frac{1}{4A}\left(1 + \sqrt{1 + \frac{4A}{3}}\right)$$

$$\widehat{\beta} = \frac{\overline{x}}{\widehat{\alpha}}$$

$$A = \ln(\overline{x}) - \frac{\sum \ln(x)}{n}$$

n is the number of precipitation observations. Let $t = x/\beta$, putting it in G(x)

$$G(x) = \frac{1}{\Gamma(\widehat{\alpha})} \int_0^x t^{\widehat{\alpha}-1} e^{-t} dt \tag{11.3}$$

For $x = 0$, the gamma function is undefined; therefore, the cumulative probability becomes

$$H(x) = q + (1 - q)G(x) \tag{11.4}$$

Here q is the value when the probability at $x = 0$.

To find the SPI value, the cumulative probability function requires to change into a standard normal cumulative distribution function having a mean value of zero and variance of one. Then SPI is evaluated using the Abramowitz and Stegun's (1965) approximation as follows:

$$\mathrm{SPI} = -\left(t - \frac{C_0 + C_1 t + C_2 t^2}{1 + d_1 t + d_2 t + d_3 t^3}\right) \text{ for } 0 < H(x) < 0.5 \tag{11.5}$$

$$\mathrm{SPI} = +\left(t - \frac{C_0 + C_1 t + C_2 t^2}{1 + d_1 t + d_2 t + d_3 t^3}\right) \text{ for } 0.5 < H(x) < 1.0 \tag{11.6}$$

where

$$t = \sqrt{\ln\left(\frac{1}{(H(x))^2}\right)} \text{ for } 0 < H(x) < 0.5$$

$$t = \sqrt{\ln\left(\frac{1}{(1.0 - H(x))^2}\right)} \text{ for } 0.5 < H(x) < 1.0$$

$$C_0 = 2.515517, C_1 = 0.802853, C_2 = 0.010328$$

$$d_1 = 1.432788, d_2 = 1.89269 \text{ and } d_3 = 0.001308$$

The sum of all SPI values within the drought duration is called magnitude of drought, and its division with duration of drought is called its intensity.

The SPI has been employed in this study at 3-, 6- and 12-month time scales by using the software RSTUDIO version 4.1.2. The graphs and tables are prepared using MS-EXCEL 2019.

Results and Discussion

The analysis has been carried out for all seven districts of the Bundelkhand region in Uttar Pradesh. Standardized Precipitation Index (SPI was calculated for all districts at three time scales of 3, 6 and 12 months to assess the intensity of meteorological drought. The values of SPI have been categorized into moderate, severe and extreme droughts. The spatio-temporal extension of drought in each district has been analysed. The calculated SPI values for all seven districts have been shown in Fig. 11.1.

The SPI values computed for each district of the Bundelkhand region of Uttar Pradesh at different time scales of 3, 6 and 12 months show variation in drought duration and occurrence. SPI3 values show a higher frequency of drought events, while SPI6 and SPI12 values show a less frequency of drought events. Although the length of SPI3 is much shorter compared to 6 and 12 months. The index exhibits a little delay in response as the time scale expands, suggesting a diminishing influence on precipitation with each successive month. Consequently, there will be a decrease in the occurrence of prolonged droughts.

Every time scale values depict different interpretations in reference to different water resources. SPI3 values show the soil moisture conditions as the monsoon period in Bundelkhand remains for 3 months (mid-June to mid-September). Water stress and soil moisture for cultivation can be accurately studied by using SPI3 values. However, stream flow and groundwater take more time to get impacted by a rainfall deficit, so SPI6 and SPI12 give a clearer picture of these conditions.

A three-month SPI was computed and shows frequent drought events in all seven districts. Jhansi shows the highest occurrence of drought, while Hamirpur has the least among all seven districts. On the other hand, Chitrakoot faces 81 months of total drought from 1980 to 2020, the highest average intensity. Table 11.2 shows the 3-month-based SPI characteristics.

Evaluation of drought characteristics, including drought intensity and duration, revealed that the highest frequency of drought with moderate to extreme intensity takes place in the period of June to September. There are an average of 25 events of drought in the monsoon period (June to September) in this region. The year 2010 has the longest duration (5 months) of a drought period after 1992–1993. Moderate intensity of droughts is more frequent in this region in comparison to severe and extreme intensity. The highest intensity (−3.19) of drought was faced by Hamirpur district in 2010.

Similar analysis was performed at 6- and 12-month time scales, and it was revealed that with increasing time scale, drought duration and severity have increased, while frequency has decreased (Table 11.3).

In the 6-month SPI calculation, it was revealed that there have been many drought events in the past 41 years. Some of the drought events last for a whole year. The drought events in 1987, 1992, 1993, 2001, 2009 and 2010 prevailed for a whole year. The average duration of the drought period is 5 months. Every district of Bundelkhand is found to be drought prone and has faced drought in similar years.

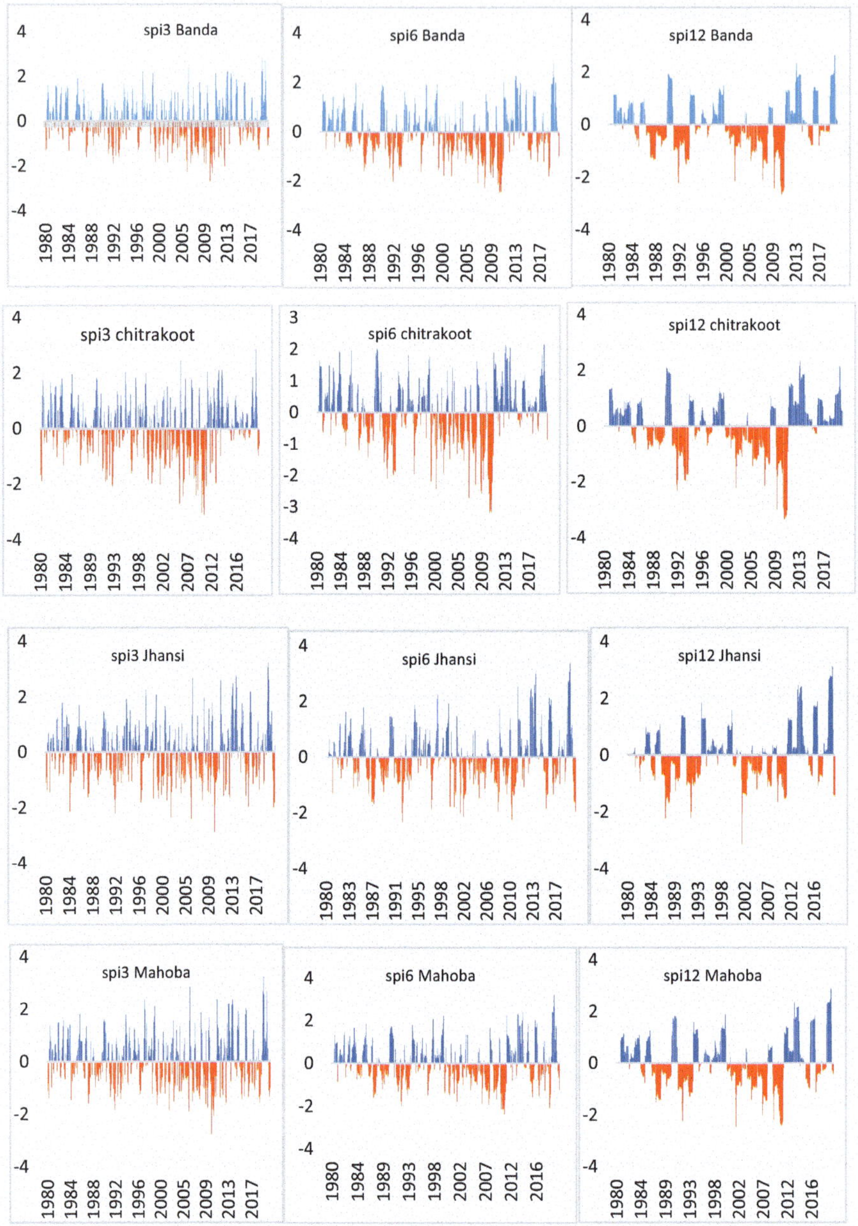

Fig. 11.1 Evaluation of 3-, 6- and 12-month SPI of seven districts of Bundelkhand, Uttar Pradesh, for the reference period (1980–2020)

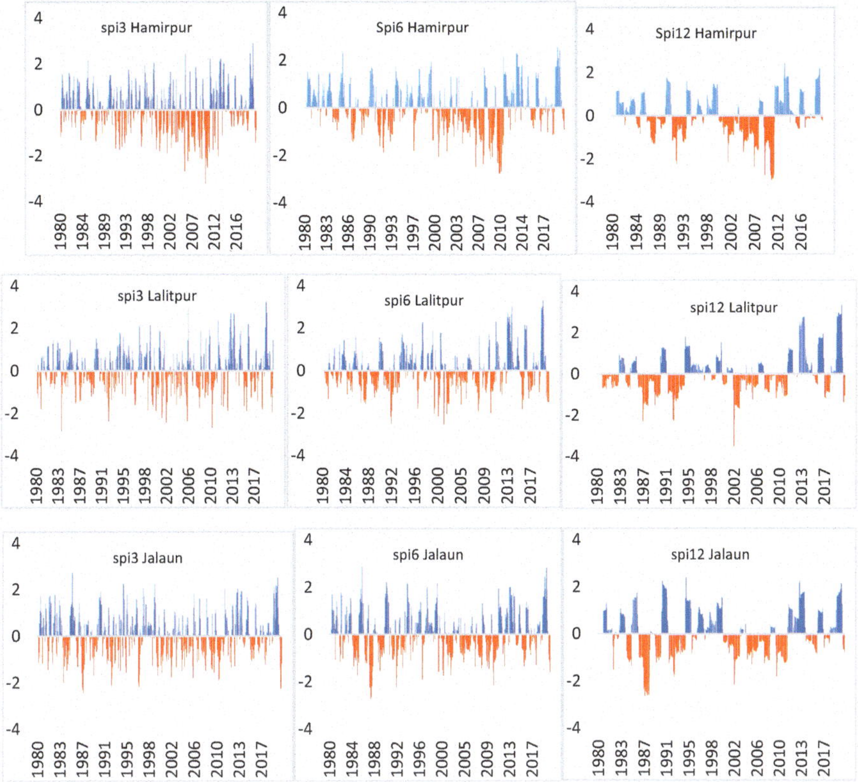

Fig. 11.1 (continued)

Table 11.2 Three-month-based SPI drought characteristics of all seven districts in the Bundelkhand region, Uttar Pradesh (1980–2020)

Districts	Number of drought events	Total duration of drought (months)	Average intensity
BANDA	24	77	−1.75
CHITRAKOOT	25	81	−1.81
MAHOBA	27	77	−1.75
JALAUN	29	72	−1.72
JHANSI	30	72	−1.76
HAMIRPUR	22	79	−1.75
LALITPUR	29	71	−1.79

The monsoon periods have more rain deficits, while the winter season also has low rainfall, which leads to drought conditions (Table 11.4).

Twelve-month SPI values also depict a similar pattern of drought conditions. The frequency of drought events drops significantly, while duration lasts for 9–12 months. Moderate-intensity droughts are more prevalent for longer duration,

Table 11.3 Six-month-based SPI drought characteristics of all seven districts in the Bundelkhand region, Uttar Pradesh (1980–2020)

Districts	Number of drought events	Total duration of drought (months)	Average intensity
BANDA	21	68	−1.73
CHITRAKOOT	19	67	−1.84
MAHOBA	20	70	−1.7
JALAUN	25	64	−1.42
JHANSI	21	59	−1.47
HAMIRPUR	20	77	−1.49
LALITPUR	20	57	−1.45

Table 11.4 Twelve-month-based SPI drought characteristics of all seven districts in the Bundelkhand region, Uttar Pradesh (1980–2020)

Districts	Number of drought events	Total duration of drought (months)	Average intensity
BANDA	13	66	−1.77
CHITRAKOOT	12	72	−1.97
MAHOBA	13	58	−1.74
JALAUN	19	65	−1.75
JHANSI	16	60	−1.72
HAMIRPUR	14	78	−1.81
LALITPUR	14	59	−1.75

while extreme-intensity drought events last for 1 or 2 months. Chitrakoot and Hamirpur face most intense drought with a long duration, while Jalaun has the highest number of drought events. It shows that districts lying in the north part (Jalaun and Jhansi) have more drought events. Districts in the western part get a longer duration of drought with high intensity and severity.

Conclusion

The spatio-temporal analysis of drought events helps in studying variations in drought intensity, characteristics of drought, frequency, occurrence, its spatio-temporal extension and its beginning and withdrawal of the drought (Thomas et al., 2015a, b). The study presents a spatio-temporal pattern of drought in seven districts of the Bundelkhand region, Uttar Pradesh. Standardized Precipitation Index (SPI) has been employed to study the drought characteristics. Three different time scales—3-, 6- and 12-month SPI—have been used for the analysis of frequency, intensity and duration of drought events. Given that the research area experiences a three-month monsoon season, it is crucial to calculate the 3-month Standardised Precipitation Index (SPI) in order to evaluate soil moisture deficiency. The result shows a higher frequency of drought events among seven districts. Jhansi, Lalitpur and Jalaun have faced more frequent droughts, while Hamirpur has the least dry spells. The moderate- and severe-intensity drought are found more common. The evaluation of 6-month SPI examined the drought for every 6-month period, which depicts frequency decreases, while intensity and duration of drought events have increased. In this case, Jalaun, Lalitpur, Banda and Jhansi have more drought events with longer durations. Though every district faces drought in the same years, the intensity and duration vary. The analysis exposed that each district of Bundelkhand region, Uttar Pradesh, is drought prone with varying intensities. The rainfall in this region is erratic, which pushes it further on the edge of a drought problem. Agriculture is the main occupation in this region, mainly fed by rainfall, but the frequent occurrence of drought led to the economic crisis. The study of SPI helps in understanding the intensity and duration of drought, which further helps in better planning for ongoing drought events.

References

Anctil, F., Larouche, W., & Viau, A. A. (2002). Exploration of the standardized precipitation index with regional analysis. *Canadian Journal of Soil Science, 82*(1), 115–125.

Bonaccorso, B., Bordi, I., Cancelliere, A., Rossi, G., & Sutera, A. (2003). Spatial variability of drought: An analysis of the SPI in Sicily. *Water Resources Management, 17*, 273–296.

Census of India. (2011). "Population census." The Registrar General & Census Commissioner, Ministry of Home Affairs, Government of India, New Delhi, India. http://www.censusindia.gov.in

Danandeh Mehr, A., & Vaheddoost, B. (2020). Identification of the trends associated with the SPI and SPEI indices across Ankara, Turkey. *Theoretical and Applied Climatology, 139*(3), 1531–1542.

Domonkos, P. (2003). Recent precipitation trends in Hungary in the context of larger scale climatic changes. *Natural Hazards, 29*(2), 255.

Ebi, K. L., & Bowen, K. (2016). Extreme events as sources of health vulnerability: Drought as an example. *Weather and Climate Extremes, 11*, 95–102.

Fischer, G., Shah, M., Tubiello, N., & Van Velhuizen, H. (2005). Socio-economic and climate change impacts on agriculture: An integrated assessment, 1990–2080. *Philosophical Transactions of the Royal Society B: Biological Sciences, 360*(1463), 2067–2083.

Guttman, N. B. (1999). Accepting the standardized precipitation index: A calculation algorithm 1. *JAWRA Journal of the American Water Resources Association, 35*(2), 311–322.

Hayes, M. J., Svoboda, M. D., Wiihite, D. A., & Vanyarkho, O. V. (1999). Monitoring the 1996 drought using the standardized precipitation index. *Bulletin of the American Meteorological Society, 80*(3), 429–438.

Heim, R. R. (2000). Drought indices: A review. *Drought: a global assessment, 1*, 159–167.

Ibrahim, K., Zin, W. Z. W., & Jemain, A. A. (2009). Evaluating the dry conditions in peninsular Malaysia using bivariate copula. *ANZIAM Journal, 51*, C555–C569.

Kundu, A., Dutta, D., Patel, N. R., Denis, D. M., & Chattoraj, K. K. (2021). Evaluation of socio-economic drought risk over Bundelkhand region of India using analytic hierarchy process (AHP) and geo-spatial techniques. *Journal of the Indian Society of Remote Sensing, 49*(6), 1365–1377.

Lana, X., Serra, C., & Burgueño, A. (2001). Patterns of monthly rainfall shortage and excess in terms of the standardized precipitation index for Catalonia (NE Spain). *International Journal of Climatology: A Journal of the Royal Meteorological Society, 21*(13), 1669–1691.

Machiwal, D., Mishra, A., Jha, M. K., Sharma, A., & Sisodia, S. S. (2012). Modeling short-term spatial and temporal variability of groundwater level using geostatistics and GIS. *Natural Resources Research, 21*, 117–136.

McKee, T. B., Doesken, N. J., & Kleist, J. (1993). The relationship of drought frequency and duration to time scales. *Proceedings of the 8th Conference on Applied Climatology, 17*(22), 179–183.

Miah, M. G., Abdullah, H. M., & Jeong, C. (2017). Exploring standardized precipitation evapotranspiration index for drought assessment in Bangladesh. *Environmental Monitoring and Assessment, 189*, 1–16.

Min, S. K., Kwon, W. T., Park, E. H., & Choi, Y. (2003). Spatial and temporal comparisons of droughts over Korea with East Asia. *International Journal of Climatology: A Journal of the Royal Meteorological Society, 23*(2), 223–233.

Nagarajan, R. (2009). *Drought assessment.* Capital Publishing Company, Co-published by Springer.

Palmer, W. C. (1965). *Meteorological drought* (Vol. 30). US Department of Commerce, Weather Bureau.

Patel, N. R., & Yadav, K. (2015). Monitoring spatio-temporal pattern of drought stress using integrated drought index over Bundelkhand region, India. *Natural Hazards, 77*(2), 663–677.

Rahman, M. R., & Lateh, H. (2016). Meteorological drought in Bangladesh: Assessing, analysing and hazard mapping using SPI, GIS and monthly rainfall data. *Environmental Earth Sciences, 75*, 1–20.

Singh, R. P., Roy, S., & Kogan, F. (2003). Vegetation and temperature condition indices from NOAA AVHRR data for drought monitoring over India. *International Journal of Remote Sensing, 24*(22), 4393–4402.

Thomas, T., Jaiswal, R. K., Nayak, P. C., & Ghosh, N. C. (2015a). Comprehensive evaluation of the changing drought characteristics in Bundelkhand region of Central India. *Meteorology and Atmospheric Physics, 127*(2), 163–182.

Thomas, T., Nayak, P. C., & Ghosh, N. C. (2015b). Spatiotemporal analysis of drought characteristics in the Bundelkhand region of Central India using the standardized precipitation index. *Journal of Hydrologic Engineering, 20*(11), 05015004.

Tsakiris, G., Pangalou, D., & Vangelis, H. (2007). Regional drought assessment based on the reconnaissance drought index (RDI). *Water Resources Management, 21*, 821–833.

Umran Komuscu, A. (1999). Using the SPI to analyze spatial and temporal patterns of drought in Turkey. *Drought Network News, 1994–2001*, 49.

Verma, P., & Ghosh, S. K. (2019). Trend analysis of climatic research unit temperature dataset for Gangotri glacier, India. *Dynamics of Atmospheres and Oceans, 85*, 83–97.

Wang, R., Chen, J., Chen, X., & Wang, Y. (2017). Variability of precipitation extremes and dryness/wetness over the southeast coastal region of China, 1960–2014. *International Journal of Climatology, 37*(13), 4656–4669.

Wilhite, D. A., Hayes, M. J., & Svoboda, M. D. (2000). Drought monitoring and assessment: Status and trends in the United States. *Drought and drought mitigation in Europe*, 149–160.

World Meteorological Organization. (2012). *Standardized precipitation index user guide*. World Meteorological Organization, (1090).

Chapter 12
Climate Change and Forest Fire in Eastern Himalaya: A Case Study of Sikkim and Darjeeling Himalayas of West Bengal

E. Ishwarjit Singh and Ajith Singha

Introduction

Forests are vital for the global carbon cycle, biodiversity conservation, climate change etc. Along with global organizations and civil societies, each country has a stern forest policy for the protection of existing forests as well as to increase forest coverage in their respective countries. Despite all these initiatives, global forest coverage has decreased by 3% in 16 years during the period of 1990–2015. In recent years, the rate of forest depletion has been at an alarming situation with 0.6% per year (FAO, 2015). Forest degradation is mainly due to forest fire, which are caused by both nature and anthropogenic-induced factors like climate change, rapid urbanization, infrastructural development, industries etc. It poses a great threat to all the lives on the earth, and many species are on the verge of extinction.

The area under forest in India is about 23.8% of the total geographical area. It covers almost 79 million hectares (FSI, 2011). Since ancient times, forests have been a pivotal to society and in everyday lives of people. It shapes the economy, livelihood, culture and religious practices. The most fragile ecosystem in India lies in the Himalayan region (Myers et al., 2000), which is also known as the yardstick of climate change. According to the Forest Report, 41% of the geographical area in the Indian Himalaya Region is under forest area, out of which 16.9% is under very dense forest cover, 45.4% under moderate forest and the remaining 37.7% under open forest category (FSI, 2011). The Himalayan region shows that a considerable forest area is under private (42%) followed by the area under community management (33%) and revenue department has only 25%. Pandit et al. (2007) reported on an alarming trend of deforestation in the Indian Himalaya and projected consequential extinctions of endemic *taxa* (species and subspecies) by 2100 across the broad range

E. I. Singh (✉) · A. Singha
Department Of Geography, Sikkim University, Gangtok, Sikkim, India
e-mail: eisingh@cus.ac.in

A. L. Singh et al. (eds.), *Climate Change, Vulnerabilities and Adaptation*,
https://doi.org/10.1007/978-3-031-49642-4_12

of taxonomic groups. The main degradation of forests in the region is due to rapid urbanization and forest fire both anthropogenic and natural factors. In India, annual forest fire ranges from 33% to 90% of forest areas in different states (Jaiswal et al. 2002). About 90% of the forest fires in India are caused by human intentionally or accidentally (NIDM, 2014). Thus, monitoring and management of forest fires is very important in India.

In this chapter, forest fire incidences have been studied in eastern Himalaya with climatic variabilities in the last 40 years (1981–2020) with forest fire data of 16 years (2000–2015). Eastern Himalaya in the Indian union includes Sikkim Himalaya, Darjeeling Himalaya and Arunachal Himalaya. There are three objectives in the study: first, to evaluate climatic variability in the study areas; second, to study spatio-temporal incidences of forest fire in the study area; and lastly, to analyse the impact of climate change on forest fire. The study also gives an account on the various forest types in the region, which are more likely to be affected.

Materials and Methods

The study is completely based on secondary data that were collected from different sources. To study climatic variabilities in the region, two stations from Sikkim and one station from Darjeeling had been selected. From Sikkim, one station from the urban centre, Gangtok, located at an altitude of 1650 m from MSL and another station at Lachung with an altitude of more than 3500 m from MSL, which is not influenced by any anthropogenic factors directly and covered mostly under snow, had been chosen for the study. It is necessary to take these two stations from Sikkim in order to check the impact of global climate change on the state climate and localized climate change due to urbanization. The climatic data of Gangtok was taken from IMD, Gangtok, for a period of 30 years from 1985 to 2015. Darjeeling climatic data were taken from an altitude of 1350 m from MSL. Both Darjeeling and Lachung climatic data were extracted from Merra-2, NASA, for a period of 40 years (1981–2020).

Forest fire incidence data were collected from satellite imageries, MODIS and Terra, from Bhuvan portal, ISRO. Forest fire data had been extracted from 2000 to 2015 for the period of 16 years. The incidences of forest fire had been plotted on a map with the help of ArcGIS version 10.2 to study the spatio-temporal aspect of forest fire incidences. The map from Google Earth imageries was also used to identify types of forests, which were affected by forest fire. LULC map extracted from the Bhuvan portal, ISRO was also incorporated to assess whether anthropogenic factors influence it or not. In the analysis, statistic techniques such as percentage and mean calculation were also used.

Study Area

The Eastern Himalaya comprises Sikkim, Darjeeling and Arunachal Pradesh of the Indian Union and Bhutan. In this study, Sikkim Himalaya and Darjeeling Himalaya had been taken (Fig. 12.1). Sikkim Himalaya lies in the state of Sikkim. Sikkim has four districts: East, West, North and South (Fig. 12.1). The altitude ranges from 8600 m to 300 m. It has a geographical area of 7096 sq. km, and an entire state comprising hilly terrain. The total population of the state is 610,557, and the literacy rate is 81.4% (Census, 2011). The total forest and tree cover of the state is 3377 sq. km, which is 47.59% of the geographical area. The area under protected forest constitutes 30.70% of the geographical area of the state. There are six types of forest in the state—Tropical moist Deciduous, Sub-Tropical Broad leave, Montane Wet Temperate, Himalayan Moist Temperate, Sub Alpine Forest and Moist Alpine Forest (Department of Forest, 2018).

Darjeeling Himalaya is situated between 26.31 and 27.13 degree latitude North and 87.59 and 88.53 degree longitude East. It includes the Darjeeling and Kalimpong districts of West Bengal. Darjeeling has four subdivisions: Darjeeling, Kurseong, Mirik and Silliguri. The altitude ranges from 2327 m on the north west to 60 m from MSL in the plain area on the south. The subdivision of Kalimpong under

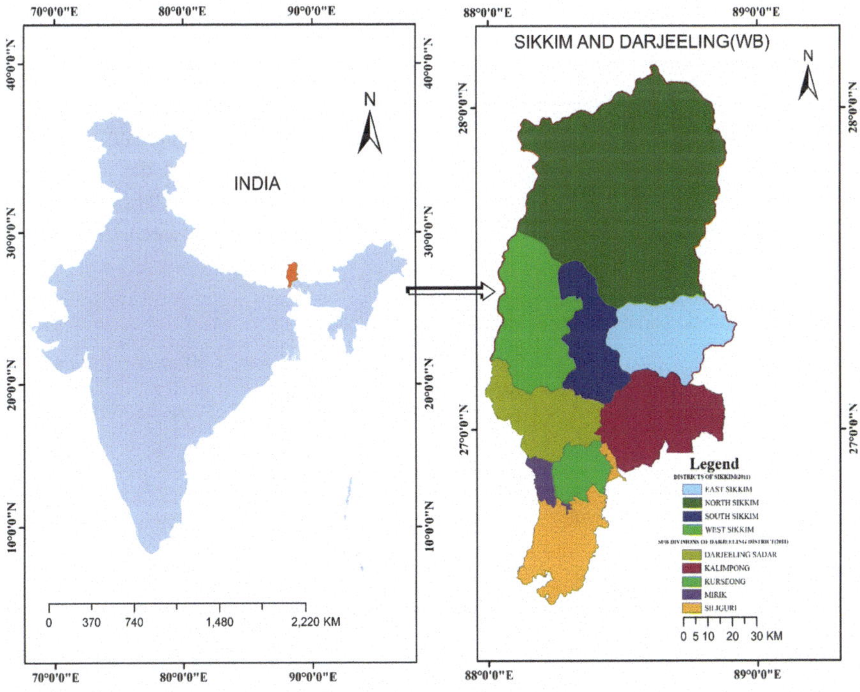

Fig. 12.1 Location map of Sikkim and Darjeeling and Kalimpong district, West Bengal

Darjeeling district became a full-fledged district on 14 February 2017 (Darjeeling Municipality, 2020).

Results and Discussion

Climatic Variability

(i) Sikkim Himalaya

Table 12.1 shows that the 1985–1995 mean decadal maximum and minimum temperatures at Gangtok were 22.37 °C and 12.20 °C, respectively. The average decadal amount of rainfall and number of rainy days were 3557.51 mm and 164.5 days, respectively. These figures had been slightly increased in the subsequent decadal period 1996–2005 where mean maximum and minimum temperatures were 22.41 °C and 13.30 °C, respectively. In this decade also observed more rainfall in terms of the amount and number of rainy days, i.e. 4359.29 mm and 183.2 days, respectively. The nature of temperature increase from the previous decade was mainly found in mean minimum temperature, i.e. +1.1 °C, whereas maximum temperature was raised by only +0.04 °C. The amount of rainfall increased from the previous decade was 602.64 mm, and number of rainy days also increased by 18.7 days in the decadal average. In the subsequent decadal period from 2006 to 2015, mean maximum and minimum temperatures were also observed more warming as 22.42 °C and 14.20 °C, respectively. The rate of increase in temperature from the previous decade is less when compared with the previous decade. The increase from the previous decade in mean minimum and maximum temperatures was +0.9 °C and + 0.01 °C, respectively. In terms of rainfall, both the number of rainy days and the amount were found lesser than the previous decadal period by – 7.2 days and – 602.64 mm, respectively. The decadal increase between 1985–1995

Table 12.1 Climatic variability at Gangtok, Sikkim (1985–2015)

Year	Rainy days	Rainfall (mm/yr)	T-min (°C)	T-max (°C)	T-mean (°C)
1985–1995 mean	164.5	3557.51	12.20	22.37	17.28
1996–2005 mean	183.2	4359.29	13.30	22.41	17.85
2006–2015 mean	170.0	3756.65	14.20	22.42	18.31
Decadal change (1985–1995 and 1996–2005)	+18.7	+801.78	+1.1	+0.04	+0.53
Decadal Change (1996–2005 and 2006–2015)	−7.2	−602.64	+0.9	+0.01	+0.46
Decadal Change(1985–1995 and 2006–2015)	+4.5	+199.14	+2	+0.05	+1.03
Grand mean	172.56	3891.15	13.23	22.40	17.81

Source: Indian Meteorological Department, Gangtok, Sikkim (2017)

Table 12.2 Climatic variability at Lachung, Sikkim (1981–2020)

Decadal year	Average Temp_Max (°C)	Average Temp_min (°C)	Total rainfall (mm)	Winter (mm)
1981–1990	13.2546	−2.83683	8775.51	207.85
1991–2000	13.60642	−2.91892	6708.09	221.51
2001–2010	13.71117	−2.35067	8710.63	183.92
2011–2020	13.26058	−2.34233	13245.49	269.9
Decadal from 1981–1990 to 1991–2000	+0.03564	+0.08209	−2067.42	+13.66
Decadal period from 1991–2000 to 2001–2010	+0.10475	−0.56825	+2002.54	−37.59
Decadal period from 2001–2010 to 2011–2020	−0.45059	−0.00834	+4534.37	+85.08
Decadal period 1981–1990 to 2001–2010	+0.45657	−0.4945	−64.88	−23.93
Decadal period 1981–1990 to 2011–2020	+0.00598	−0.4945	+4469.98	+62.03

Source: MERRA-2, NASA, 2022

and 2006–2015 in maximum and minimum temperature were +0.05 °C and + 2 °C, respectively, whereas for rainfall, more number of rainy days and more amount, i.e. +4.5 days and + 199.41 mm, respectively, in the later decade. The average temperature, average maximum and average minimum temperature between 1985 and 2015 were 17.81 °C, 22.40 °C and 19.12 °C, respectively. In 30 years, increase in temperature was profound at minimum and slightly at maximum. It shows clearly that climate change in the state in the last 30 years has caused a constant rise in temperature. This substantial increase in temperature is due to the rapid urbanization of Gangtok. But the amount of rainfall and the number of rainy days were observed to be highly fluctuating.

Table 12.2 shows the climatic variabilities from 1981 to 2020 at Lachung station located at an altitude of more than 3500 m from MSL where there is least direct human implication. For the decade of 1981–1990, the average maximum and minimum temperatures were 13.25 °C and – 2.83 °C, respectively. The total amount of precipitation received in this decade was 8775.51 mm. The total winter precipitation in the decade was 207.85 mm. In the following decade from 1991 to 2000, the average maximum and minimum temperatures were 13.60 °C and – 2.9 °C, respectively. The decadal average maximum and minimum temperatures were more than the previous decade by 0.05 °C in maximum and 0.082 °C in minimum. The total amount of precipitation in the decade was 6708.09 mm, which was much lower than the previous decade with an amount of 2067.42 mm. But the decadal total winter precipitation was 221.51 mm, which was more 13.66 mm than from the previous decade. In the decade of 2001–2010, the average maximum temperature was 13.7 °C, which was warmer than the previous decade by 0.10 °C and average minimum temperature was −2.35 °C which was colder from the previous decade by

0.56 °C. The total decadal precipitation was 8710.63 mm, which was much higher from the previous decade. But winter precipitation was found less than the previous decade. In the recent decade, i.e. 2011–2020, it was found colder than previous decade but experienced heavier rainfall than any decade in the last 40 years. During this decade, the average maximum temperature was 13.26 °C, which was colder than the previous decade by 0.45 °C and minimum average temperature was −2.34 °C, which was also colder by 0.008 °C from the previous decade. But in terms of precipitation, this decade received highest in 40 years, which was 13245.49 mm. It was around 4534.37 mm more than the previous decades. As well as in winter received precipitation about 269.9 mm, which was more 85.08 mm than the previous decade.

In the decadal comparison between 1981–1990 and 2001–2010, the average maximum temperature was higher by 0.45 °C but average minimum temperature was lower by 0.49 °C in 2001–2010. Summer experienced warmer and winter was more chilling than the previous decade. The total amount of precipitation received during 2001–2010 was lesser by 64.88 mm and winter precipitation also lesser by 23.93 mm. This shows that 2001–2010 was a warmer decade and more dryer too. Between 1981–1991 and 2011–2020, the maximum temperature was slightly increased by 0.0059 °C, whereas the minimum temperature declined by 0.49 °C in 2011–2020. This shows that global warming is also impacting Sikkim, which is warmer in summer and colder in winter. More melting of ice is likely to take place that accelerate ice recession.

From the study of climatic variabilities of two stations in Sikkim, i.e. Lachung and Gangtok, there is a clear indication of the impact of urbanization in Gangtok. At Lachung, the average maximum temperature in 40 years was increased slightly by 0.0059 °C and the minimum temperature was decreased by 0.49 °C which seems global influence. Whereas, in Gangtok, both maximum and minimum temperatures were found increasing, but increase in average minimum temperature was profound with 2 °C in three decades. Overall temperature at Gangtok increased by 1 °C in 30 years.

(ii) Darjeeling Himalaya

Table 12.3 shows the climatic variabilities at Darjeeling in the last 40 years from 1981 to 2020. In the decade of 1981–1990, the average maximum and minimum temperatures were 26.49 °C and 11.38 °C, respectively. The average amount of rainfall was 1297.72 mm, and total amount of rain during winter in the decade was 287.51 mm. In 1991–2000, the average maximum and minimum temperatures were 26.96 °C and 11.48 °C, respectively. The average rainfall was 992.752 mm, and total winter rainfall was 261 mm. It was found that 1991–2000 was warmer and drier than the previous decade. The increase in average maximum and minimum temperatures were 0.467 °C and 0.09858 °C, respectively from the previous decade. The amount of rainfall decreased by 304.9 mm on average. In winter, total amount of rain also found slightly decreased with 261.98 mm from previous decade. In the decade of 2001–2010, the average maximum and minimum temperature were 26.45 °C and 11.88 °C, respectively. The average decadal rainfall was 1313.043 mm, which was

Table 12.3 Climatic variability at Darjeeling, West Bengal (1981–2020)

Decadal year	Temp_Max (°C)	Temp_min (°C)	Rainfall (mm)	Winter
1981–1990	26.49892	11.38667	1297.726	287.51
1991–2000	26.965	11.48525	992.752	261.98
2001–2010	26.45942	11.88367	1313.043	231.49
2011–2020	26.24175	12.16125	1776.766	270.11
Decadal from 1981–1990 to 1991–2000	+0.467	+0.09858	−304.974	−25.53
Decadal period from 1991–2000 to 2001–2010	−0.506	+0.3984	+320.29	−30.49
Decadal period from 2001–2010 to 2011–2020	−0.218	+0.27758	+463.723	+38.62
Decadal period 1981–1990 to 2001–2010	−0.039	+0.497	+15.317	−56.02
Decadal period 1981–1990 to 2011–2020	−0.257	+0.7752	+479.034	−17.39

Source: MERRA-2, NASA, 2020

more rain than the previous decade by 320.29 mm, whereas winter rain decreased by 30.49 mm in total from the previous decade. This decade experienced a wetter and milder summer, whereas in the winter, it was warmer and drier than the previous decade. In 2011–2020, the average maximum and minimum temperatures were 26.24 °C and 12.16 °C, respectively. The decadal average and total rainfall in winter were 1776.766 mm and 270.11 mm, respectively. This decade had the lowest maximum average temperature, whereas minimum average temperature was recorded highest in the last 40 years. It means that the rise in temperature is profoundly marked in the minimum temperature. The minimum temperature has been rising continuously for the last 40 years, whereas the maximum temperature has been found to be very fluctuated. In the decadal comparison between 1981–1990, 2001–2010 and 2011–2020, it was found that 2001–2010 decade had a lower average maximum temperature of 0.039 °C and same in 2011–2020 with 0.257 °C when compared with 1981–1990. But average minimum temperature was recorded warmer in both subsequent decades than 1981–1990, i.e. 0.497 °C in 2001–2010 and 0.77 °C in 2011–2020. Same also recorded in rainfall: 15.317 mm more in 2001–2010 and 479 mm in 2011–2020 in decadal average rainfall, whereas winter rainfall were decreased by 56.02 mm in 2001–2010 and 17.39 mm in 2011–2020 from 1981 to 1990.

Forest Fire Incidences

(i) Sikkim Himalaya

The total number of forest fire incidences that occurred in the state for the last 16 years (2000–2015) was 120 (Table 12.4). Maximum forest fire incidences took place in 2006 with 41.50%, followed by 2014 with 14.15%. In these years, winter rain was very less. The maximum forest fire incidence took place in South district, i.e. 40 (33.32%), followed by East district with 30 (25.0%), North Sikkim with 26 (21.6%) and the least incidence in West district with 24 (20%). Forest fire incidences were confined in four months: January, February, March and April. Because in these months, most vegetation in sub-tropical deciduous forests become

Table 12.4 Forest fire incidence in Sikkim Himalaya (2000–2015)

Year	West district	South district	East district	North district	Total (%)
2000	–	–	–	–	Nil
2001	2 (Apr)	1 (Apr)	1 (Apr)	–	4 (3.3%)
2002	2 (Jan)	1 (May)	–		3 (2.5%)
2003	–	1 (Apr)	–	–	1 (0.8%)
2004	–	–	–	–	Nil
2005	2 (Mar)	1 (Mar), 1 (Dec)		2 (Dec)	6 (5%)
2006	2 (Jan), 5 (Mar)	2 (Mar), 2 (Apr)	14 (Jan)	19 (Jan)	44 (41.50%)
2007	–	–	–	–	Nil
2008	1 (Apr)	–	3 (Feb)	1 (Jan)	5 (4.7%)
2009	–	8 (Mar)	1 (Mar)	1 (Jan)	10 (9.43%)
2010	4 (Feb)	–	3 (Feb)	1 (Mar)	8 (7.5%)
2011	1 (Feb)	4 (Mar), 1 (Apr)	1 (Feb)	–	7 (6.6%)
2012	1 (Apr)	5 (Mar), 1 (Apr)	1 (Mar), 1 (Nov)	–	9 (8.4%)
2013	1 (Mar)	2 (Mar)	–	–	3 (2.8%)
2014	1 (Apr)	6 (Apr), 1 (Jan)	5 (Apr)	2 (Mar)	15 (14.15%)
2015	2 (Mar)	3 (Mar)	–	–	5 (4.7%)
Total	24 (20%)	40 (33.33%)	30 (25%)	26 (21.6%)	120
January	4 (16.6%)	1 (2.5%)	14 (46.6%)	21 (80.7%)	40 (33.3%)
February	5 (20.8%)	0	7 (23.3%)	0	12 (10%)
March	10 (41%)	25 (62.5%)	2 (6.6%)	3 (11.5)	40 (33.3%)
April	5 (20.8%)	12 (30%)	6 (20%)	0	23 (19.1)
May	0	1 (2.5%)	0	0	1 (0.8%)
November	0	0	1 (3.3%)	0	1 (0.8%)
December	0	1 (2.5%)	0	1 (3.84)	2 (1.6%)

Source: Bhuvan, NRSC, ISRO (2020)

dry and shed their dry leaves, as well as grasses on the ground also become dry. With this condition, small spark from natural or anthropogenic factors leads to extensive forest fire, which accentuated by a rainless winter. The highest occurrence of forest fire incidences took place in South district with 40 (33.3%). South district is drier than any other districts in the state. North district of Sikkim is the least populated district and mostly snow cover with evergreen alpine forest along with patches of pastural land. In 2006, 19 forest fire incidences took place in the district in January itself out of a total of 26 forest fire incidences in 16 years in a few concentrated area. It may be due to the minimum rainfall in winter in the state as well as rapid development in the areas. After that, North district had experienced the least forest fire incidence. In East district also, most of the forest fires happened in month of January (46.6%) and also took place in 2006. It may be caused by land-use changes due to rapid development and urbanization that affect local hydrology and micro-climatic conditions besides dry winter. In West district, mostly forest fire happens in the month of March (62.5%). In the state as a whole, 33.3% forest fire incidences took place in the month of March in the last 16 years (2000–2015). This shows that forest fire is related to the pre-monsoon's sudden rise in temperature associated with dry plants and grasses in sub-tropical forests. The same number of forest fire incidences took place in month of January that happened in 2006, 14 incidents in East district and 19 incidents in North district. February month experienced less comparatively with these four months, i.e. 10%. It may be due to a slight shower rain from the western disturbance. But a rare forest fire occurred in North district in the month of December in 2005. Most forest fires occurred up to the elevation of 1760 m from MSL, but few occurred exceptional above 3000 m where scrubs are found. Forest fires also took place in the months of May, December and November with negligible numbers. Spatially, most of the forest fires took place in open forests located near agricultural sites. In the higher altitudes, scrubs are more vulnerable (Fig. 12.3). The location of the forest fire mostly on lee ward sites (Fig. 12.4). It was also observed that forest fire in Sikkim mostly occurred near settlement, town, agriculture and along NH 31A, which give the figure of around 80% of the forest fire incidence. This shows that forest fire in the state is from anthropogenic causes and accentuated by climate condition. Most of forest fires (75%) have been occurred in and around sub-tropical deciduous forest. In 2004, 58.5 hectares of forest land was damaged by forest fire in which 80% occurred in Sal forest. In 2005, 225 hectares of forest land had been affected mostly in ground bushes and Sal forest (Department of State Disaster and Management, 2017). In 2014, the area affected by the forest fire was reported to be about 570 hectares in which ground bushes were affected, mostly Sal, teak, bamboo, etc. It was observed that exponential increased in effected areas.

(ii) Darjeeling Himalaya

Darjeeling Himalaya includes Darjeeling and Kalimpong districts of West Bengal. Four subdivisions of Darjeeling (Darjeeling, Kurseong, Mirik and Silliguri) and Kalimpong districts have been taken for the study of forest incidence in the last 16 years (2000–2015). Silliguri subdivision of Darjeeling is almost plain and least areas under forest land. The total forest fire incidences that took place in 16 years were 247 (Table 12.5). It is more than Sikkim because the forests in the southern part mostly belonged to tropical deciduous forest and are drier in the pre-monsoon season. Sikkim forest fire incidences are also located mostly along the boundary of Kalimpong. In Darjeeling Himalaya, Kalimpong district has more forest fire incidences than Darjeeling district, with an account of 52.63%. Most forest fires occur in areas that are below 1500 m MSL. In Kalimpong district, western part of it along the boundary of Darjeeling and southern part of the district along Duar region have the maximum concentration of forest fire (Fig. 12.2). In Darjeeling district, Darjeeling subdivision has the highest forest fire incidences with an account of 18.4%, followed by Mirik subdivision, i.e. 13.93% of the region. Kurseong subdivision has 8.9% least fire incidences among hilly regions because this subdivision receives more rain than other subdivisions. Silliguri subdivision has recorded least forest fire incidences, which is 6.55% because of the small area of the forest land and mostly land under settlement and agriculture. In Darjeeling subdivision, maximum forest fire took place in the eastern and southern parts, whereas middle and northern parts have less because of the high altitude above 1700 m from MSL. The west part of the subdivision also experienced more forest fire where it was located on the rain shadow. Mirik subdivision is the smallest subdivision but has extensive forest fire because of its location on rain shadow. Darjeeling and Kalimpong districts do not have areas above 3400 m. Ninety percent of forest fire occurred below an elevation of 1760 m (Fig. 12.2). Most forest fires took place in forests located near agricultural sites (Fig. 12.3). The location of the forest fire mostly on lee ward site (Fig. 12.4).

It was observed that forest fire incidences occurred in the region for eight months: January, February, March, April, May, September, October and December. The maximum number of forests occurred in the month of March, which accounted for more than 55.8% followed by April with 20%. Among these eight months, the least forest fire occurred in the month of May because of pre-monsoon rain.

Conclusion

Forest fire incidences have been a local issue, but they are also influenced by global phenomena. Its become very complicated presently. The incidence of forest fire has been increasing in an unprecedented way in terms of frequency and extension. It is clear from the observation of temporal incidence of forest fire that pre-monsoon in which sudden rise of temperature and dryness in the region specially in tropical and subtropical deciduous are the main responsible factors by which most of the forest fire. The forest fire incidences took place in the months of March and April in Darjeeling Himalaya, whereas in Sikkim Himalaya mostly occurred in the month of January followed by March, which shows the relationship with winter rain and pre-monsoon temperature. Most of the forest fires in this region are also caused by anthropogenic factors. Forest fire affects the ground and Sal forest extensively. The area affected by forest fire in this region has increased exponentially, but the number

Table 12.5 Forest fire incidences in Darjeeling Himalaya (2000–2015)

Year	Darjeeling subdivision	Kurseong subdivision	Mirik subdivision	Silliguri subdivision	Kalimpong district	Total (%)
2000	–	–	–	–	–	Nil
2001	–	2 (Mar)	–	1 (Mar)	4 (Mar)	7 (2.47%)
2002	–	–	–	–	6 (Oct), 2 (Sep), 1 (Dec)	9 (3.6%)
2003	2 (Apr), 2 (May)	1 (Jan)	–	–	1 (Mar), 1 (Apr), 1 (Dec)	8 (3.23%)
2004	1 (Mar)	1 (Mar)	2 (Apr)	–	1 (Mar), 1 (Sep), 1 (Oct)	7 (2.83%)
2005	1 (Apr)	1 (Mar)	–	–	3 (Mar), 3 (Apr)	8 (3.6%)
2006	4 (Mar), 2 (Apr)	–	2 (Feb), 1 (Mar), 1 (Jan)	4 Mar	2 (Jan), 10 (Mar), 1 (Apr)	27 (10.9%)
2007	–	–	–	–	3 (Oct), 1 (Dec), 1 (Feb)	5 (2.02%)
2008	4 (Mar), 2 (Apr), 1 (Feb)	–	1 (Mar)	–	6 (Mar), 2 (Feb), 2 (Apr), 2 (Sep), 1 (Oct), 1 (Dec)	22 (8.9%)
2009	3 (Mar), 1 (Apr)	1 (Mar)	3 (Jan), 1 (Feb), 8 (Mar)	2 (Mar), 1 (Feb)	1 (Feb), 13 (Mar), 4 (Sep), 1 (Oct)	39 (15.78%)
2010	4 (Mar), 1 (Apr)	3 (Mar), 1 (Apr),	3 (Jan)	1 (Apr)	1 (Feb), 6 (Mar), 1 (Apr)	21 (8.5%)
2011	6 (Mar), 1 (Apr)	2 (Apr)	4 (Mar)	1 (Feb)	2 (Feb), 11 (Mar), 5 (Apr), 1 (SEP)	33 (13.36%)
2012	3 (Mar), 1 (Apr)	3 (Mar), 2 (Apr)	4 (Mar), 2 (Apr)	1 (Feb), 4 (Mar)	1 (Jan), 18 (Mar), 4 (Apr)	43 (17.4%)
2013	–	–	–	–	–	Nil
2014	6 (Apr)	5 (Apr)	2 (Apr)	1 (Apr)	4 (Apr)	18 (7.28%)
2015	–	–	–	–	–	–
Total	45 (18.4%)	22 (8.9%)	34 (13.93%)	16 (6.55%)	131 (52.63%)	247 (100%)
January	0	1 (4.5%)	7 (20.58)	0	3 (2.3%)	

(continued)

Table 12.5 (continued)

Year	Darjeeling subdivision	Kurseong subdivision	Mirik subdivision	Silliguri subdivision	Kalimpong district	Total (%)
						11 (4.45%)
February	1 (2.2%)	0	3 (8.8%)	3 (18.7%)	7 (5.38%)	14 (5.6%)
March	25 (55.5%)	11 (50%)	18 (52.9%)	11 (68.7%)	73 (56.15%)	138 (55.87%)
April	17 (37.7%)	10 (45.45%)	6 (17.64%)	2 (12.5)	21 (16.15%)	56 (22.67)
May	2 (4.4)	0	0	0	0	2 (0.8)
September	0	0	0	0	10 (7.6%)	10 (4.0%)
October	0	0	0	0	12 (9.2%)	12 (4.8%)
December	0	0	0	0	4 (3.0%)	4 (1.6%)

Source: MERRA-2, NASA

of forest fire incidence is fluctuated. As a result, precious biodiversities are lossed irreparably. It also affects to surface run off and declining soil fertility in the region.

More forest fire incidences occur in Darjeeling Himalaya than in Sikkim Himalaya because of its lower elevation and exposure to more dry and sudden temperature rises in the pre-monsoon season, accentuated by anthropogenic activities as Darjeeling Himalaya is located in the southern part. In Darjeeling Himalaya, Kalimpong also experienced more forest fire compared with Darjeeling district as it lies in a lower area. It is also clear from climatic variability in the last 40 years (1981–2020) that the temperature in this region is becoming warmer. This trend of climate change, so-called global warming, is also influenced by localized urbanization.

Forest fire has been a constant threat to the ecosystem, biodiversity, resources, properties etc. It is high time to save forests from forest fire. Changes in the attitudes and actions of individuals, stakeholders including the private sectors, and governments are instrumental in mitigating initiative programme for forest fire.

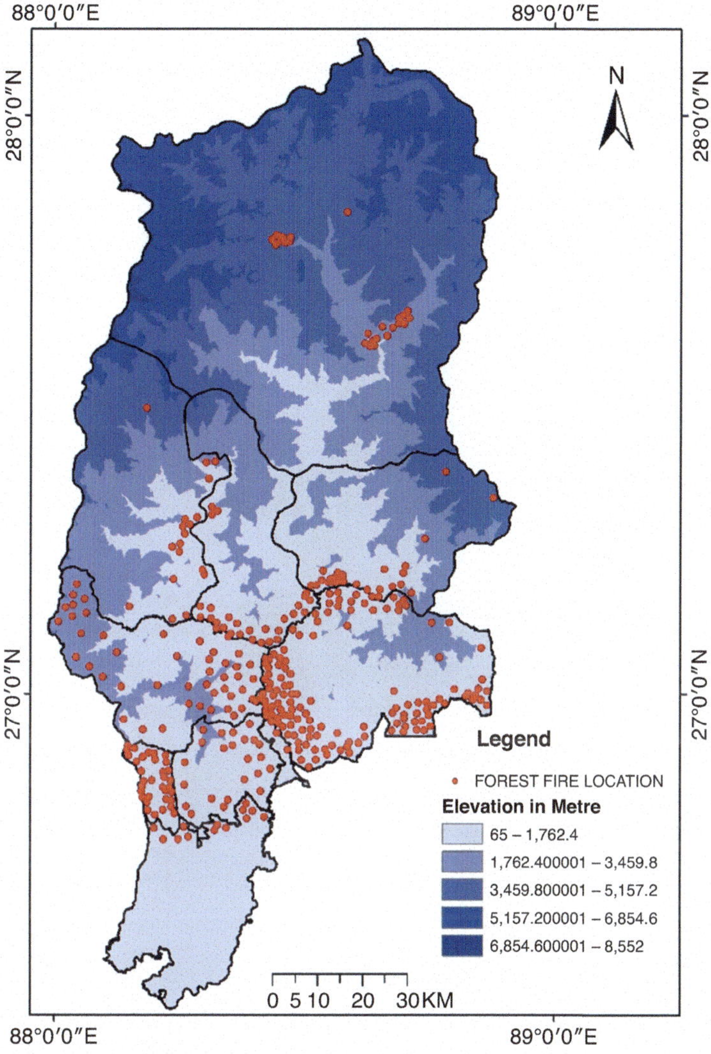

Fig. 12.2 Forest fire with elevation in Sikkim and Darjeeling Himalaya (2000–2015). (Source: ASTER DEM, ISRO 2015)

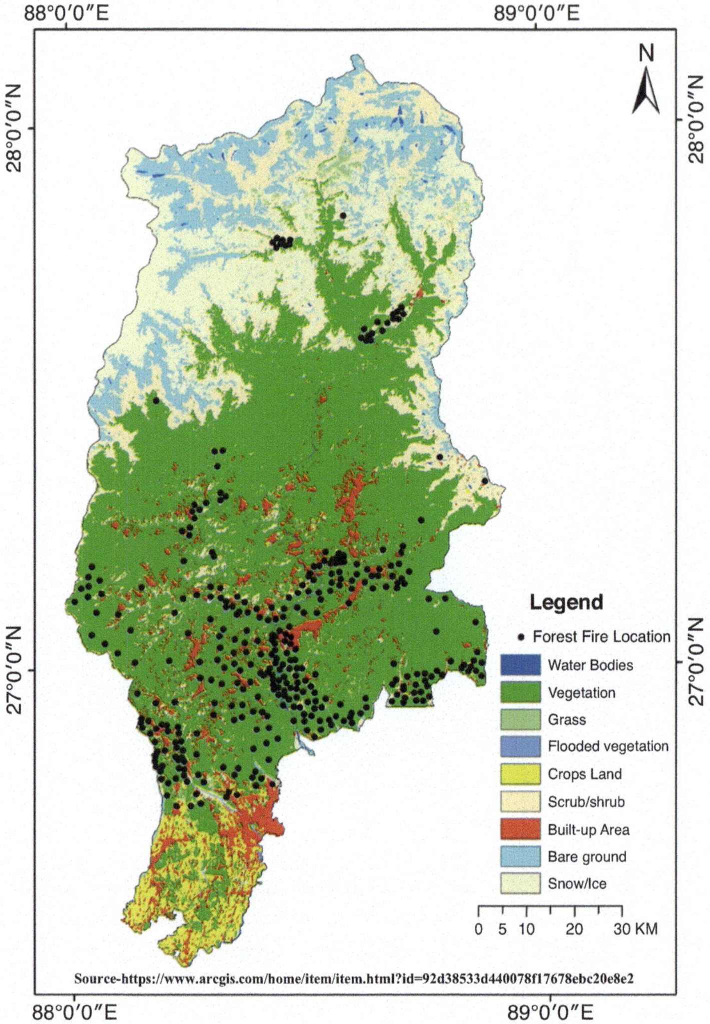

Fig. 12.3 Forest fire and different land use land covers of Sikkim and Darjeeling Himalaya (2000–2015)

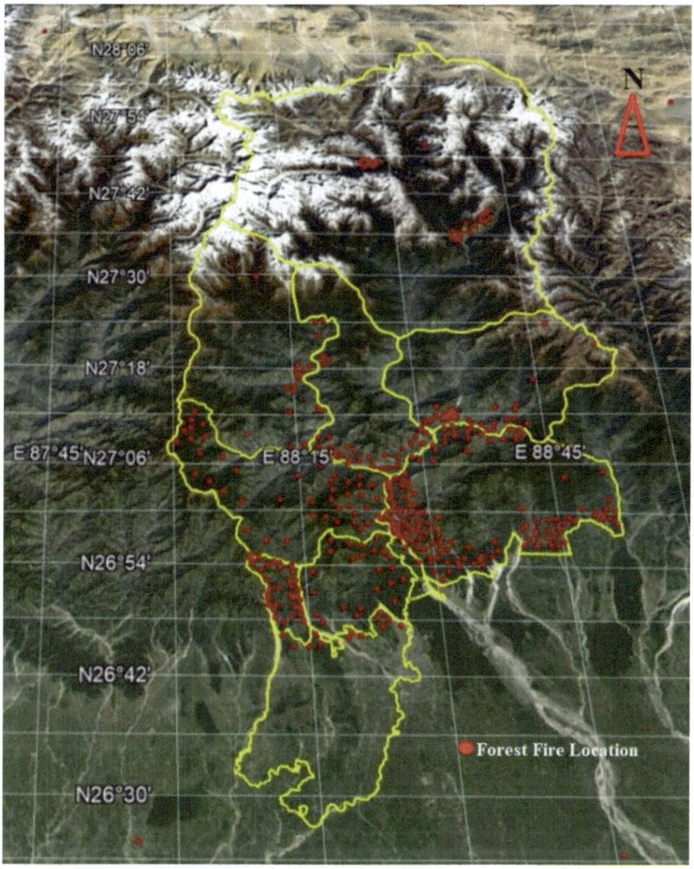

Source: Google Earth and ISRO, 2015

Fig. 12.4 Forest fire in Sikkim and Darjeeling Himalaya. (Source: Google Earth and ISRO, 2015)

References

Census of India (2011). Government of India.

Darjeeling Municipality. (2020). Government of West Bengal.

Department of Forest. (2018). *Annual Forest report 2017*. Sikkim Forest Department.

Department of State Disaster and Management (2017). Annual Report 2016, Government of Sikkim.

FAO. (2015). Global Forest Resources Assessment 2015: Main report.

Forest Survey of India (FSI). (2011). *State of Forest report, 2011*. Ministry of Environment and Forests.

https://bhuvan-app1.nrsc.gov.in/disaster/disaster.php?id=fire

https://earth.google.com/web/

https://earthdata.nasa.gov/learn/articles/new-aster-gdem

https://power.larc.nasa.gov/data-access-viewer/

IMD, 2017. Governement of India.

Jaiswal, R. K., Mukherjee, S., Raju, K. D., & Saxena, R. (2002). Forest fire risk zone mapping through satellite imagery & geographical information system. *International Journal of Applied Earth Observation and Geoinformation, Elsevier Publications, 4*, 1–10.

Myers, N., Mittermeier, R. A., Mittermeier, C. G., Da Fonseca, G. A., & Kent, J. (2000). Biodiversity hotspots for conservation priorities. *Nature, 403*, 403.

NIDM. (2014). *Forest fire report*. GOI.

Pandit, M. K., Sodhi, N. S., Koh, L. P., Bhaskar, A., & Brook, B. W. (2007). Unreported yet massive deforestation driving loss of endemic biodiversity in Indian Himalaya. *Biodiversity and Conservation, 16*, 153–163.

Chapter 13
Flood Disaster Risk Governance in Changing Climate Contexts

Gowhar Farooq Wani, Syed Towseef Ahmad, Rayees Ahmed, Abid Farooq Rather, Ajinder Walia, and Pervez Ahmed

Introduction

Flooding is a common problem in many parts of the world. It has the potential to cause massive damages, particularly in flood-prone areas of poor and developing countries such as Bangladesh, Sri Lanka, India and Pakistan. In contrast, floods wreaked havoc in parts of the wealthiest and technologically advanced countries like USA, Germany and Belgium in 2021, exposing their existing flood control mechanisms, vulnerabilities of critical infrastructure, effectiveness of early warning systems and response to flooding in a way to safeguard public and private infrastructure as well as carry safer evacuations of people (Fekete & Sandholz, 2021). In 2020, flooding was observed as a common disaster (201 flood events out of 389 total disaster events), with Asia experiencing most of it (EM-DAT, 2020; CRED, 2021).

Cities are home to more than half of the world's population including the urban poor, which are exposed to flooding. In the future, the incidence of floods may increase in frequency and magnitude as a result of many factors like climate change (Carmin et al., 2013; McCarthy et al., 2001), unplanned development, rapid urbanization, socio-economic problems and bad governance. Climate change is impacting everyone and every region. Since the beginning of pre-industrial period, average global temperatures have increased by about 1 °C (Masson-Delmotte et al., 2021; Bajracharya et al., 2008; Mool et al., 2011). Kulkarni et al. (2013) and Sun et al. (2017) reported a significant increase in surface temperatures in the Himalayan region over the last 60 years. The annual mean surface air temperature in the

G. F. Wani (✉) · S. T. Ahmad · R. Ahmed · A. F. Rather · P. Ahmed
Department of Geography and Disaster Management, School of Earth and Environmental Sciences, University of Kashmir, Srinagar, India

A. Walia
Department of Governance and Inclusive Disaster Risk Reduction, National Institute of Disaster Management (NIDM), Ministry of Home Affairs, Government of India, New Delhi, India

A. L. Singh et al. (eds.), *Climate Change, Vulnerabilities and Adaptation*,
https://doi.org/10.1007/978-3-031-49642-4_13

Himalayan region increased at a rate of 0.1 °C per decade between 1901 and 2014 (Ren et al., 2017). During the early twentieth century, the Himalayan warming trend was 0.16 °C per decade, which eventually doubled to 0.32 °C per decade (Yan & Liu., 2014).

Since past 108 years, from 1901 to 2009, the mean annual temperature in India has increased by 0.56 °C (Attri & Tyagi, 2010), with an average temperature predicted to rise by 3.5–5.5 °C by the end of the twenty-first century (Lal, 2002). The region will be negatively affected if the warming trend persists.

Recently, the Intergovernmental Panel on Climate Change (IPCC) in its Special Report on Extreme Events and Disasters (SREX) predicted that there will be a further increase in the frequency of extreme weather events in twenty-first century, including the increase in intensity of floods. Climate change causes different changes in different regions, such as faster warming and flooding (Masson-Delmotte et al., 2021), which disproportionately affects socially vulnerable people living in the flood-prone areas. The impacts are, however, expected to be amplified in urban centres, which are commonly understood as areas that are warmer than the surroundings (Masson-Delmotte et al., 2021) and prone to flooding because the water flows to low-lying areas and into water bodies, causing increased runoff rather than seeping into the ground as a result of pavements and abundant impervious surfaces.

In recent times, many areas around the globe have been developing as urban at a faster rate, originating urban development as a problem with change in land use. It is believed that urbanization may increase the frequency and intensity of floods and expose communities to multiple flood hazards (Konrad, 2003). Nevertheless, flood exposure is a critical aspect to understand and manage flood risk. This has become more convincing after the introduction of two voluntary and non-binding agreements: the proactive approach 'Hyogo Framework for Action 2005–2015 (HFA)' and the people-centred approach 'Sendai Framework for Disaster Risk Reduction 2015–2030 (SFDRR)'. In this backdrop, the regional or city planners are tasked to prepare developmental strategies with maximum gains by incorporating disaster risk into policy and planning at the local level. In the past, flood risk management was primarily concerned with the distribution of relief and humanitarian aid to affected populations, which was later expanded to include prevention and mitigation measures through technological interventions such as the construction of dikes, protection walls, and dredging of water bodies. However, as knowledge has advanced and threats have increased, flooding is now being studied as a complex chain of climatic, hydrological, and social factors, with a clear emphasis on preparedness, adaptation, and resilience aspects of flood risk management. With one of the strategic priorities of the SFDRR 2015–2030 being 'strengthening disaster risk governance for disaster risk management' (UNISDR, 2015), the efforts to advance disaster risk governance theory to improve disaster risk reduction practices are gaining momentum globally (Shi et al., 2010)

Disaster Risk Governance: Concept and Meaning

Disaster risk governance has been a centre of attention since the HFA 2005–2015, which advocates for good governance and disaster risk reduction and terms them as mutually supportive, but its origins can be traced back to 3200 BC, with evidence from Iraq indicating that members of the Asipu social group were advised on how to manage disasters (Coppola, 2015). The approach to dealing with disasters in place for centuries was hazard specific and relief centric; for example, actions were taken to reduce the negative impacts of floods and then separately for fires or other hazards by providing relief. Recognizing that this approach was time consuming and required significant resources; however, accomplishments were made on a variety of fronts, including disaster awareness, capacity building and the involvement of international agencies to take responsibility, set up committees and develop policies and legal frameworks under the umbrella of the United Nations to guide all relevant stakeholders in administrative and development spheres on disaster risk reduction. In the nineteenth century, for example, planning committees were set up by the British and Indian governments to manage drought impacts facing the region (Coppola, 2015). The response phase of disaster management was prioritized, and disasters were mostly viewed as natural calamities rather than human induced.

After a long period of ignoring it, disaster risk governance, or simply risk governance, shapes under the umbrella of governance. According to the United Nations Development Programme (UNDP), governance is 'the exercise of authority (politically, economically and administratively) for managing country affairs at the various levels. Encompassing and transcending government, as well as enabling all individuals and groups to express themselves, exercise their legal rights, meet obligations and reduce disparities through the mechanisms, processes and institutional frameworks contained within this'. Good and supportive governance is essential for building disaster-conscious and-resilient societies by establishing adequate and appropriate (top-and ground-level) governance arrangements, strengthening institutional mechanisms and developing policy and legal frameworks, enhancing international cooperation, coordinating and overseeing disaster risk reduction at the global, national and regional levels. Because it is the vulnerabilities of populations and other elements at risk that turn a natural hazard into a disaster (Wani et al., 2022). The Sendai Framework for Disaster Risk Reduction 2015–2030 includes risk governance as the second of four priorities, as 'strengthening of disaster risk governance to manage risk'.

A Paradigm Shift in Disaster Risk Reduction Policy at the Global Level

Change in Approach: From Reactive to Proactive

In recent times, disaster management has been a subject attracting attention from everyone including common people to policymakers and researchers around the globe. The shift, which was inevitable, progressed from a culture of reaction to a culture of prevention. More precisely, from a reactive, relief-centric approach to a more holistic and proactive (action and result oriented) preparedness, prevention and mitigation approach. Rightly so, given that the global community, especially the poor and developing countries, have been adversely impacted and suffering substantially in terms of human losses and economic costs, proving a setback to development. The distribution of relief and humanitarian aid to disaster-affected communities through the involvement of local and global humanitarian organizations is a well-known, age-old dimension of disaster management that is predominantly observed in developing countries and is thought to have existed since the dawn of global cooperation (Kamidohzono et al., 2015).

Declaration of 1990s as the International Decade for Disaster Risk Reduction

The General Assembly of the United Nations in its 44th session declared the 1990s as the International Decade for Natural Disaster Risk Reduction, which effectively started on January 1, 1990 (IDNDR, 1994). The main objective of declaring 1990s as a decade for natural disaster risk reduction was to reduce human loss, damages to assets, social and economic disruption as a result of natural disasters such as drought, earthquake, and other adversities of natural origin, particularly in the poor and developing countries and regions of the world through international cooperation (UNISDR, 2012a). The need for such a decade was felt with the goal of reducing disaster losses by gathering various stakeholders, experts, professionals, policy and decision makers from various backgrounds, cultures and nationalities under one umbrella organization, namely the United Nations, with moral authority and a strong commitment to addressing disaster-related issues confronting the global community (Lechat, 1990).

In addition, the developing countries were witnessing more devastation and were somehow neglected at the international forums when it came to managing disasters as a global concern. The world community, particularly comprised of developed nations, later recognized that reducing disaster risk has positive impacts on all people, with a special focus on developing countries, because the poor countries housing millions of vulnerable people (in any case) were not in a position to respond to or withstand disasters on its own.

In a nutshell, IDNDR's main goals were as follows (IDNDR, 1994; Lechat, 1990):

- To effectively mitigate the effects of natural disasters by building the capacity of each country, with a focus on developing countries, by providing them with special assistance in assessing disaster damage potential, developing early warning systems and disaster-resilient infrastructure.
- To develop appropriate guidelines and future strategies for applying scientific knowledge and technical expertise while considering the various cultures and economic structures of the countries.
- To plug crucial knowledge gaps and alleviate human suffering by promoting scientific and engineering enterprises.
- To facilitate the dissemination of technological information.
- To develop certain measures on critical aspects of disaster management such as assessment, early warning, prevention and mitigation through programmes like transfer of technology, education and training, and project demonstration, which are disaster and area specific and to evaluate the effectiveness of such programmes

International Strategy for Disaster Reduction and Hyogo Framework for Action 2005–2015

In a significant development, a global framework, namely the International Strategy for Disaster Reduction (ISDR), was adopted in 1999 in Geneva, Switzerland, with the vision 'to enable all communities/nations to become resilient to all types of hazards and reduce associated risks' (UNISDR, 2012b). This was within the United Nations system and successor to the 1990s International Disaster Reduction Decade (Fig. 13.1). ISDR's implementation was facilitated by the United Nations Office for Disaster Reduction (UNISDR). Following that, the breakthroughs in disaster management continued, with the world community taking disasters losses seriously while attempting to manage them effectively through the adoption of the Hyogo Framework for Action (HFA) 2005–2015: Building the Resilience of Nations and Communities to Disasters (UNISDR, 2005).

This feat of managing disasters through the constant and systematic efforts was accomplished at the United Nations General Assembly (UN/GA) second World Conference on Disaster Risk Reduction (DRR) from 18 to 22 January 2005 in Kobe-Hyogo, Japan, after very carefully reviewing the Yokohama Strategy for a Safer World and its Plan of Action, which was adopted during the first World Conference on Natural Disasters from 23 to 27 May 1994 in Yokohama Japan.

The HFA was a 10-year plan that first identified the ways of and detailed the measures for building resilience of communities required across different sectors (global and local) to reduce disaster losses, vulnerabilities, risks from hazards by bringing together different stakeholders (national and international, government and

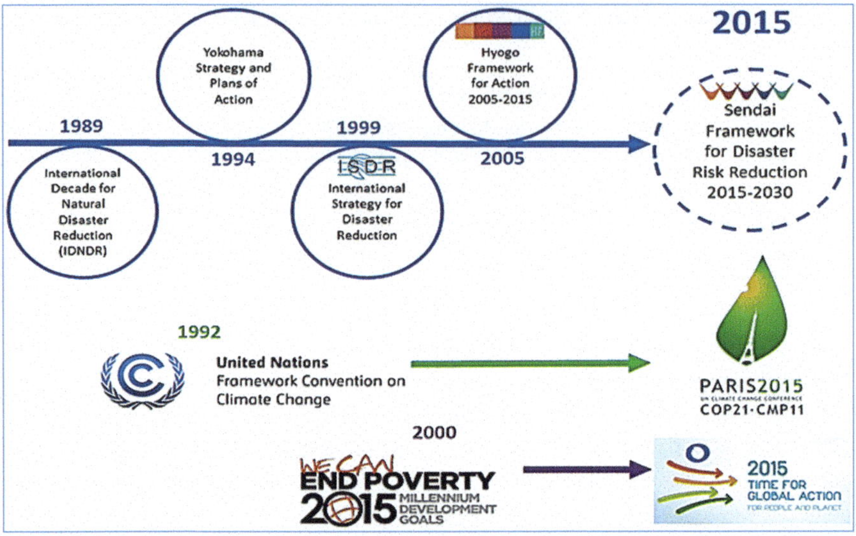

Fig. 13.1 Global commitments towards disaster risk reduction from 1989 to 2015. [Adapted from Andrew Maskrey's Presentation (UNISDR, 2015)]

non-government, experts, etc.) under one umbrella through international cooperation and coordination (UNISDR, 2005).

The HFA had set the following five priorities for action:

(1) Make Disaster Risk Reduction (DRR) a priority at all levels (global and local) with robust institutional arrangements/networks in place to support its implementation.
(2) Know the disaster risk by identifying, assessing, and monitoring procedures, and enhance early warning system.
(3) Use knowledge, education and innovation to better understand disasters and develop a culture of safety and resilience.
(4) Reduce the underlying disaster risk factors.
(5) Improve disaster preparedness at all levels to ensure an effective disaster response.

The United Nations Framework Conventionon Climate Change (UNFCCC), established on principles to promote climate change cooperation and regulate greenhouse gases, adopted the Kyoto Protocol in 1997. The Kyoto Protocol with its first commitment period from 2000 to 2012, ran parallel to the Eight Millennium Development Goals (MDGs), which were intended to be achieved by 2015. These initiatives operated concurrently with the Hyogo Framework of Action. These words of action (policies, frameworks) worked out by various people from varied backgrounds represented the global community's concerted efforts to address issues of global concern and emerging challenges as a result of the complex interactions of natural phenomena with social and economic conditions, compounded by anthropogenic interference. The synergies between these policies and frameworks were not clearly defined, whether it was reducing global hunger or poverty, gender equality or women's empowerment, environmental sustainability, or climate change and management of disasters.

From Disaster Management to Disaster Risk Management

The pre-2015 efforts at the global level (Fig. 13.1) were majorly concentrated on some important phases of disaster management and particularly focused on preparedness, response and recovery. However, a shift was observed since the inception of HFA 2005–2015, which to a great extent focused on reducing losses from disasters by managing disaster risk. Although it started with just providing relief to the affected communities in the eventuality of a disaster—disaster management—the global works prioritized disaster risk management.

The expected outcome of prioritizing disaster risk management was to substantially reduce damages and losses from disasters by reducing existing risks or preventing new risks through the application of DRR policies and strategies (corrective, prospective, compensatory and community-based approaches) at all levels. This approach of managing risk gained momentum in the following years, i.e. post-2015 era, and was mainly guided by the Sendai Framework for Disaster Risk Reduction 2015–2030 (SFDRR). A successor instrument to HFA, the SFDRR was adopted at the third World Conference on Disaster Risk Reduction on 18 March 2015 in Sendai, Japan, after threadbare international-level consultations of multi-stakeholders and concerted inter-governmental negotiations (UNISDR, 2015).

Innovation is a key in disaster research and this is believed to be a defining factor in the introduction of SFDRR after reviewing HFA, as well as to ensure the continuity of work and progress made during the HFA period. Promoting engagement at all levels (community, institutional and societal) and across the sectors is among the guiding principles of SFDRR for the prevention and reduction of disaster risk. Both natural and human-made hazards and its associated risks as well as health resilience finds a mention in SFDRR envisaging and broadening its scope while clearly recognizing the global and regional platforms for disaster risk reduction. The action priorities give a strong emphasis on understanding risk and investing in disaster risk reduction as well as building preparedness, however, 'strengthening of disaster risk governance to manage disaster risk' as a priority was a major highlight, something never seen before.

Sendai Framework for DRR has the following four priorities of action (UNISDR, 2015):

(1) Understanding disaster risk
(2) Strengthening disaster risk governance to manage disaster risk
(3) Investing in DRR to improve resilience

(4) Building preparedness for better response and building back better for safer and resilient infrastructure

With the advances in disaster risk management theory and knowledge, as well as innovation in science and technology, it is believed that money spent on aspects/ activities in the pre-disaster phase, such as preparedness and mitigation, will save huge bucks during the post-disaster phase, such as response and relief, as well as mobilizing resources and encouraging multi-stakeholder participation and community involvement will prepare ways for effective and efficient disaster risk management.

Understanding Disaster Risk Governance in Kashmir Context

At the global level, there have been consistent and calibrated efforts to mitigate the negative effects of disasters. The policies and legal frameworks that provide strategies and actions were broad in nature and needed to be tailored to the conditions of specific locations while keeping in mind the hazards and risks that were context specific. In developing countries, institutional mechanisms were non-existent, and human resources were inadequately equipped to develop and implement risk-informed policies and plans. The multi-hazard-prone Kashmir region with a history of flood disasters found itself in a similar situation. Following the 2005 Kashmir earthquake, there were dramatic changes in the approach towards disaster management, with more concerted efforts made on policy, planning, education and research fronts to recognize risks from impending disasters as well as lessen their impacts by focusing on all phases of disaster management and particularly pre-disaster phase for a safe and disaster-resilient Kashmir.

In 2005, after the enactment of Disaster Management Act in India, many national and local-level authorities and committees were formed to ensure effective disaster management. The UT of Jammu and Kashmir (an erstwhile state of India) envisaged a plan document, the first of its kind, a separate, 'Disaster Management Policy in 2011', which was later revised in 2017, after learning lessons from the 2014 Kashmir flood. The policy document advocated for a proactive, integrated and holistic, multi-hazard approach to disaster risk reduction by providing guiding principles on all phases of disaster management, establishing disaster management authorities at different levels, and encouraging multi-stakeholder role and participation in disaster risk reduction efforts for a safe & resilient future.

The 2014 floods in Kashmir were historic, causing widespread devastation and highlighting various existing vulnerabilities and loopholes in disaster planning and development aspects. The community's role was exceptional during the disaster response & recovery phase, motivating local planners/authorities to provide guidance on community-based disaster management through the State Disaster Management Plan in 2017. Which also focused on the establishment of a comprehensive multi-hazard early warning system and a state-level emergency operation centre in the region. Following the twin disasters of 2005 and 2014, as well as the region's warming climate as understood from various research works considering the local climatic regime (Romshoo et al., 2020; Ahmed et al., 2021; Ahsan et al., 2021; Ahmad et al., 2021), the field of disaster management has received some serious attention on various fronts, including risk governance (Table 13.1).

Conclusion

The poor and vulnerable populations across the world, especially in developing countries, are at the greatest risk from disasters. Climate variability and change, combined with multiple environmental, social and economic vulnerabilities of elements at risk, are increasing overall disaster risks for all communities worldwide. The scientific community now widely accepts the fact that disasters or severe impacts of hazards are not 'acts of god' but rather natural hazards that turn into disasters as a result of human actions or decisions. In the disaster management context, individuals, societies and institutions have been taking initiatives since time immemorial to deal with the effects of disasters after facing losses in terms of death and destruction.

The ongoing disaster risk reduction process has evolved tremendously from the traditional approach of managing single hazards by focusing on humanitarian relief to a more action-oriented and effective all-hazard approach with a focus on preparedness, prevention and mitigation. The ultimate goal is to reduce disaster risks for all individuals, particularly the poor and vulnerable populations in developing countries, in order to ensure a safe and resilient future. At the international level, this has been accomplished by recognizing disaster risk governance as a key factor in disaster risk reduction, strengthening it through initiatives that provide platforms for multi-stakeholder involvement (public and private authorities, media, civil society and non-governmental organizations), and coordinating and guiding on how to reduce disaster risks at the community, regional and national levels.

Good governance is essential for fostering a resilient culture and a disaster-conscious society. Which is poorly understood in countries and regions that lack strong political will, technical expertise, sufficient human resources, institutional mechanisms and processes. Kashmir Valley, a multi-hazard-prone region, is one such example. Following the twin disasters of the 2005 earthquake and the 2014 flood, the region has been on a path of developing and implementing plans and policies to ensure effective disaster risk management. The implementation of these risk-informed programmes, plans and policies has been slow due to inherent challenges and issues, including a lack of political will. With the increased frequency and intensity of disasters, there is a growing interest in strengthening disaster risk governance in order to make disaster risk management a top priority at all levels and across all sectors by building capacities of societies and institutions.

Table 13.1 Insights into historical flood events in Kashmir Valley, India

Year of occurrence	Description of the flood event/extent of damage	Evidence
1838	A major flooding known as the "Great Flood" forced the residents to use boats owned by them to rescue themselves and move to safety.	Bates (1873) and Digby (1890)
1841	A major flooding in the region caused i. Loss of lives ii. Sweeping of security personnels iii. Damage to important city bridges	Bates (1873) and Lawrence (1895)
1865	Flood in the month of August resulted in damage to the crops	Ince (1888) and Digby (1890)
1869	Heavy rains caused damage to standing crops	Bewell (1875) and Digby (1890)
1871	Complete inundation of the entire state giving resemblance to a large lake and caused damage to crops	Khoihami (1885), Ram (1895) and Rai (2004)
1885	Compound event: Earthquake-induced landslides triggered flooding in the low-lying areas surrounding the Jhelum River.	Jones (1885) and Neve (1885)
1893	Continuous rains over two days, Lawrence described as a "Great Calamity" in his book, followed by famine. 2225 houses damaged, 225,426 acres of crop land submerged, 329 cattle killed	Climo (1893), Ganga Ram (1928) and Mehran (2015)
1903	Entire Srinagar city was inundated and resembled a big lake. 7000 houses damaged, including 773 on Dal Lake. Besides that, a total of 83 villages were affected; among them, 26 have lost their whole Kharif crops.	Saleh et al., (2017) provides more sources.
1950	A major flood that resulted in loss of 100 human lives as well as damage to 5000 houses.	Ballesteros-Cánovas et al., (2020)
1957	Severe flooding across Kashmir Valley caused damages to crops and public property. It led to famine as a compounding effect. 600 villages were reportedly inundated with economic losses approximating 4 crore and 2 lakh Indian rupees.	Kelman et al., (2018) and Ballesteros-Cánovas et al., (2020)
1959	Incessant rains over four days triggered floods in the Jhelum floodplain and its low-lying areas.	Jammu and Kashmir state archives, Council Records, Ref 15842
1992	A major flood across Jammu and Kashmir and parts of Pakistan affected thousands of people and caused multiple casualties.	Kelman et al., (2018) and Bhat et al., (2019)
2014	Unprecedented rainfall in the first week of September as a result of an interaction between western disturbances and monsoonal currents caused flooding in the Jhelum floodplain. It caught people unawares and caused large-scale devastation and affected millions of people in India and Pakistan administered Kashmir. The causalities were low due to the heroic efforts of young people who braved floodwaters to remove trapped people to safety before the external help came. Many houses in low-lying areas collapsed due to the persistence of water for more than two weeks. Standing crops were also destroyed.	Venugopal and Yasir (2017)

Understanding that disaster risk management is both an individual and collective responsibility, the role of government and other relevant stakeholders is to guide everyone to make disaster reduction efforts, with a particular emphasis on community-centred disaster risk management activities in all phases of a disaster. The progress in disaster risk reduction policy and planning (especially post-2005) under the auspices of the United Nations is a new beginning, but this must be developed, adapted and implemented in letter and spirit to location-specific requirements at national and regional levels. Yet a lot is to be done to achieve a safe and resilient future for everyone.

References

Ahmad, S. T., Ahmed, R., Wani, G. F., Sharma, P., & Ahmed, P. (2021). Glacier changes in Sind basin (1990–2018) of North-western Himalayas using earth observation data. *Modeling Earth Systems and Environment*, 1–13.

Ahmed, R., Ahmad, S. T., Wani, G. F., Ahmed, P., Mir, A. A., & Singh, A. (2021). Analysis of landuse and landcover changes in Kashmir valley, India—A review. *Geo Journal, 1-13*, 4391. https://doi.org/10.1007/s10708-021-10465-8

Ahsan, S., Bhat, M. S., Alam, A., Ahmed, N., Farooq, H., & Ahmad, B. (2021). Assessment of trends in climatic extremes from observational data in the Kashmir basin. *NW Himalaya. Environmental Monitoring and Assessment, 193*(10), 1–18.

Attri, S. D., & Tyagi, A. (2010). *Climate profile of India*. Environment Monitoring and Research Center: India Meteorology Department.

Bajracharya, S. R., Mool, P. K., & Shrestha, B. R. (2008). *Global climate change and melting of Himalayan glaciers* (pp. 28–46). Melting glaciers and rising sea levels: Impacts and implications.

Ballesteros-Cánovas, J. A., Koul, T., Bashir, A., Del Pozo, J. M. B., Allen, S., Guillet, S., ... & Stoffel, M. (2020). Recent flood hazards in Kashmir put into context with millennium-long historical and tree-ring records. *Science of The Total Environment, 722*, 137875.

Bates, C. E. (1873). *A gazetteer of Kashmir*. Light & Life Publishers.

Bellew, H. W. (1875). *Kashmir and Kashghar: A Narrative of the Journey of the Embassy to Kashghar in 1873–74*. Trübner & Company.

Bhat, M. S., Ahmad, B., Alam, A., Farooq, H., & Ahmad, S. (2019). Flood hazard assessment of the Kashmir Valley using historical hydrology. *Journal of Flood Risk Management, 12*, e12521.

Carmin, J., Dodman, D., & Chu, E. (2013). Urban climate adaptation and leadership: From conceptual understanding to practical action.

Climo, W. H. (1893). The floods in Kashmir and Srinagar a new focus of endemic cholera. *British Medical Journal, 2*(1702), 396.

Coppola, D. P. (2015). Hazards. *Introduction to international disaster management*, 40.

CRED (Centre for Research for Epidemiology for Disaster) (2021). *Disaster Year in Review 2020: Global Trends and Perspectives*. https://cred.be/sites/default/files/CredCrunch62.pdf. Accessed 15 Sept 2021.

Digby, W. (1890). *Condemned unheard: The Government of India and HH the Maharaja of Kashmir: a Letter to the Rt.* Hon. Sir Ughtred Kay-Shuttleworth. Indian Political Agency.

EM-DAT. (2020). *Disaster profiles*. https://www.emdat.be/emdat_db/. Accessed 10 Sept 2021.

Fekete, A., & Sandholz, S. (2021). Here comes the flood, but not failure? Lessons to learn after the heavy rain and pluvial floods in Germany 2021. *Water, 13*(21), 3016.

IDNDR (International Decade for Natural Disaster Reduction). (1994). Yokohama strategy and plan of action for a safer world. Guidelines for natural disaster prevention, preparedness and mitigation. http://www.unisdr.org/we/inform/publications/8241. Accessed 20 Oct 2021.

Ince, J. (1888). *Ince's Kashmir handbook: A guide for visitors, re-written and much enlarged by Joshua Duke*. Thacker, Spink and Co.

Jammu and Kashmir state archives, Council Records, Ref 15842

Jones, E. J. (1885). Notes on the Kashmir Earthquake of 30th May 1885. *Records of the Geological Survey of India, 18*(3), 153–156.

Kamidohzono, S. G., Gomez, O. A., & Mine, Y. (2015). *Embracing human security: New directions of Japan's ODA for the 21st century*. JICA Research Institute Technical Report no. 94. http://jicari.jica.go.jp/publication/assets/JICA-RI_WP_No.94.pdf. Accessed 10 Oct 2021

Kelman, I., Field, J., Suri, K., & Bhat, G. M. (2018). Disaster diplomacy in Jammu and Kashmir. *International Journal of Disaster Risk Reduction, 31*, 1132–1140.

Khoihami, P.G.H. (1885). *Tarikh-i-Hassan*. Persian manuscript. Vols. I–II, folios 511 and 263. Srinagar, India: Research & Publication Division, Srinagar Library, University of Kashmir.

Konrad, C. P. (2003). *Effects of urban development on floods*.

Kulkarni, A., Patwardhan, S., Kumar, K. K., Ashok, K., & Krishnan, R. (2013). Projected climate change in the Hindu Kush–Himalayan region by using the high-resolution regional climate model PRECIS. Mountain Research and Development, 33(2), 142–151.

Lal, M. (2002). *Possible impacts of global climate change on water availability in India. Report to Global Environment and Energy in the 21st Century* (pp. 5–25). Indian Institute of Technology.

Lawrence, W. R. (1895). *The valley of Kashmir*. H. Frowde.

Lechat, M. F. (1990). *The international decade for natural disaster reduction: background and objectives*. Blackwell.

Masson-Delmotte, V., Zhai, P., Pirani, A., Connors, S. L., Péan, C., Berger, S., ... & Zhou, B. (2021). Climate change 2021: the physical science basis. *Contribution of working group I to the sixth assessment report of the intergovernmental panel on climate change, 2*.

McCarthy, J. J., Canziani, O. F., Leary, N. A., Dokken, D. J., & White, K. S. (Eds.). (2001). *Climate change 2001: impacts, adaptation, and vulnerability: contribution of Working Group II to the third assessment report of the Intergovernmental Panel on Climate Change (Vol. 2)*. Cambridge University Press.

Mehran, D. (2015). Food shortages in Kashmir: Response of society. *Journal of Central Asian Studies, 22*.

Mool, P. K., Maskey, P. R., Koirala, A., Joshi, S. P., Wu, L., Shrestha, A. B., et al. (2011). Glacial Lakes and Glacial Lake Outburst Floods in Nepal. *Icimod*, 1–109.

Neve, A. (1885). The late earthquake in Kashmir. *Lancet, 126*(3236), Article 455.

Rai, M. (2004). *Hindu rulers, Muslim subjects: Islam, rights, and the history of Kashmir*. Princeton University Press.

Ram, P. B. (1895). Report on the administration of the Jammu and Kashmir State for 1892–93. .

Ram, G. (1928). *Floods in Jammu and Kashmir. Jammu and Kashmir government flood management report*. Published by the State Government.

Ren, Y. Y., Ren, G. Y., Sun, X. B., Shrestha, A. B., You, Q. L., Zhan, Y. J., et al. (2017). Observed changes in surface air temperature and precipitation in the Hindu Kush Himalayan region over the last 100-plus years. *Advances in Climate Change Research, 8*(3), 148–156.

Romshoo, S. A., Bashir, J., & Rashid, I. (2020). Twenty-first century-end climate scenario of Jammu and Kashmir Himalaya, India, using ensemble climate models. *Climatic Change, 162*(3), 1473–1491.

Saleh, S. F., Rather, F. F., & Jabbar, M. J. (2017). Floods and mitigation techniques with reference to Kashmir. *International Journal of Engineering and Computer Science, 7*(4), 6359–6363.

Shi, P., Li, N., Ye, Q., Dong, W., Han, G., & Fang, W. (2010). Research on integrated disaster risk governance in the context of global environmental change. *International Journal of Disaster Risk Science, 1*(1), 17–23.

Sun, X. B., Ren, G. Y., Shrestha, A. B., Ren, Y. Y., You, Q. L., Zhan, Y. J., et al. (2017). Changes in extreme temperature events over the Hindu Kush Himalaya during 1961–2015. *Advances in Climate Change Research, 8*(3), 157–165.
UNISDR (United Nations International Strategy for Disaster Reduction). (2005). *Hyogo framework for action 2005–2015: Building the resilience of nations and communities to disasters.* www.unisdr.org/files/1037hyogoframeworkforactionenglish.pdf. Accessed on 14 Sept 2021.
UNISDR (United Nations International Strategy for Disaster Reduction). (2012a). *Milestones in the history of disaster risk reduction—yearly archives.* http://www.unisdr.org/who-we-are/history. Accessed on Sept 2021.
UNISDR (United Nations International Strategy for Disaster Reduction). (2012b). *Milestones in the history of disaster risk reduction—yearly archives.* http://www.unisdr.org/who-we-are/history. Accessed 03 Jan 2022.
UNISDR (United Nations International Strategy for Disaster Reduction). (2015). *Sendai framework for disaster risk reduction 2015–2030.* http://www.wcdrr.org/uploads/Sendai_Framework_for_Disaster_Risk_Reduction_2015-2030.pdf. Accessed on 10 Jan 2022.
Venugopal, R., & Yasir, S. (2017). The politics of natural disasters in protracted conflict: the 2014 flood in Kashmir. *Oxford Development Studies, 45*(4), 424–442.
Wani, G. F., Ahmed, R., Ahmad, S. T., Singh, A., Walia, A., Ahmed, P., ... & Mir, R. A. (2022). Local perspectives and motivations of people living in flood-prone areas of Srinagar city, India. *International Journal of Disaster Risk Reduction, 82*, 103354.
Yan, L., & Liu, X. (2014). Has climatic warming over the Tibetan Plateau paused or continued in recent years. *Earth Ocean and Atmospheric Science, 1*(1), 13–28.

Chapter 14
Modelling Potential Zones of Gangotri Glacier Using GIS and ML in the Wake of Physico-Climatic Factors

Zainab Khan, Mohd Mohsin, Uzma Ajmal, and Ateeque Ahmad

Introduction

Glaciers on Earth are important components of the climate system. Changes in glaciers are indicators of climate change (Forsberg et al., 2017), and they contribute to and sustain significant river systems on the face of the earth (Bolch et al., 2012; Verma et al., 2021). In total, 15–28% of run off is contributed by glaciers (Liljedahl et al., 2017), and thus their existence is viable for rivers and streams (Kong et al., 2019).

Himalayas are the largest adobe of ice and snow outside poles, and various mighty glaciers straddle in the Himalayan Valleys (Ramsankaran et al., 2021), covering an area of 33,000 km^2 (Bahuguna, 2003). These glaciers hold great ecological importance for the Indian sub-continent, as rivers draining the sub-continent acquire a considerable proportion of their recharge from them (Thayyen & Gergan, 2010; Jones et al., 2018). However, the past few decades have not been very favourable for glaciers due to climate change (Clark et al., 2002; Pörtner et al., 2022). It is claimed that colder areas are getting hot faster than the rest of the earth's average (Arndt and Schembri, 2015), rendering glaciers especially vulnerable. The Himalayan range recorded the second fastest warming on the earth after the poles, leaving glacial bodies highly susceptible to rapid meltdown (Banerjee & Shankar, 2013; Kargel et al., 2011). Several studies are strongly suggestive of glacier recession under rising temperatures in the Kumaun Himalayas (Bisht et al., 2018; Singh et al., 2018) and Garhwal Himalayas (Chaujar, 2009).

Z. Khan (✉) · U. Ajmal · A. Ahmad
Department of Geography, Faculty of Science, Aligarh Muslim University, Aligarh, India

M. Mohsin
Department of Civil engineering, Faculty of Engineering and Technology, Zakir Husain College of Engineering, Aligarh Muslim University, Aligarh, India

A. L. Singh et al. (eds.), *Climate Change, Vulnerabilities and Adaptation*,
https://doi.org/10.1007/978-3-031-49642-4_14

Gangotri is one such glacier resting in the Garhwal Himalayas with strong signs of retreat (Bhambri et al., 2012) and has been the focus of researchers due to its hydrological, fluvial and economic significance (Arora & Malhotra, 2023). Ambinakudige, 2010, applied optical remote sensing images to support the recession of Gangotri while Varugu & Rao, 2016, exploited the SAR dataset for assessing Gangotri. Shi et al. (2023) used dug deeper using sedimentary biological structure to delve into the ecology of these glaciers. The glacial variability in the context of space and time is also assessed (Joshi et al., 2020). In studies, sub-alpine tree extension was used to support the crunching Gangotri glacier over the past century. Most of these scholars attributed the declining glaciers to climate change. Scholars such as Mitra et al. (2009), Ambinakudige (2010), and Bhambri et al. (2012), also associate the retreat with warming temperatures; however, the role of physical characteristics of the catchment remained uncredited in the context of Gangotri glacier. Paxman et al. (2017) argue that underlying topography plays a key role in the dynamics of overlying ice mass in a mountain range. The influence of underlying physiography is believed to have an effect even on the ice sheets with gentler and uniform aspects and altitude (Gassen et al. 2015), and thus, zones of ablation and accumulation in a glacier are not the sole outcome of temperature, but other physiological characteristics (Yu et al., 2013). This sort of research work is limited to the context of Gangotri or Himalayan region for that matter. Singh et al. (2017) have attempted to associate glacial retreat with morphological zones; however, their impact on the glaciers was not studied. Local physiological construct of the glacier catchment is in situ factors. Climate change or global temperature rise are ex site factors influencing a glacier. Both in situ and ex situ factors and their complex interrelationship in association with climatic factors play a great part in glacial dynamics, and therefore, these causative factors must be considered in holistic glacial studies. Arguably, a catchment represents the fundamental unit of hydrological studies that is defined by underlying topography (Jarvis, 2012); therefore, it is rational to study glacial dynamics at the catchment level.

In this study, for the first time, an attempt is made to explain the complex interrelationships of physico-climatic factors using Random Forest Regression (RFR), their impact on Gangotri and tried to model the potential zone of ablation, equilibrium, and accumulation at catchment level. The aims to explain the impact of the topographic factors as well as the temperature on the glacial dynamics. The study also seeks to quantify the areal changes in the foresaid potential zones of the glacier by temporal comparison.

Study Area

Gangotri is an alpine glacier nestling in the central Himalayas, falling in the district of Uttarkashi, Uttarakhand (Fig. 14.1). Being one of the largest glacial deposits in the Himalayas (Naithani et al., 2001), Gangotri pertains massive hydrological and ecological significance. It is one of the major tributary glaciers of the river Ganges.

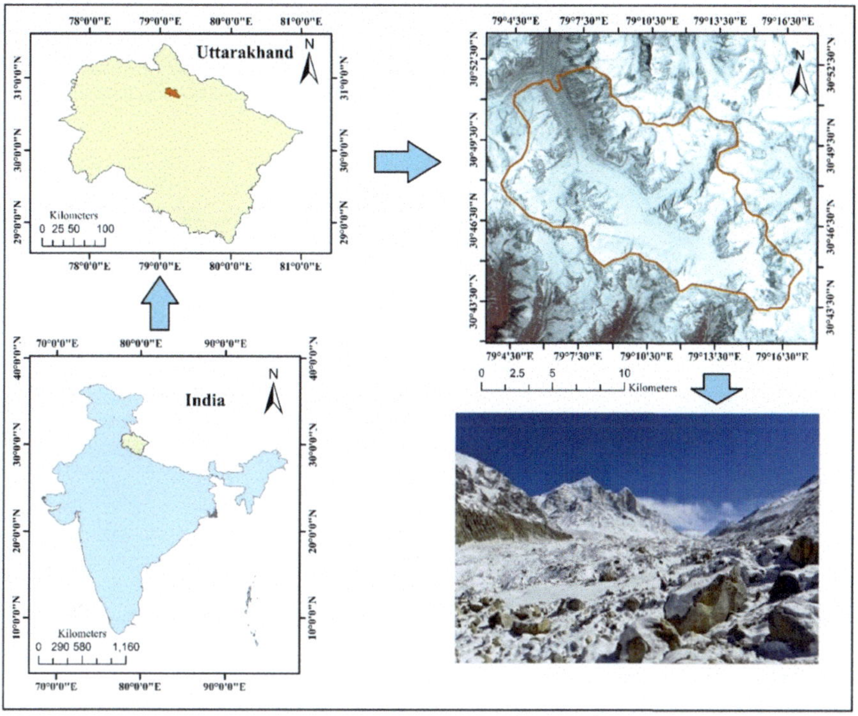

Fig. 14.1 Study area

Apart from that, Gangotri also has great mythological and religious importance in Hindu culture. The study area, i.e. catchment of Gangotri—the fundamental hydrological unit, is processed in a GIS environment. The processed glacial catchment covers 184 km^2. There are massive variations in the altitude due to uneven topography. The altitude varies from 4474 m at its lowest point to 7085 m at its highest point across the catchment. The terrain of the catchment is full of undulation. This undulating terrain is also responsible for changing aspects over short distances causing uneven reception of the Sun's energy flux. Being curved by glacial ice, Gangotri sits in a U-shaped valley with a variant slope that ranges from 0° in the central part to 82° along the ridges. Hypsometric curve (Fig. 14.2) suggests that roughly 50% area of the catchment lies above 6500 m and roughly 95% area is above 5000 m; therefore, it can be implied that almost all of the glacial catchment is above the snow-line (Ray, 2009).

A strong influence of aspect, slope and elevation can be visually observed in the maps. Aspect affects insolation and therefore determines temperature, slope determines movement under gravity and environmental lapse rate of temperature is controlled by elevation.

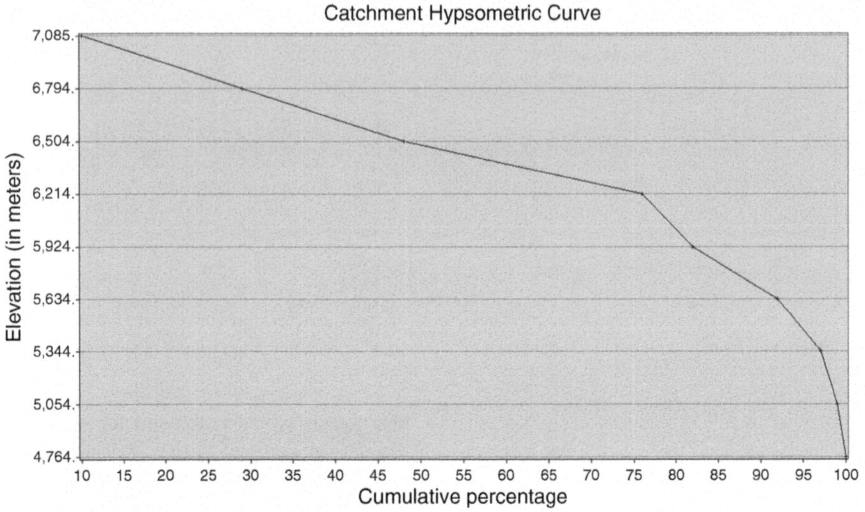

Fig. 14.2 Hypsometric curve of the catchment

Database and Methods

Database (Table 14.1)

Table 14.1 Data sources

Satellite	Sensor	Bands used	Cloud cover	Source	Maps generated	Methods
Landsat 5	TM	B2, B5, B6	0%	https://earthexplorer.usgs.gov/	NDSI	NDSI (B2-B5/B2 + B5)
					LST	Mono window algorithm (band 6)
Landsat 8	OLI and TIRS	B3, B6, B10, B11	0%	https://earthexplorer.usgs.gov/	NDSI	NDSI (B3-B6/B3 + B6)
					LST	Split-window algorithm (bands 10 and 11)
Terra	ASTER	ASTER DEM	N.A.	https://earthexplorer.usgs.gov/	Slope	Calculated in GIS environment
					Elevation	ASTER DEM
					Insolation	Estimated in GIS environment

Rationale of Selection

Elevation

There is an inverse relationship between altitude, i.e. the mean height from sea level and temperature (Aigang et al., 2009); therefore, it seems fair to acknowledge altitude as one of the most important factors affecting the distribution of snow in a region. Altitude is the lone responsible factor for the accumulation of snow-clad regions in low latitudes in the form of alpine glaciers (Braithwaite, 2008). Increasing altitude provides favourable climatic conditions, i.e. decreasing temperatures and increasing precipitation directly leading to the formation of glaciers (Křížek & Mida, 2013). It has been found that glaciers situated at higher altitudes show less susceptibility towards depletion, i.e. their rate of retreat is lower, while glaciers at lower altitudes are retreating at faster rates. A study based on the Chandra basin in Himalayan glaciers has found that out of 18% of water loss of the total basin from the year 1984–2012, about 67% of the loss was reported from smaller and lower altitude glaciers (Tawde et al., 2017). The data for the altitude of the Gangotri basin has been obtained from the DEM (digital elevation model) and ASTER (Advanced Spaceborne Thermal Emission and Reflection Radiometer) project by USA and Japan in 2009. A digital elevation model (DEM) is a digital raster representation of ground surface topography or terrain. Each raster cell (or pixel) has a value corresponding to its altitude above sea level. ASTER's GDEM was created by stereo correlation of more than 1.2 million individual ASTER stereo scenes contained in the archive. The GDEM had 1 arc-second latitude and longitude postings (~30 m) and a vertical accuracy of approximately 10 m (Abrams et al., 2020).

Except obtaining, altitude raster, DEM dataset is used in terrain analysis and for extracting other terrain parameters such as slope and aspect, and modelling water flow and catchment modelling.

Slope

The *slope* or *gradient* of a line is a number that describes both the *direction* and the *steepness* of the line. The direction of a slope is significant as it offers information about the duration of incoming solar radiation. In the northern hemisphere, south-facing slopes receive solar radiation for longer duration in comparison to north-facing slopes. Therefore, the south-facing slopes are more prone to melting due to longer exposure to solar radiation (Wegmann et al., 1998). The steepness of slope determines the movement of glaciers under gravity (Evans, 2018). Slope is an important factor controlling the speed of glacier movement, together with their temperature, and amount of meltwater at the base of the glacier. Snow cover is expected to be low on steeper slopes. Therefore, slope of a glacier is expected to be an important factor behind glacier retreat, as the rate of retreat is directly controlled by the slope of the glacier (Falaschi et al., 2017). It has been estimated that in the glaciers of the same climatic zones, different rates of retreat/advancement of glaciers can be explained by the mean slope and size of the glacier. In this study, slope in the glacier catchment area has been extracted from the ASTER DEM using GIS tools and techniques.

Insolation

Insolation, short for *incoming solar radiation*, is the incident energy of the Sun on any celestial body in a unit of area for a given time. One unit is Watt-hour/meter-square (WH/m^2). In the context of Earth, not all the Sun's energy that strikes the earth actually reaches the surface. 30% of it gets reflected back by the atmosphere into space. Insolation is responsible for maintaining the temperature, and more insolation means higher temperature (Lund, 1968). Therefore, insolation is another factor affecting glacier cover. However, insolation is controlled by the topography of the surface and the interaction between slope, aspect and solar geometry, which decide the angle of incoming solar radiation. Insolation is also affected by any nearby higher landscape, which is responsible for shading the glaciers. The diffused radiation received from the sky is also affected by the nearby landscape, as it controls the proportion of the sky visible from the glacier (Bertoldi et al., 2010). Since Himalayan terrain is very undulating and has steep slopes (Kumar et al., 2021), the incoming solar radiation is greatly affected by the location of glaciers with respect to slope and aspect. The incoming solar radiation in this study has been calculated in a GIS environment. The tool calculates incoming solar radiation by using methods from the hemispherical view shed algorithm developed by Hetrick et al., 1993; Rich et al., 1995; Fu & Rich, 1999, 2002). The first step of calculation of incoming solar radiation involves the calculation of global radiation, i.e. total radiation of a particular area. Furthermore, the calculation of direct solar radiation is repeated for every feature or location of the topographic surface to generate an insolation map of the region.

$$\mathrm{Global}_{\mathrm{tot}} = \mathrm{Dirt}_{\mathrm{tot}} + \mathrm{Dif}_{\mathrm{tot}} \tag{14.1}$$

where

$\mathrm{Global}_{\mathrm{tot}}$ = total global radiation
$\mathrm{Dir}_{\mathrm{tot}}$ = total direct radiation of all sun map sectors
$\mathrm{Dif}_{\mathrm{tot}}$ = diffuse radiation of all sky map sectors

Total direct insolation ($\mathrm{Dir}_{\mathrm{tot}}$) for a given point is the sum total of the direct insolation ($\mathrm{Dir}_{\theta,\alpha}$) from all sun map sectors, which is calculated using the following equation:

$$\mathrm{Dir}_{\mathrm{tot}} = \Sigma\, \mathrm{Dir}_{\theta,\alpha} \tag{14.2}$$

The direct insolation from the sun map sector, i.e. ($\mathrm{Dir}_{\theta,\alpha}$) with a zenith angle of θ and azimuth angle of α is calculated using the following equation:

$$\mathrm{Dir}_{\theta,\alpha} = S_{\mathrm{Const}}{}^{*}\beta^{m(m)*}\mathrm{SunDur}_{\theta,\alpha}{}^{*}\mathrm{SunGap}_{\theta,\alpha}{}^{*}\,cos\left(\mathrm{AngIn}_{\theta,\alpha}\right) \tag{14.3}$$

where

$S_{\mathrm{Const}} =$ 1367 W/m^2
$\beta =$ atmospheric transmissivity
$m(\theta) =$ relative optical path length
$\mathrm{SunDur}_{\theta,\alpha} =$ time duration represented by the sky sector
$\mathrm{SunGap}_{\theta,\alpha} =$ gap fraction for the sun map sector
$\mathrm{AngIn}_{\theta,\alpha} =$ angle of incidence between the centroid of the sky sector and the normal

Land Surface Temperature

Temperature is the most important element of climate. Land surface temperature (LST) is a robust remote sensing-based method for the estimation of the temperature of the land surface. The impact of LST over glaciers is a well-recognized fact. Surface temperature is one of the most important parameters for estimating the effect of climatic change on glaciers. LST has a vast application in the study of glaciers. Brabyn & Stichbury, 2020, used LST to estimate the rate of thaw of ice sheet, while Mortimer et al., 2016, used MODIS LST data set to literally procure the temperature of the glacier surface. Wu et al., 2015, and Baral et al., 2020, also adopted LST to obtain glacier surface temperature. In this study, an attempt has been made to estimate surface temperatures from Landsat 5 ETM and Landsat 8 TIRS for Gangotri Glacier. Mono window algorithm using band 6 has been used for ETM, while the split window algorithm using bands 10 and 11 has been adopted for TIRS data. Mono Window algorithm has been used for LST calculation. The processes involved in the retrieval of LST are as follows:

Calculation of TOA (Top of Atmospheric) spectral radiance.

$$\mathrm{TOA}(L) = M_L{}^{*}(Q_{\mathrm{cal}} + A_L) \tag{14.4}$$

where

$M_{\mathrm{L}} =$ band-specific multiplicative rescaling factor from the metadata of downloaded imagery (RADIANCE_MULT_BAND_**x**, where x is the band number)
$Q_{\mathrm{cal}} =$ corresponds to band 10
$A_{\mathrm{L}} =$ band-specific additive rescaling factor from the metadata of downloaded imagery (RADIANCE_ADD_BAND_x, where x is the band number)

TOA to Brightness Temperature conversion utilized the following formula:

$$BT = (K_2/(ln(K_1/L) + 1)) - 273.15) \quad (14.5)$$

where

K_1 = band-specific thermal conversion constant from the metadata of downloaded imagery (K1_CONSTANT_BAND_**x**, where x is the thermal band number).
K_2 = band-specific thermal conversion constant from the metadata of downloaded imagery (K2_CONSTANT_BAND_**x**, where x is the thermal band number).

Calculation of NDVI utilized the following formula:

$$\mathrm{NDVI} = (\mathrm{NIR–RED})/(\mathrm{NIR} + \mathrm{RED}) \quad (14.6)$$

The formula for Landsat 5:

$$\mathrm{NDVI} = (\mathrm{Band\ 4–Band\ 3})/(\mathrm{Band\ 4} + \mathrm{Band3}) \quad (14.7)$$

where Band 4 is a near-infrared band and Band 3 is a visible red band in Landsat 5.

The formula for Landsat 8:

$$\mathrm{NDVI} = (\mathrm{Band\ 5–Band\ 4})/(\mathrm{Band\ 5} + \mathrm{Band4}) \quad (14.8)$$

where Band 5 is a near-infrared band and Band 5 is a visible red band in Landsat 8.

Calculation of the NDVI is important because, subsequently, the proportion of vegetation (P_v), which is highly related to the NDVI, and emissivity (ε), which is related to the P_v, must be calculated.

Calculation of the proportion of vegetation P_v utilized the following formula:

$$P_v = \mathrm{Square}((\mathrm{NDVI–NDVI_{min}})/(\mathrm{NDVI_{max}–NDVI_{min}})) \quad (14.9)$$

Calculation of emissivity ε:

$$\varepsilon = 0.004^* P_v + 0.986 \quad (14.10)$$

It is essential to calculate the land surface emissivity LSE(ε) in order to estimate LST, since the LSE is a proportionality factor that scales blackbody radiance (Planck's law) to predict emitted radiance, and it is the efficiency of transmitting thermal energy across the surface into the atmosphere (Jiménez-Muñoz & Sobrino, 2006). The land surface emissivity LSE (ε) is calculated as proposed by Sobrino and Jiménez-Muñoz et al., 2014.

Calculation of LST:

$$\mathrm{LST} = BT/(1 + \lambda^{*}(BT/\rho) * Ln(e)$$

where

BT = brightness (at satellite temperature)
λ = Wavelength of emitted radiance (0.00115)
ρ = 14,380 (Constant)
e = Emissivity

In order to deal with seasonal variation in temperature, taking mean annual temperature is a safe method. Therefore, in this study, LST is calculated on a quarterly basis. The member LST rasters for each reference year, thus, represent the mean LST value of all four seasons, i.e. summer, spring, autumn and winter.

Normalized Difference Snow Index

Normalized Difference Snow Index (NDSI) developed by Hall et al., 1995, is a remote-sensing method to map snow-cover area. It is a widely accepted method to study glacier dynamics and glacier expanse (Khan, 2019; Kulkarni et al., 2002; Yao et al., 2010). Here, one should acknowledge the optical properties and thermodynamics of ice and snow, having an albedo of 0.5–0.7, respectively, meaning that these surfaces reflect 50–70% of insolation (NSIDC, n.d.). Higher intrinsic albedo of snow and ice results in more radiative heat loss (Demenocal & Rind, 1993). While studying glaciers, Rose et al., 2017, explained the albedo-feedback of the snow, which accounted responsible for affecting the zonal energy budget. More snow-covered glacier catchment means a greater amount of radiative heat loss. Therefore, it is justified to assess the snow cover by the means of the NDSI method for an integrated catchment-level study of Gangotri. NDSI is a measure of the relative magnitude of the reflectance difference between visible (green) and shortwave infrared (SWIR). It controls the variance of two bands (one in the short-wave infrared and another one in the visible parts of the spectrum), which is suitable for snow mapping. The snow absorbs most of the shortwave radiance from the sun, but the cloud does not. Thus, the NDSI can effectively differentiate snow and cloud and, therefore, can be used, subsidiarity, in glacier monitoring.

$$\mathrm{NDSI} = (\mathrm{GREEN{-}SWIR})/(\mathrm{GREEN} + \mathrm{SWIR}) \tag{14.11}$$

Formula for Landsat 5:

$$\mathrm{NDSI} = (\mathrm{Band2{-}BAND5})/(\mathrm{BAND2} + \mathrm{BAND5}) \tag{14.12}$$

where Band 2 is a visible green band and Band 5 is short wave infrared band in Landsat 5.

Formula for Landsat 8:

$$\text{NDSI} = (\text{Band3–BAND6})/(\text{BAND3} + \text{BAND6}) \quad (14.13)$$

where Band 3 is a visible green band and Band 6 is short-wave infrared band in Landsat 8.

In order to capture the correct amount of snow in the catchment, a quarterly mean NDIS is calculated for each reference year. The member NDSI raster, thus, represents the mean NDSI value of all four seasons, i.e. summer, spring, autumn and winter.

Methods

Principle Component Analysis

Principle component analysis (PCA) is a mathematical tool based on linear algebra that is used to reduce the dimensionality of a dataset. The dimensionality reduction is done by reduction of directions, i.e. principal component (PC) (Ringnér, 2008). In PCA, each PC is identified as a linear combination of variables, whereas the first principal component presents maximum variation (Wold et al., 1987). PCA can effectively be employed to procure the variables to analyse a phenomenon at an optimal level (Sarkar & Chouhan, 2021). In the process of data analysis, PCA is adopted to better discriminate the data (Adler & Golany, 2002) and to select the variables that are not equally important in explaining the given phenomena (Das et al., 2021). Lencioni et al., 2007, used PCA for ecological studies of glacier. Walsh & Butler, 1997, used PCA weights to monitor glacier debris flow for morphometric analysis using slope, aspect, sun angle and elevation.

Application of PCA: In this study, we require weights for the predictor variables for running fuzzy product algorithms over the rasters representing predictor variables. Eigenvalues of PCs are assigned as weights for predictor rasters in fuzzy product. For that purpose, it is most suitable to run PCA over all the pixels of predictor rasters to obtain eigenvalues that represent the whole raster rather than a sample of its pixels. To calculate eigenvalues of the predictor rasters i.e., elevation, slope, Insolation, LST and NDSI, a suitable GIS environment is utilized.

Random Forest Regression (RFR)

Within the last few decades, various predictive techniques have been used in civil and environmental engineering applications (Gislason et al., 2006; Pal et al. 2013; Were et al. 2015). In this research, we refer to RFR, which is an extension of a technique developed by Breiman, 2001, called random forest (RF) that has outstanding performance with regard to predicting error. Having tree-structured predictors, injected with randomness, makes it an exceptionally robust predictor. It is a supervised machine learning method that uses ensemble learning (Bakshi, 2020) for regression by boosting and bagging and generates significantly unbiased estimations of generalized errors (Chakure, 2019). The decision trees randomly create training data and test data. The test data used to test the fitting of the model and the error is usually called 'out of bag error' or OOB. OOB error is expressed as:

$$\mathrm{MSE_{OOB}} = n^{-1} \sum_{1}^{n} (yi - \widehat{y}OOBi)^2 \tag{14.14}$$

where' $\widehat{y}$OOBi is the mean of OOB predictions for ith observation.

Mean square residual (MSE) is another method of cross-validation. The lower the value, the better the fit of data. However, a value below 0.05 is safe for the validity of the results. MSR is mathematically expressed as:

$$y - \widehat{y} = \widehat{\epsilon},^2 \tag{14.15}$$

where y is observed value and $\widehat{y}$ and percent of variance explained are computed as:

$$1 - \frac{\mathrm{MSEOOB}^2}{\sigma_y^2} \tag{14.16}$$

where $\sigma 2_y$ is computed as a divisor with n (not as n-1).

RFR has been very popular in various scientific investigations of numerical relationships. It has been applied to remotely sensed data as well (Zhou et al., 2016), while Wangchuk & Bolch, 2020, applied this method to sentinel datasets for glacial mapping. RFR method has also been used in automatic glacier rock detection by Brenning, 2009. RF is a binary stepwise regressor that can explain mutual non-linear relationship (Brenning, 2009).

Application of RFR: In this study, the rationale to utilize RFR is to delve into the complexity of various factors that affect glacial dynamics. For modelling the potential zones of accumulation, equilibrium or ablation, it is of utmost importance that that we perform RFR that give Gini-impurity for all predictor variables and regress the values with explained variation. Thus, RFR is applied to predictor variables, i.e. elevation, slope, Insolation, LST and NDSI for validating their adoption for modelling the zones of glacial catchment. Pixel values of 251 random pixels (less than 10% of population according to central limit theorem) of the predictor variables are extracted using random sampling without replacement for the ML. RFR is performed using *R* version 3.6. cforest package is utilized for performing RFR in RStudio. Before running the algorithm, multicollinearity is visually checked for the reference years for all the variables. To train the data, 70% of the sample are selected, and ntree is tuned to 170. ntree 170 gives minimal test error (Figs. 14.5 and 14.6) and optimizes the accuracy for present modelling. For validation of results of RF MSE is used.

Fuzzy Overlay

Fuzzy overlay is based on Zadeh's (1972) fuzzy theory (Hsieh & Chen, 1999) that allows the analysis of a phenomenon belonging to multiple sets. It explains the relationship between multiple members of the set. Unlike Boolean, it transforms the data onto a scale of 0 to 1 using the theory of partial truth (Kaur & Mahajan, 2015). 0 is given to the data points that are absolute non-members and 1 is given to the data points that are absolute members; however, there are various methods to perform the computation, ranging from fuzzy Sum, fuzzy And, fuzzy Or, fuzzy Gaussian and fuzzy product.

Application of Fuzzy Product: We, in this study, have utilized fuzzy products that multiply each of the values for all member rasters (DEM, slope, insolation, LST, NDSI) for each cell in a GIS environment. The eigenvalues (Tables 14.2 and 14.3) calculated through PCA are supplied as weights for fuzzy products for each corresponding year. The returned raster is then classified into accumulation, equilibrium and ablation zones. The threshold for classification of the returned easter for the zone of accumulation is 0–0.2; for the zone of equilibrium, the range is 0.2–0.36, and for zone of ablation, it is 0.36–0.99.

In this study, the reference period denotes a 20-year difference. Therefore, we argue that it is a short period of time for orogenic elevation increase or associated slope change; however, small avalanches have not been taken into account. We also assume that the insolation has remained constant over the 20 years. Thus, rasters of DEM, slope and insolation are kept uniform for both years. In each operation, NDSI and LST rasters are supplied accordingly.

Table 14.2 Accumulative Eigen values, 1995

1995 – Percent and accumulative eigenvalues				
PC Layer		Eigen value	Percent of Eigen values	Accumulative of Eigen values
Insolation	1	0.04047	49.5870	49.5870
LST	2	0.02231	27.3319	76.9189
Slope	3	0.01151	14.1070	91.0260
DEM	4	0.00540	6.6178	97.6437
NDSI	5	0.00192	2.3563	100.0000

Table 14.3 Accumulative Eigen values, 2015

2015 – Percent and accumulative eigenvalues				
PC Layer		Eigen value	Percent of Eigen values	Accumulative of Eigen values
Insolation	1	0.04143	48.3052	48.3052
LST	2	0.02315	26.9871	75.2923
Slope	3	0.01275	14.8612	90.1535
DEM	4	0.00476	5.5554	95.7089
NDSI	5	0.00368	4.2911	100.0000

Results

PCA

All causative factors are important in the modelling of the glacial zones of Gangotri, with varying degrees of importance. The potential zones of the ablation loosely coincide with LST variation across the catchment. However, the eigenvalues of PCA are highest for Insolation followed by LST, i.e. 50% and 27%, respectively. Slope and DEM account for only 14% and 7% of eigenvalues for the year 1995. For the year 2015, the order of variables remains constant, but percentage eigenvalue contribution changes where insolation contributes 48% followed by LST with 14%. Slope and DEM count 14% and 7% of eigenvalues, respectively. NDSI eigenvalues increased from 2% to 4% approximately from 1995 to 2015. All these values are case specific, and we propound no generalization.

Collinearity in Data

The visual assessment of the scatter plots for the reference years represents peculiar associations in the variables (Figs. 14.3 and 14.4). There is a strong negative correlation between NDSI and LST. The respective figures represent a strong positive correlation in slope and elevetion of the catchment area. The scatter plot is also suggestive of a weak negative association between LST. The association of other variables is not explainable through the scatter plots.

RFR

The RFR for 1995 regresses the values of predictor variables at an MSE (Eq. 14.15) of 0.019 and the variance explained is 41.16%. The values of the mean of MSE remain 0.020 and the variance explained 41.4% for 2015. Error plots for RF through minimal error at 170 ntree for both reference years, i.e. 1995 and 2015 (Figs. 14.5 and 14.6).

Fuzzy Product

The member rasters (*map a, map b, map c, map d, map e, map f, map g*) are together (Fig. 14.7) supplied to the GIS environment for running fuzzy products with corresponding weights. The returned rasters produce interesting results. The result demonstrates a vast increase in the potential ablation zone of the glacial catchment

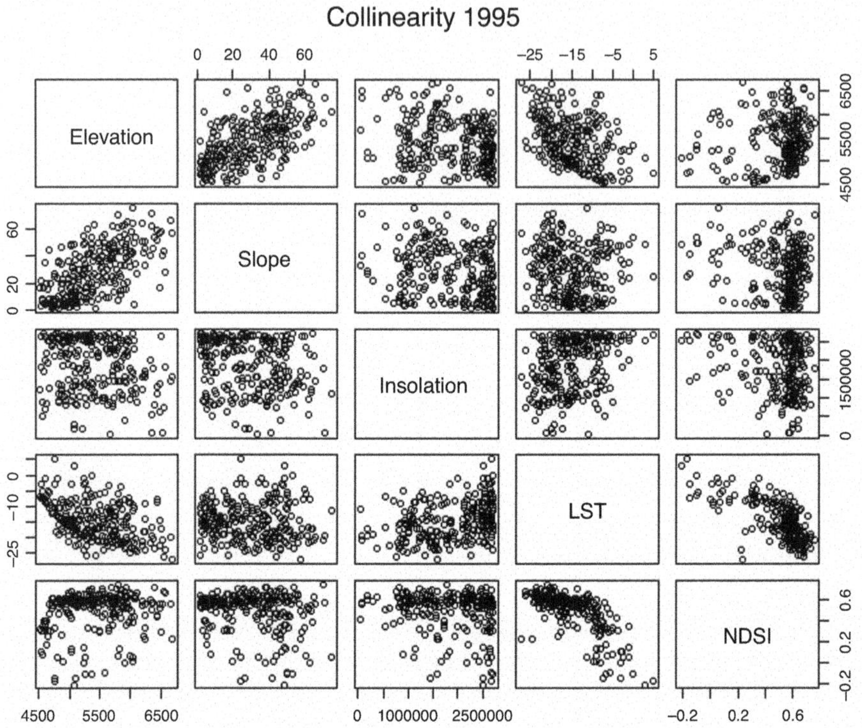

Fig. 14.3 Collinearity of data, 1995

over the period of 1995 and 2015 (map h and map j). The increase in the potential ablation zone is more prominent on the south-facing slope, whereas the potential equilibrium zone and potential accumulation zone seem to have depleted much more in lower elevation areas. The higher latitudes and north-facing slopes represent minimal changes.

The study propounds a vast increase in the areal extent of the potential ablation zone over the period of 1995 and 2015, expanding over threefold of what it used to be in 1995. The areal increase in the potential ablation zone has taken place at the expense of the potential accumulation zone and potential equilibrium zones (Fig. 14.8).

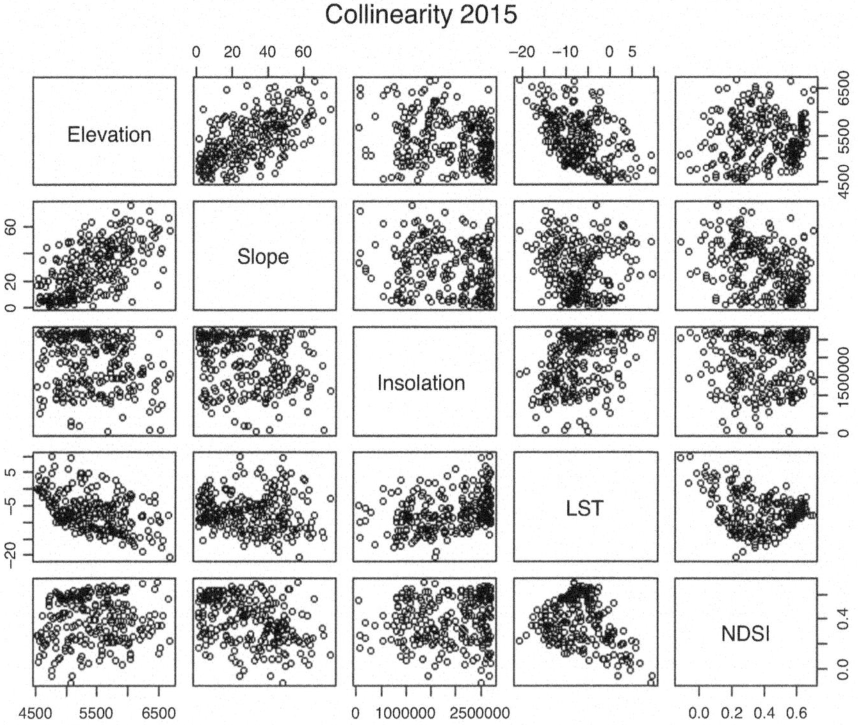

Fig. 14.4 Collinearity of data, 2015

Fig. 14.5 Error Plot, 1995

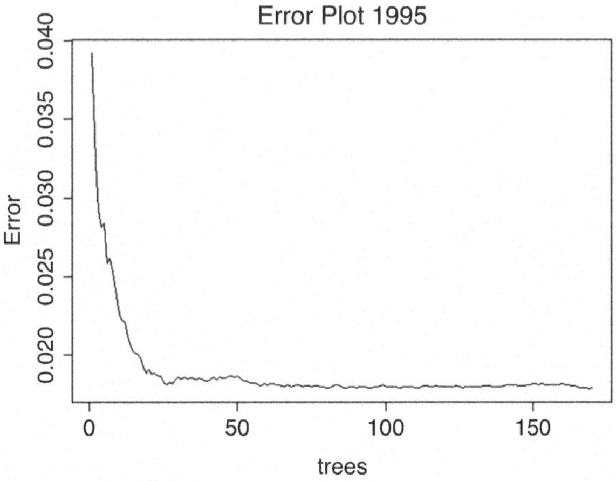

Fig. 14.6 Error Plot, 2015

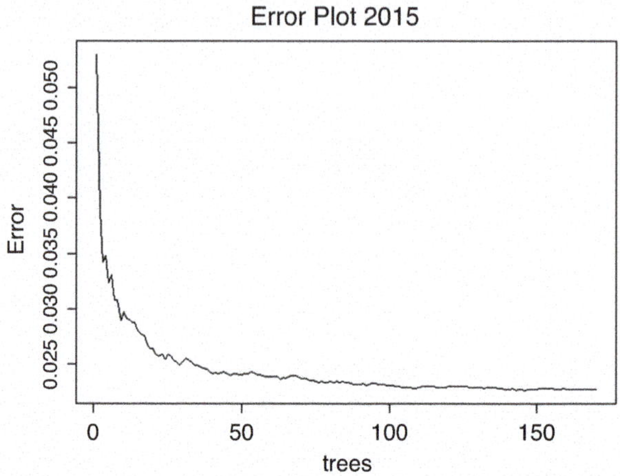

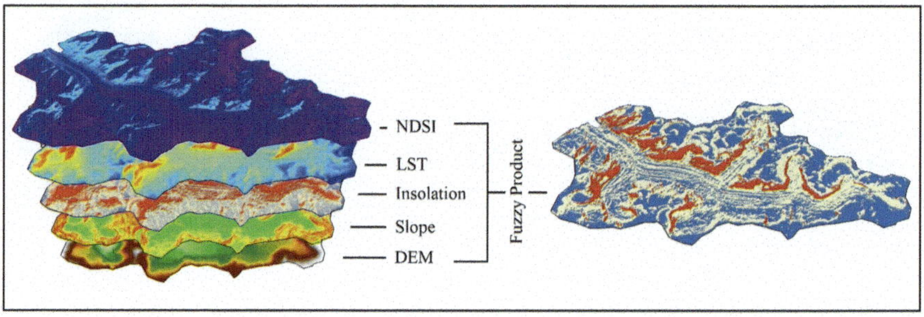

Fig. 14.7 Input raster and resultant raster

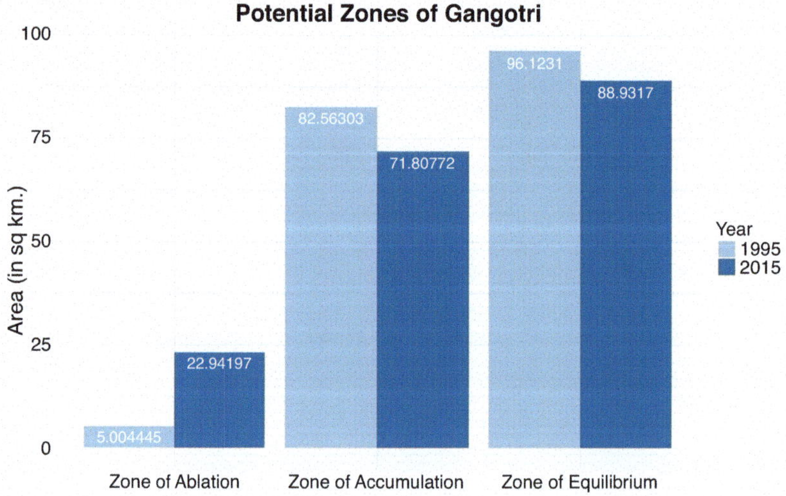

Fig. 14.8 Area under different potential zones of Gangotri

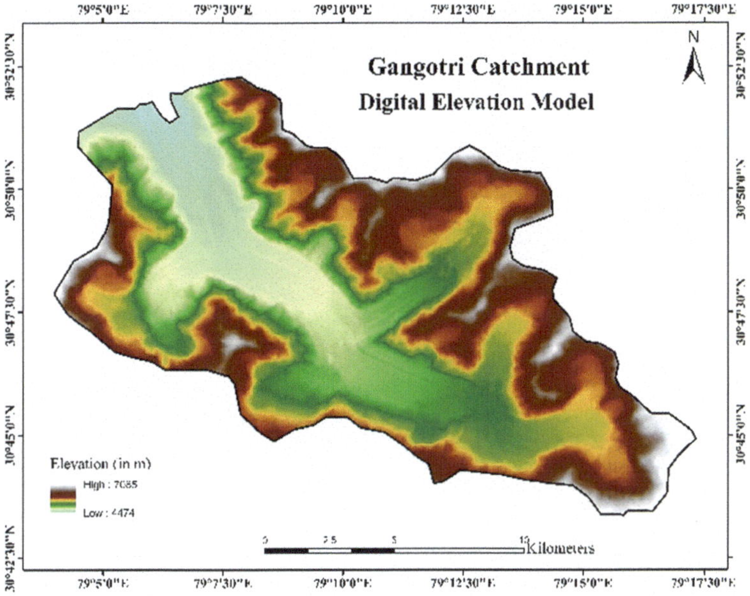

Map a: Digital elevation model

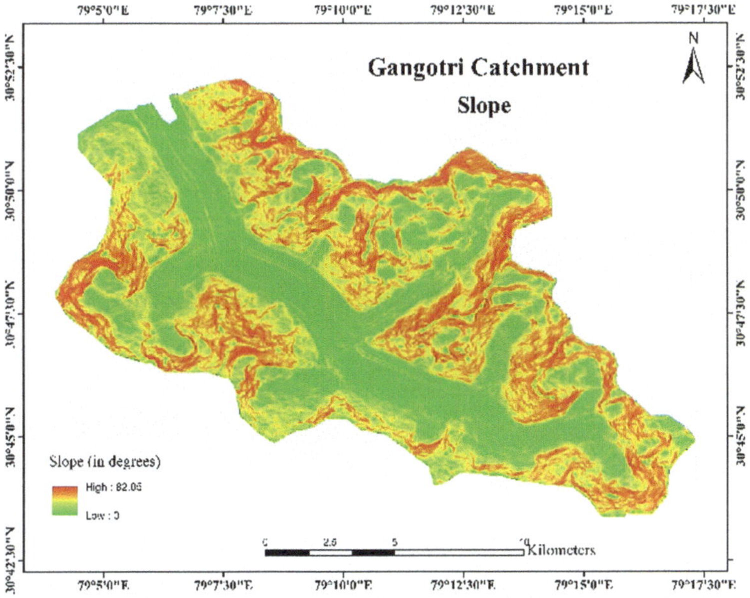

Map b: Slope

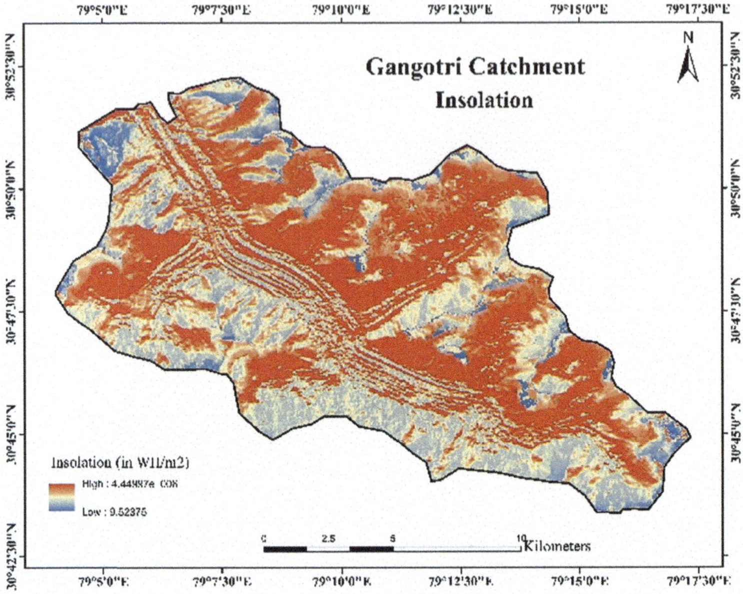

Map c: Insolation

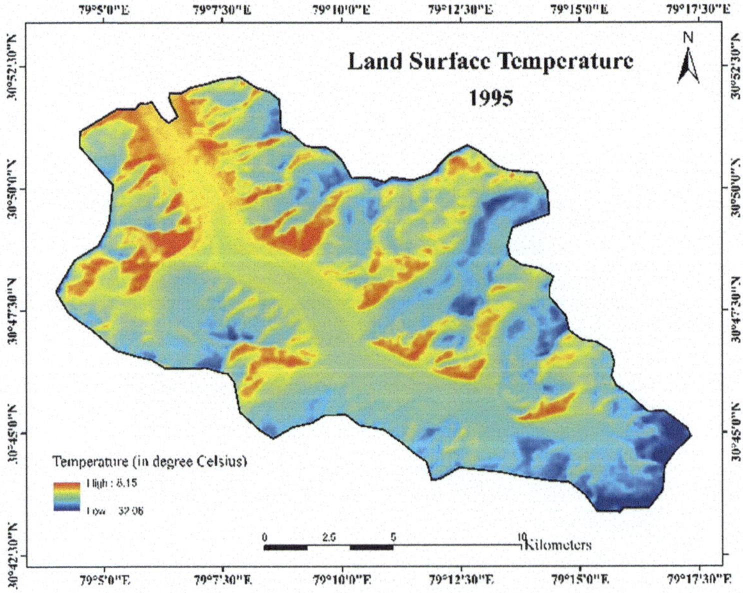

Map d: Land surface temperature,1995

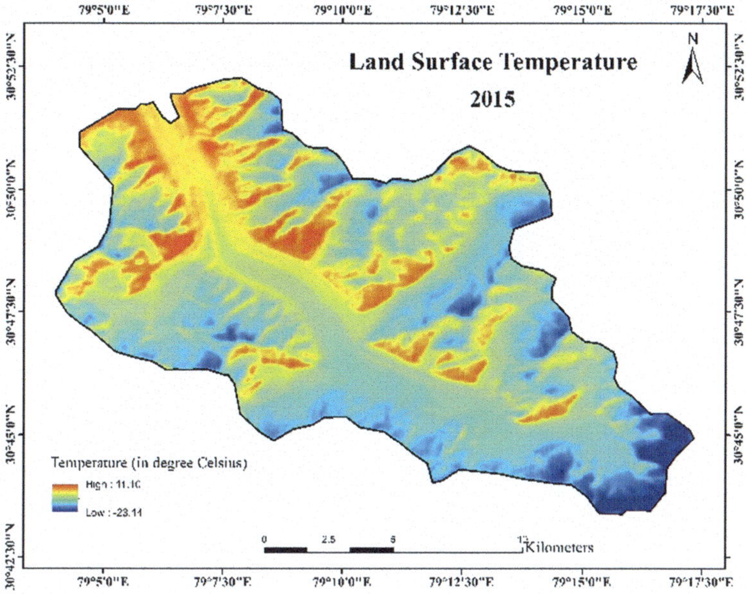

Map e: Land surface temperature, 2015

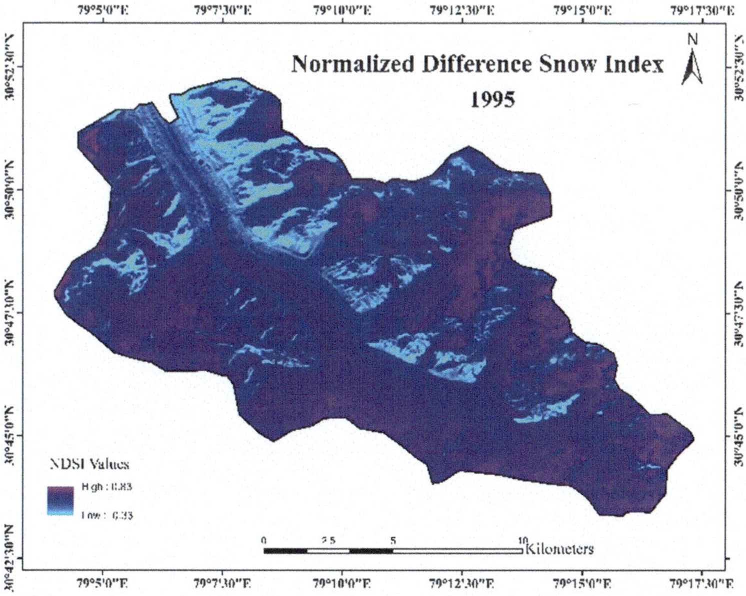

Map f: NDSI, 1995

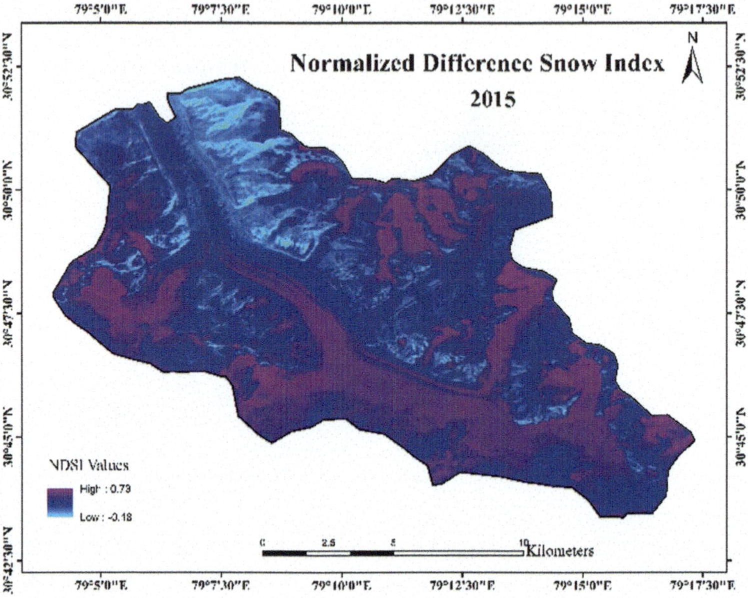

Map g: NDSI, 2015

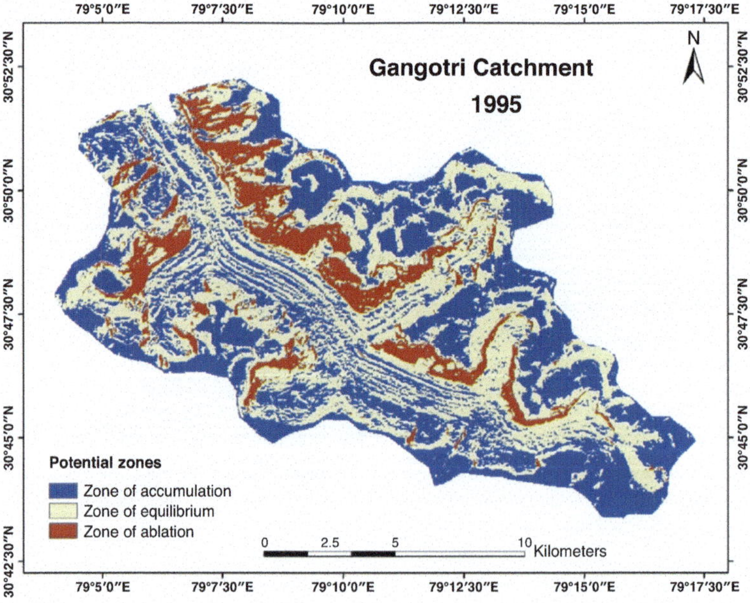

map h: Potential zones of gracial dynamics, 1995

Map h: Potential zones of gracial dynamics, 1995

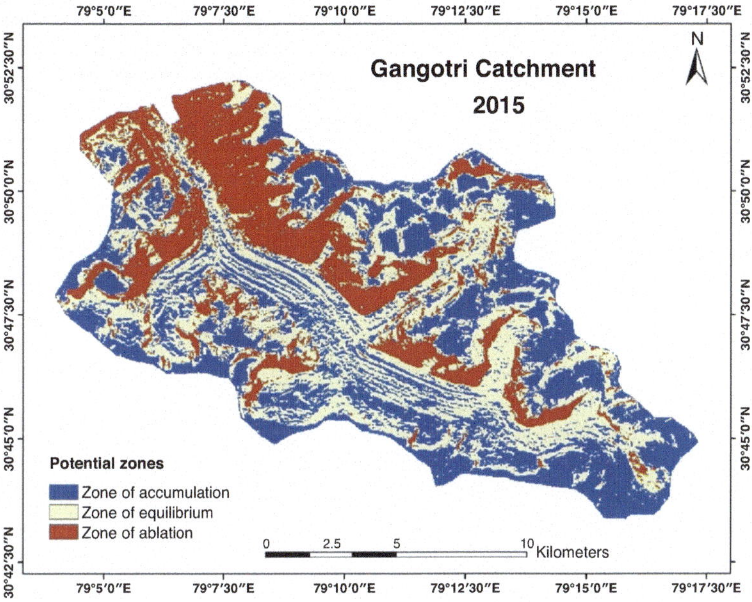

map i: Potential zones of gracial dynamics, 2015

Map i: Potential zones of gracial dynamics, 2015

Discussion and Conclusion

Gangotri glacier is receding in space over time, and a large glacial mass has been claimed by rising temperatures. The accumulation zone is retreating, and the ablation zone is increasing in the area (Haritashya et al., 2006). It appears that an increase in the ablation zone is very crucial for the sustenance of a glacier (Oerlemans, 1991) and a very strong negative sign for the health of the glacier (Agrawal et al., 2018). This study confirms the increase in the potential ablation zone of Gangotri from 1995 and 2015 at the expense of potential zones of equilibrium and accumulation. However, it is noteworthy that the zoning glacier is based on the potential outcomes of member rasters, which does not represent the true ablation, equilibrium or accumulation zone of Gangotri.

Fuzzy overlay operations have been used for demarcating hierarchical zones of complex geographical phenomena. Through GIS, input of causative rasters can be re-scaled and integrated generating a single raster that can be reclassified in multiple zones. Unlike weighted overlay, this method defines possibility rather than probability, and therefore, the membership of a causative raster is immune to assigned weights (Ahmed et al., 2014). Scholars have gone to great lengths to propound the significance of these methods for mapping suitable zones or susceptible zones for a specific phenomenon (Dimri et al., 2007; Kayastha et al., 2013; Kumar et al., 2013).

For instance, the fuzzy AHP method was used by Kumar et al., 2021, to demarcate potential zones of groundwater. The application of fuzzy overlay combined with dimensionality reduction methods is limited for glacial studies. Most of the previous GIS-based inventory research focuses on the glacial retreat in Gangotri (Singh et al., 2017; Ghosh 2017; Sparavigna, 2017) where due importance is not given to the zonal dynamics. The role of topographical causative factors is not modelled at large by scholars like Ding (2010), Rai et al. (2017). In this study, we combine the fuzzy product method with PCA to demarcate the potential zones of glacial mass balance along with the statistically significant result of RFR. We have to acknowledge that the modelled potential zones of glaciers are estimations and have limited capacity to represent the true processes and mass balance of the glaciers.

The objective of this chapter is to model and analyse the dynamics of Gangotri in the wake of underlying physiological factors in association with climatic factor, i.e. temperature. The study is one of its kind where for the first time role topography is modelled for glacial zonation of Gangotri catchment using the fuzzy product overly method, and the mutual relationship of various predictive variables is assessed with ML. Glacial zones are essential sub-units of a glacier, and each one is important to understand the glacial mass balance. This study puts forth the spatial extent of these zones for Gangotri and compares it with two time periods and, thus, highlights the gravity of climate change by areal quantification of the potential for glacial mass loss.

This study concludes that the Gangotri glacier is experiencing climate-imposed vulnerability and is prone to further loss of ice mass. The areal increase in the potential zone of ablation in the catchment poses higher risks of glacial losses against reduced possibility of replenishment from fresh snow at the catchment level. Being an alpine glacier, the dynamics are complex in association with topographical heterogeneity. Glacial zones coincide with slope, and physiology seems to have a key role in glacial dynamics and zonation. Hence, the study strongly suggests the inclusion of physiological factors in holistic glacial studies. Gangotri crucially contributes to India's fluvial system and, therefore, its thorough analyses are of key importance, and quantification of risks and vulnerabilities of the glacier both in situ and ex situ causative factors are worth delving.

Fuzzy product combined with PCA eigenvalues produces reasonable results for the potential glacier zone modelling and, hence, can be safely used for glacial studies in the future. Regression methods are useful to explain the mutual relationship of the various predictive variables. This study is a stand-alone piece of research that opens up a new arena for alpine glacial studies.

References

Abrams, M., Crippen, R., & Fujisada, H. (2020). ASTER global digital elevation model (GDEM) and ASTER global water body dataset (ASTWBD). *Remote Sensing, 12*(7), 1156.

Adler, N., & Golany, B. (2002). Including principal component weights to improve discrimination in data envelopment analysis. *Journal of the Operational Research Society, 53*(9), 985–991.

Agrawal, A., Thayyen, R. J., & Dimri, A. P. (2018). Mass-balance modelling of Gangotri glacier. *Geological Society, London, Special Publications, 462*(1), 99–117.

Ahmed, M. F., Rogers, J. D., & Ismail, E. H. (2014). A regional level preliminary landslide susceptibility study of the upper Indus river basin. *European Journal of Remote Sensing, 47*(1), 343–373.

Aigang, L., Tianming, W., Shichang, K., & Deqian, P. (2009). On the relationship between latitude and altitude temperature effects. In *2009 International Conference on Environmental Science and Information Application Technology* (Vol. 2, pp. 55–58). IEEE.

Ambinakudige, S. (2010). A study of the Gangotri glacier retreat in the Himalayas using Landsat satellite images. *International Journal of Geoinformatics, 6*(3), 7.

Arndt, E., & Schembri, P. J. (2015). Common traits associated with establishment and spread of Lessepsian fishes in the Mediterranean Sea. *Marine Biology, 162*, 2141–2153.

Arora, M., & Malhotra, J. (2023). Prevalent climate variables during ablation season around Gangotri glacier. In *Climate change and environmental impacts: Past, present and future perspective* (pp. 205–214). Springer International Publishing.

Bahuguna, I. M. (2003). Satellite stereo data analysis in snow and glaciated region. In *Training document: Course on remote sensing for glaciological studies; Manali, India* (pp. 85–99). SAC/RESA/MWRG/ESHD/TR.

Bakshi, C. (2020). *Random forest regression.* Retrieved 23 February 2022, from https://levelup.gitconnected.com/random-forest-regression-209c0f354c84

Banerjee, A., & Shankar, R. (2013). On the response of Himalayan glaciers to climate change. *Journal of Glaciology, 59*(215), 480–490.

Baral, P., Haq, M. A., & Yaragal, S. (2020). Assessment of rock glaciers and permafrost distribution in Uttarakhand, India. *Permafrost and Periglacial Processes, 31*(1), 31–56.

Bertoldi, G., Notarnicola, C., Leitinger, G., Endrizzi, S., Zebisch, M., Della Chiesa, S., & Tappeiner, U. (2010). Topographical and ecohydrological controls on land surface temperature in an alpine catchment. *Ecohydrology: Ecosystems, Land and Water Process Interactions, Ecohydrogeomorphology, 3*(2), 189–204.

Bhambri, R., Bolch, T., & Chaujar, R. K. (2012). Frontal recession of Gangotri glacier, Garhwal Himalayas, from 1965 to 2006, measured through high-resolution remote sensing data. *Current Science, 102*(3), 489–494.

Bisht, K., Joshi, Y., Upadhyay, S., & Metha, P. (2018). Recession of Milam glacier, Kumaun Himalaya, observed via lichenometric dating of moraines. *Journal of the Geological Society of India, 92*(2), 173–176.

Bolch, T., Kulkarni, A., Kääb, A., Huggel, C., Paul, F., Cogley, J. G., et al. (2012). The state and fate of Himalayan glaciers. *Science, 336*(6079), 310–314.

Brabyn, L., & Stichbury, G. (2020). Calculating the surface melt rate of Antarctic glaciers using satellite-derived temperatures and stream flows. *Environmental Monitoring and Assessment, 192*(7), 1–14.

Braithwaite, R. J. (2008). Temperature and precipitation climate at the equilibrium-line altitude of glaciers expressed by the degree-day factor for melting snow. *Journal of Glaciology, 54*(186), 437–444.

Breiman, L. (2001). Random forests. *Machine Learning, 45*, 5–32.

Brenning, A. (2009). Benchmarking classifiers to optimally integrate terrain analysis and multispectral remote sensing in automatic rock glacier detection. *Remote Sensing of Environment, 113*(1), 239–247.

Chakure, A. (2019). *Random forest and its implementation.* Retrieved 23 February 2022, from https://medium.com/swlh/random-forest-and-its-implementation-71824ced454f

Chaujar, R. K. (2009). Climate change and its impact on the Himalayan glaciers–a case study on the Chorabari glacier, Garhwal Himalaya, India. *Current Science*, 703–708.

Clark, P. U., Pisias, N. G., Stocker, T. F., & Weaver, A. J. (2002). The role of the thermohaline circulation in abrupt climate change. *Nature, 415*(6874), 863–869.

Das, M., Das, A., Giri, B., Sarkar, R., & Saha, S. (2021). Habitat vulnerability in slum areas of India–What we learnt from COVID-19? International Journal of Disaster Risk Reduction, 65, 102553.

Demenocal, P. B., & Rind, D. (1993). Sensitivity of Asian and African climate to variations in seasonal insolation, glacial ice cover, sea surface temperature, and Asian orography. *Journal of Geophysical Research: Atmospheres, 98*(D4), 7265–7287.

Dimri, S., Lakhera, R. C., & Sati, S. (2007). Fuzzy-based method for landslide hazard assessment in active seismic zone of Himalaya. *Landslides, 4*(2), 101–111.

Ding, J. (2010). Retreating glaciers of the Himalayas: A case study of gangotri glacier using 1990–2009 satellite images.

Evans, D. J. (2018). *Glaciation: A very short introduction*. Oxford University Press.

Falaschi, D., Bolch, T., Rastner, P., Lenzano, M. G., Lenzano, L., & VECCHIO, A. L., & Moragues, S. (2017). Mass changes of alpine glaciers at the eastern margin of the northern and southern Patagonian Icefields between 2000 and 2012. *Journal of Glaciology, 63*(238), 258–272.

Forsberg, R., Sørensen, L., & Simonsen, S. (2017). Greenland and Antarctica ice sheet mass changes and effects on global sea level. *Integrative Study of the Mean Sea Level and Its Components*, 91–106.

Fu, P., & Rich, P. M. (1999). Design and implementation of the solar analyst: An ArcView extension for modeling solar radiation at landscape scales. In *proceedings of the nineteenth annual ESRI user conference* (Vol. 1, pp. 1–31). USA: San Diego.

Fu, P., & Rich, P. M. (2002). A geometric solar radiation model with applications in agriculture and forestry. *Computers and Electronics in Agriculture, 37*(1–3), 25–35.

Gasson, E., DeConto, R., & Pollard, D. (2015). Antarctic bedrock topography uncertainty and ice sheet stability. *Geophysical Research Letters, 42*(13), 5372–5377.

Ghosh, P. (2017). Remote sensing and GIS analysis of Gaumukh snout retreat and ice loss estimation at Gangotri glacier, during 1962–2015. *Global Journal of Current Research, 5*(3), 113–119.

Gislason, P. O., Benediktsson, J. A., & Sveinsson, J. R. (2006). Random forests for land cover classification. *Pattern Recognition Letters, 27*(4), 294–300.

Hall, D. K., Riggs, G. A., & Salomonson, V. V. (1995). Development of methods for mapping global snow cover using moderate resolution imaging spectroradiometer data. *Remote Sensing of Environment, 54*(2), 127–140.

Haritashya, U. K., Singh, P., Kumar, N., & Gupta, R. P. (2006). Suspended sediment from the Gangotri glacier: Quantification, variability and associations with discharge and air temperature. *Journal of Hydrology, 321*(1–4), 116–130.

Hetrick, W. A., Rich, P. M., Barnes, F. J., & Weiss, S. B. (1993). GIS-based solar radiation flux models. In *ACSM ASPRS annual convention* (Vol. 3, pp. 132–132). American Soc Photogrammetry & Remote Sensing+ Amer Cong On.

Hsieh, C. H., & Chen, S. H. (1999). A model and algorithm of fuzzy product positioning. *Information Sciences, 121*(1–2), 61–82.

Jarvis, W. T. (2012). Integrating groundwater boundary matters into catchment management. In *The dilemma of boundaries* (pp. 161–176). Springer.

Jiménez-Muñoz, J. C., & Sobrino, J. A. (2006). Error sources on the land surface temperature retrieved from thermal infrared single channel remote sensing data. *International Journal of Remote Sensing, 27*(5), 999–1014.

Jones, D. B., Harrison, S., Anderson, K., Selley, H. L., Wood, J. L., & Betts, R. A. (2018). The distribution and hydrological significance of rock glaciers in the Nepalese Himalaya. *Global and planetary change, 160*, 123–142.T, D. (2015). Climate change rule of thumb: Cold "things" warming faster than warm things | NOAA Climate.gov. Retrieved 31 January 2022, from https://

www.climate.gov/news-features/blogs/beyond-data/climate-change-rule-thumb-cold-things-warming-faster-warm-things

Joshi, S., Khobragade, S., & Kumar, S. (2020). Spatio-temporal variability in glacier melt contribution in Bhagirathi river discharge in the headwater region of Himalaya.

Kargel, J. S., Cogley, J. G., Leonard, G. J., Haritashya, U., & Byers, A. (2011). Himalayan glaciers: The big picture is a montage. *Proceedings of the National Academy of Sciences, 108*(36), 14709–14710.

Kaur, J., & Mahajan, M. (2015). Hybrid of fuzzy logic and random Walker method for medical image segmentation. *IJ Image, Graphics and Signal Processing, 7*(2), 23.

Kayastha, P., Bijukchhen, S. M., Dhital, M. R., & De Smedt, F. (2013). GIS based landslide susceptibility mapping using a fuzzy logic approach: A case study from Ghurmi-Dhad Khola area, eastern Nepal. *Journal of the Geological Society of India, 82*(3), 249–261.

Khan, Z. (2019). *Tracing dynamics of snow cover in Pithoragarh district using geo-spatial techniques*. Aligarh Muslim University Aligarh, India.

Kong, Y., Wang, K., Pu, T., & Shi, X. (2019). Nonmonsoon precipitation dominates groundwater recharge beneath a monsoon-affected glacier in Tibetan plateau. *Journal of Geophysical Research: Atmospheres, 124*(20), 10913–10930.

Křížek, M., & Mida, P. (2013). The influence of aspect and altitude on the size, shape and spatial distribution of glacial cirques in the high Tatras (Slovakia, Poland). *Geomorphology, 198*, 57–68.

Kulkarni, A. V., Srinivasulu, J., Manjul, S. S., & Mathur, P. (2002). Field based spectral reflectance studies to develop NDSI method for snow cover monitoring. *Journal of the Indian Society of Remote Sensing, 30*(1), 73–80.

Kumar, U., Kumar, B., & Mallick, N. (2013). Groundwater prospects zonation based on RS and GIS using fuzzy algebra in Khoh River watershed, Pauri-Garhwal District, Uttarakhand, India. *Global Perspectives on Geography (GPG), 1*(3), 37–45.

Kumar, D., Singh, A., & Israil, M. (2021). Necessity of terrain correction in Magnetotelluric data recorded from Garhwal Himalayan region. *India. Geosciences, 11*(11), 482.

Lencioni, V., Maiolini, B., Marziali, L., Lek, S., & Rossaro, B. (2007). Macroinvertebrate assemblages in glacial stream systems: A comparison of linear multivariate methods with artificial neural networks. *Ecological Modelling, 203*(1–2), 119–131.

Liljedahl, A. K., Gädeke, A., O'Neel, S., Gatesman, T. A., & Douglas, T. A. (2017). Glacierized headwater streams as aquifer recharge corridors, subarctic Alaska. *Geophysical Research Letters, 44*(13), 6876–6885.

Lund, I. A. (1968). Relationships between insolation and other surface weather observations at Blue Hill, Massachusetts. *Solar Energy, 12*(1), 95–106.

Mitra, A., Banerjee, K., Sengupta, K., & Gangopadhyay, A. (2009). Pulse of climate change in Indian Suindarbans: A myth or reality? *National Academy Science Letters (India), 32*(1), 19.

Mortimer, C. A., Sharp, M., & Wouters, B. (2016). Glacier surface temperatures in the Canadian high Arctic, 2000–2015. *Journal of Glaciology, 62*(235), 963–975.

Naithani, A. K., Nainwal, H. C., Sati, K. K., & Prasad, C. (2001). Geomorphological evidences of retreat of the Gangotri glacier and its characteristics. *Current Science*, 87–94.

Oerlemans, J. (1991). The mass balance of the Greenland ice sheet: Sensitivity to climate change as revealed by energy-balance modelling. *The Holocene, 1*(1), 40–48.

Pal, M., Singh, N. K., & Tiwari, N. K. (2013). Pier scour modelling using random forest regression. *ISH Journal of Hydraulic Engineering, 19*(2), 69–75.

Paxman, G. J., Jamieson, S. S., Ferraccioli, F., Bentley, M. J., Forsberg, R., Ross, N., et al. (2017). Uplift and tilting of the Shackleton range in East Antarctica driven by glacial erosion and normal faulting. *Journal of Geophysical Research: Solid Earth, 122*(3), 2390–2408.

Pörtner, H. O., Roberts, D. C., Adams, H., Adler, C., Aldunce, P., Ali, E., et al. (2022). *Climate change 2022: Impacts, adaptation and vulnerability* (p. 3056). IPCC.

Ramsankaran, R. A. A. J., Navinkumar, P. J., Dashora, A., & Kulkarni, A. V. (2021). UAV-based survey of glaciers in Himalayas: Challenges and recommendations. *Journal of the Indian Society of Remote Sensing, 49*(5), 1171–1187.

Rai, P. K., Mishra, V. N., Singh, S., Prasad, R., & Nathawat, M. S. (2017). Remote sensing-based study for evaluating the changes in glacial area: a case study from Himachal Pradesh, India. *Earth Systems and Environment, 1*, 1–13.

Ray, K. (2009). *Snowline recedes on the Himalayas by 400 mts*. Retrieved 1 March 2022, from https://www.deccanherald.com/content/23732/snowline-recedes-himalayas-400-mts.html

Rich, P. M., Hetrick, W. A., & Saving, S. C. (1995). *Modeling topographic influences on solar radiation: A manual for the SOLARFLUX model* (no. LA-12989-M). Los Alamos National lab. (LANL), Los Alamos, NM (United States).

Ringnér, M. (2008). What is principal component analysis? *Nature Biotechnology, 26*(3), 303–304.

Rose, B. E., Cronin, T. W., & Bitz, C. M. (2017). Ice caps and ice belts: The effects of obliquity on ice – Albedo feedback. *The Astrophysical Journal, 846*(1), 28.

Sarkar, A., & Chouhan, P. (2021). COVID-19: District level vulnerability assessment in India. *Clinical Epidemiology and Global Health, 9*, 204–215.

Shi, S., Xu, H., Shui, Y., Liu, D., Xie, Q., Zhang, J., et al. (2023). Sedimentary organic molecular compositions reveal the influence of glacier retreat on ecology on the Tibetan plateau. *Science of the Total Environment, 163629*, 163629.

Singh, D. S., Tangri, A. K., Kumar, D., Dubey, C. A., & Bali, R. (2017). Pattern of retreat and related morphological zones of Gangotri glacier, Garhwal Himalaya, India. *Quaternary International, 444*, 172–181.

Singh, D. K., Thakur, P. K., Naithani, B. P., & Kaushik, S. (2018). Temporal monitoring of glacier change in Dhauliganga basin, Kumaun Himalaya using geo-spatial techniques. *ISPRS Annals of Photogrammetry, Remote Sensing & Spatial Information Sciences, 4*(5), 203.

Sobrino, J. A., & Jiménez-Muñoz, J. C. (2014). Minimum configuration of thermal infrared bands for land surface temperature and emissivity estimation in the context of potential future missions. *Remote Sensing of Environment, 148*, 158–167.

Sparavigna, A. C. (2017). The retreat of the terminus of Gangotri glacier in Google earth images.

Tawde, S. A., Kulkarni, A. V., & Bala, G. (2017). An estimate of glacier mass balance for the Chandra basin, western Himalaya, for the period 1984–2012. *Annals of Glaciology, 58*(75pt2), 99–109.

Thayyen, R. J., & Gergan, J. T. (2010). Role of glaciers in watershed hydrology: A preliminary study of a "Himalayan catchment". *The Cryosphere, 4*(1), 115–128.

Varugu, B. K., & Rao, Y. S. (2016, May). Glacier retreat monitoring from SAR coherence images: Application to Gangotri glacier. In *Land surface and cryosphere remote sensing III* (Vol. 9877, p. 987715). International Society for Optics and Photonics.

Verma, A., Tiwari, S. K., Kumar, A., Sain, K., Rai, S. K., & Kumari, S. (2021). Assessment of water recharge source of geothermal systems in Garhwal Himalaya (India). *Arabian Journal of Geosciences, 14*(22), 1–18.

Walsh, S. J., & Butler, D. R. (1997). Morphometric and multispectral image analysis of debris flows for natural hazard assessment. *Geocarto International, 12*(1), 59–70.

Wangchuk, S., & Bolch, T. (2020). Mapping of glacial lakes using Sentinel-1 and Sentinel-2 data and a random forest classifier: Strengths and challenges. *Science of Remote Sensing, 2*, 100008.

Wegmann, M., Gudmundsson, G. H., & Haeberli, W. (1998). Permafrost changes in rock walls and the retreat of alpine glaciers: A thermal modelling approach. *Permafrost and Periglacial Processes, 9*(1), 23–33.

Were, K., Bui, D. T., Dick, Ø. B., & Singh, B. R. (2015). A comparative assessment of support vector regression, artificial neural networks, and random forests for predicting and mapping soil organic carbon stocks across an Afromontane landscape. *Ecological Indicators, 52*, 394–403.

Wold, S., Geladi, P., Esbensen, K., & Öhman, J. (1987). Multi-way principal components-and PLS-analysis. *Journal of Chemometrics, 1*(1), 41–56.

Wu, Y., Wang, N., He, J., & Jiang, X. (2015). Estimating mountain glacier surface temperatures from Landsat-ETM+ thermal infrared data: A case study of Qiyi glacier, China. *Remote Sensing of Environment, 163*, 286–295.

Yao, X., Zhu, Y., Tian, Y., Feng, W., & Cao, W. (2010). Exploring hyperspectral bands and estimation indices for leaf nitrogen accumulation in wheat. *International Journal of Applied Earth Observation and Geoinformation, 12*(2), 89–100.

Yu, W., Yao, T., Kang, S., Pu, J., Yang, W., Gao, T., et al. (2013). Different region climate regimes and topography affect the changes in area and mass balance of glaciers on the north and south slopes of the same glacierized massif (the west Nyainqentanglha range, Tibetan Plateau). *Journal of Hydrology, 495*, 64–73.

Zadeh, L. A. (1972). A fuzzy-set-theoretic interpretation of linguistic hedges.

Zhou, X., Zhu, X., Dong, Z., & Guo, W. (2016). Estimation of biomass in wheat using random forest regression algorithm and remote sensing data. *The Crop Journal, 4*(3), 212–219.

Part III
Livelihood Adaptation to Climate Change: Insight for Socio-Ecological Sustainability

Chapter 15
Analysing the Household-Level Adaptation to Flood Hazard in Bhagirathi Sub-basin of West Bengal, India: A Concern for Mitigation

Sufia Rehman and Haroon Sajjad

Introduction

Flood is the most disastrous phenomenon that may not only cause socio-economic vulnerabilities but also result in prolonged mental distress. Increasing climate change is making the flooding new normal globally. The global annual temperature anomaly is around 0.98 °C, while the monthly anomaly is about 0.91 °C. These rising temperatures are likely to enhance the evaporation over the surface and increase the retention capacity of air to hold more moisture, which will lead to frequent, intense precipitation even for a period of time (UNEP, 2021). Intergovernmental Panel on Climate Change (IPCC) in AR6 with high confidence highlighted that warmer climate may lead to flooding and drought incidence whose intensity would depend on the regional climatic conditions, i.e. mid-latitude storms and monsoon (IPCC, 2021). Global Climate Risk Index 2020 indicated that Japan, Philippines, Germany, Madagascar, India and Sri Lanka are the most disaster-affected nations, where India ranks 5, incurring a share of 0.36% disaster losses of the GDP. The major fatalities in India were found from landslide-induced flooding (Eckstein et al., 2019). Around 15,082 deaths were reported globally (NDRCC, 2021) where maximum deaths (6171) were reported from floods followed by storms (1742), landslides (514), earthquakes (196) and wildfires (70). Flash floods have also caused severe implications in northern and coastal India (Eckstein et al., 2019).

S. Rehman
Department of Civil Engineering, National Institute of Technology, Imphal, Manipur, India

H. Sajjad (✉)
Department of Geography, Faculty of Sciences, Jamia Millia Islamia, New Delhi, India
e-mail: hsajjad@jmi.ac.in

A. L. Singh et al. (eds.), *Climate Change, Vulnerabilities and Adaptation*,
https://doi.org/10.1007/978-3-031-49642-4_15

Vulnerability analysis is an integral part of flood risk assessment. Vulnerability reflects a situation where a system is exposed and sensitive to climate change-induced implications and tries to cope with them (Rehman et al., 2019). This coping mechanism or ability to resist the changes driven by climate change is called adaptation. Adaptation has now become an essential aspect of climate change analysis. Scientific assessments and policy progress on understanding varied dimensions of adaptation have increased recently (Swart et al., 2014). Adaptation at the local level may help in modulating risk and increasing resilience to prevent the climate change-induced disaster implications. IPCC special report (IPCC, 2012) defines adaptation as *'In human systems, the process of adjustment to actual or expected climate and its effects, in order to moderate harm or exploit beneficial opportunities. In natural systems, the process of adjustment to actual climate and its effects; human intervention may facilitate adjustment to expected climate'*. Substantial attention has been given to the factors that determine the ability of a region or community to adapt to the effects resulting from disaster-induced vulnerability and risk (Smit & Pilifosova, 2003; Crane et al., 2011; Chersich & Wright, 2019). Ullah et al. (2018) examined the climate change-induced vulnerability and varied adaptation strategies by the farming community in Khyber Pakhtunkhwa, Pakistan. De Silva and Kawasaki (2018) investigated the flood- and drought-induced socio-economic vulnerability of rural Sri Lankan community. Their findings indicated that rural households are largely dependent on natural resources for their livelihood, which is largely affected by flood and drought. Singha et al. (2018) analysed the flood erosion-induced livelihood vulnerability along the Ganges riparian corridor of India. They suggested expansion of cultural and urban landscape and unplanned manmade resources combined with physical vulnerability are leading to social and ecological losses.

The coastal areas are largely affected by multi-hazard shocks, which raises the vulnerability of the coastal communities. Inadequate planning, preparedness and management of land also raises the vulnerability of the community to disaster events. In such areas, adaptation and resilience are required for lessening the risk and vulnerability caused by hydrometeorological disasters (Buchori et al., 2018). Resilience reflects the ability of the community to absorb and recover the effects of disasters in an efficient manner (IPCC, 2012). Bhagirathi sub-basin is an important hydrological unit of India, lying over the states of Jharkhand, Bihar and West Bengal. The southern part of the sub-basin is undermined by various disasters including flooding, cyclones, storm surges and coastal erosion (Sahana et al., 2020; Rehman et al., 2021). River flooding is a big concern in West Bengal, Bihar and some adjacent parts of Jharkhand. The riverine topography, changing land use, massive discharge from barrages and dams and extreme rainfall are some of the causes of flooding (Tripathi et al., 2020; Ghosh & Kar, 2018). Owing to these implications, this chapter examined the household-level adaptation to flood in the Bhagirathi sub-basin. The findings of the analysis may aid in formulating effective measures for obtaining sustainable adaptation to floods.

Study Area

Bhagirathi sub-basin is located between 86° 7′–89° 28′ E and 21° 39′–26° 56′ N latitudes (Fig. 15.1) extending over the three states (administrative division of country) of India, namely Jharkhand, Bihar and West Bengal comprising 35 districts (administrative division of state). A total of 75 watersheds are located in the study area (Hasan & Rai, 2020). This hydrological unit is formed by the three important rivers of India, namely *Ganga*, *Meghna* and *Hooghly*. The sub-basin is marked by diversified topography, microclimate and vegetation types. The study area is also rich in water resources including 77,873 surface bodies, 12 reservoirs and 12 dams. Rainfall ranges between 1000 and 1500 mm with maximum showers during the monsoon season (June–September). The study area experiences tropical savanna and hot Mediterranean-type climate. Around 9.3 billion reside in the sub-basin (Census of India, 2011). The population is largely dependent upon natural resources and agriculture for their livelihoods. Moreover, animal rearing, honey collection and fishing are also prominent among the coastal communities. The relief variation, extreme weather events and riverine topography have increased the vulnerability of the communities in the sub-basin (Sahana et al., 2021). Flood has been evident since several decades causing severe devastation to life and property. Thus, a close examination of household-level adaptation to flood hazard is essential for enhancing the resistance of the local communities.

Methods

Sampling Criteria

A household-level survey was conducted to derive the database for analysing the socio-economic vulnerability to floods in the sub-basin during June and August 2019. These 2 months witnessed huge monsoonal rainfall and thus selected as study period to get personal observation on the intensity of rainfall. The sample design adopted for the study consisted of two stages (Fig. 15.2). In the first stage, selection of villages was made from 12 drainage systems. From each drainage basin, 2 villages (one located near the river and one away from the river) were selected randomly, and thus, a total of 24 villages were selected from the sub-basin. The second stage consisted of selection of households. The selected villages were classified into three categories (high, medium and low) based on population. From each category, six households were selected randomly. In this way, a total of 432 households were surveyed in the study area. Perception of the households on various adaptation strategies including effectiveness of early warning system, adaptation by community during and post flood strategies, protection strategies and role of stakeholders in respective villages was observed. During the field visit, a close disclosure of the social and economic conditions of the households was also observed to determine flood-induced implications (Table 15.1).

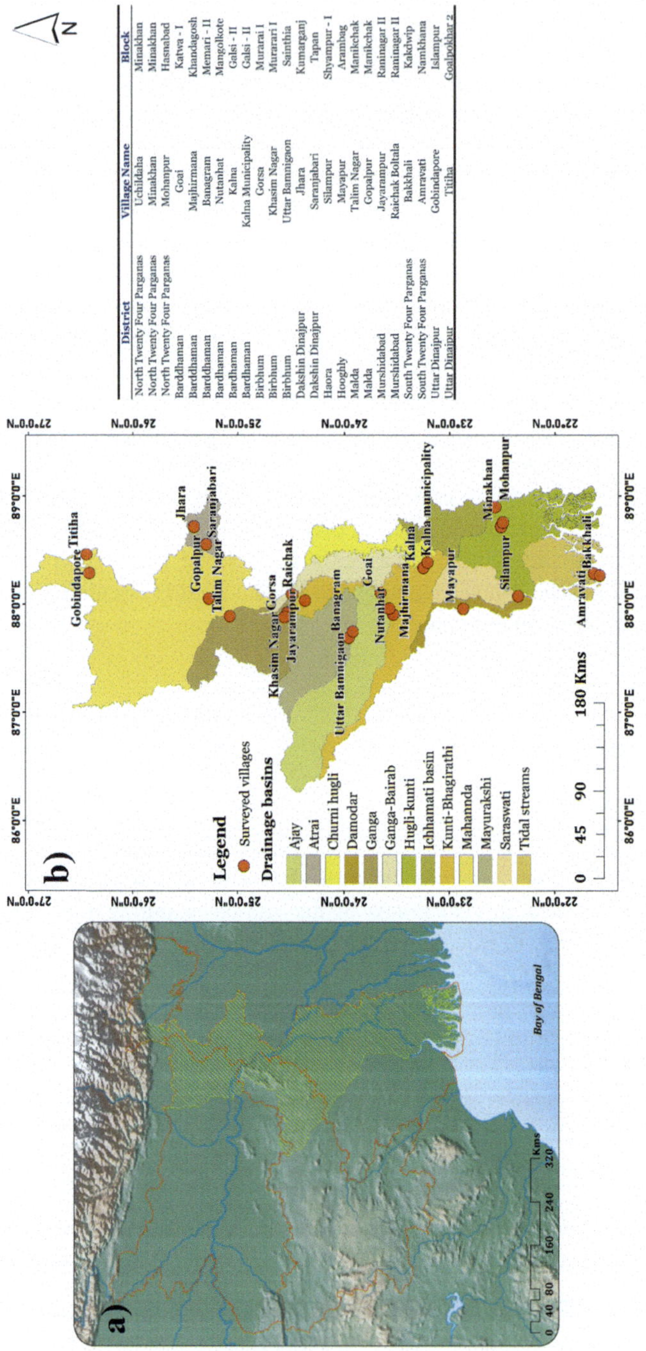

Fig. 15.1 Location of the sub-basin: **(a)** sub-basin in India and **(b)** surveyed villages identified among different drainage basins

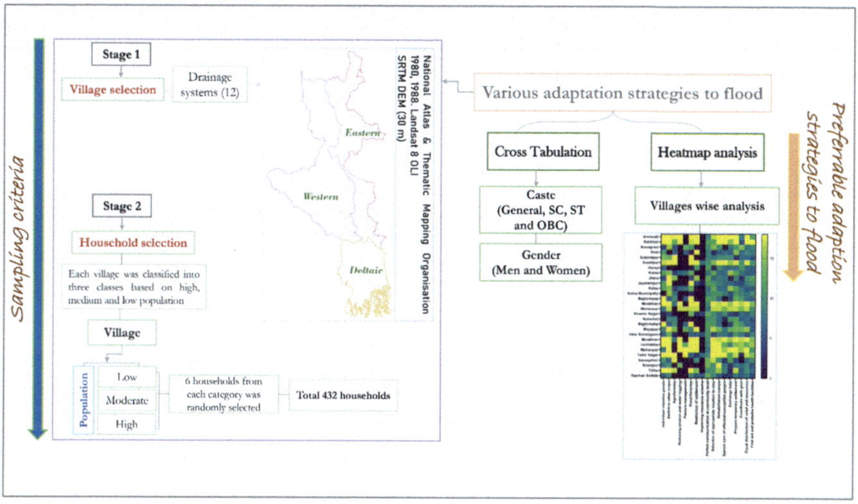

Fig. 15.2 Flowchart of methodology

Table 15.1 Descriptive statistics of housing characteristics in the sampled households

Housing characteristics		Percentage	Mean	First quantile	Third quantile	Std.
Housing type	Muddy house	27.8	25	22.07	31.02	14.31
	Semi-cemented house	28.4				
	Fully cemented house	4.9				
	Grass/bamboo/ others	38.9				
Housing facilities	Electricity	61.6	33.33	19.18	44.52	25.86
	Drinking water	11.0				
	Latrine	27.4				
Combustion materials	Firewood and chips	0.2	20	6.71	40.52	20.21
	Dung cake	43.1				
	Kerosene	9.4				
	LPG	40.5				
	Others	6.7				

Std. standard deviation

Cross Tabulation

After examining the various adaptation strategies, a cross-tabulation analysis was carried out to ascertain the effective adaptation methods by communities in general, during and post flood strategies particularly preferred by gender and caste groups (Table 15.2). A heatmap was also prepared to ascertain the preferable adaptation methods in the surveyed villages (Fig. 15.2).

Results

Of the total sampled households in the study area, around 85% were males while 15% were females. Average age of the respondents was found 44.7 years. Majority of the sampled households (62%) belonged to Hinduism followed by Islam (35%) and Christianity (3%). Nearly 41% of the sampled households belonged to general category followed by other backward castes (27%), scheduled castes (26%) and scheduled tribes (7%). About 73% of the head of the households have completed the primary education followed by secondary (20%) and graduate (5%), while only 2% were post-graduates. Most of the sampled households around 53% were having nucleated family system followed by joint family (45%) and extended family systems (2%). Only 1% of them were found living single. Most of them were married (91%), while 6% were unmarried and 3% were widows. The descriptive statistical results demonstrated that over 39% of houses were made of grass, bamboos and other locally available resources in the sub-basin. Table 15.1 also displays that more than one-fourth of the houses are still made of mud. Though the houses are not cemented, about two-thirds of houses are having electricity facilities.

The adaptation among the sampled households was analysed as strategies adopted by local communities, effectiveness of early warning system, during and post flood strategies, measures for flood protection and role of stakeholders.

Perception of Adaptation Strategies Adopted by Local Communities

The most preferred strategy adopted by the sampled households was switching over other crops (63%) followed by pond fisheries (47%), crop insurance (43%), agroforestry (33%) and restricting construction in flood risk areas (32%). Pasture management (40%) and reducing soil erosion and waterlogging (31%) were less adapted strategies by the sampled households (Fig. 15.3). Improving insurance schemes (45%) was the least adapted by the sampled respondents. Plate 15.1 represents the fishing ponds in various villages.

Table 15.2 Gender- and caste-wise preferred adaptation strategies

Adaptation strategies		Gender				Caste							
		Men		Women		GEN		OBC		SC		ST	
		Yes	No	Yes	No	Yes	No	Yes	No	Yes	No	Yes	No
Adaptation by communities	Individual retention ponds	57.2	42.8	66.7	33.3	64.4	35.6	53.4	46.6	50.5	49.5	56.7	43.3
	Switch to other crops	72.4	27.6	76.4	23.6	84.6	15.4	61.2	38.8	66.7	33.3	73.3	26.7
	Agroforestry	48.2	51.8	38.1	61.9	54.9	45.1	36.2	63.8	37.8	62.2	73.3	26.7
	Reducing erosion and waterlogging	21.4	78.6	50.8	49.2	29.7	70.3	21.6	78.4	23.4	76.6	26.7	73.3
	Pasture management	22.8	77.2	17.5	82.5	8.6	91.4	35.3	64.7	29.7	70.3	20	80
	Pond fisheries	59.6	40.4	58.7	41.3	71.4	28.6	51.7	48.3	50.5	49.5	53.3	46.7
	Restriction of settlement	48.2	51.8	63.5	36.5	58.3	41.7	39.7	60.3	45	55	66.7	33.3
	Improving crop insurance scheme	7	93	9.5	90.5	10.9	89.1	6	94	4.5	95.5	3.3	96.7
During flood strategies	Perfect communication at community level	58.3	41.7	69.8	30.2	58.9	41.1	53.4	46.6	66.7	33.3	66.7	33.3
	Selection of appropriate location to stay	65	35	74.6	25.4	65.1	34.9	55.2	44.8	73	27	93.3	6.7
	Rehabilitation camps	65.6	34.4	76.2	23.8	70.9	29.1	62.9	37.1	68.5	31.5	56.7	43.3
	Special care of affected/susceptible people	50.4	49.6	61.9	38.1	56.6	43.4	52.6	47.4	44.1	55.9	53.3	46.7
Post-flood strategies	Exchange help	50.4	49.6	55.6	44.4	47.4	52.6	43.1	56.9	62.2	37.8	63.3	36.7
	Prepare temporary settlement	65.9	34.1	68.3	31.7	61.7	38.3	66.4	33.6	75.7	24.3	56.7	43.3
	Coordinate with govt	46.6	53.4	54	46	38.9	61.1	53.4	46.6	56.8	43.2	43.3	56.7
	Equal distribution of relief and resources	58.8	41.2	66.7	33.3	63.4	36.6	56.9	43.1	56.8	43.2	63.3	36.7
	First aid and probable health facilities	51.2	48.8	57.1	42.9	55.4	44.6	53.4	46.6	41.4	58.6	66.7	33.3

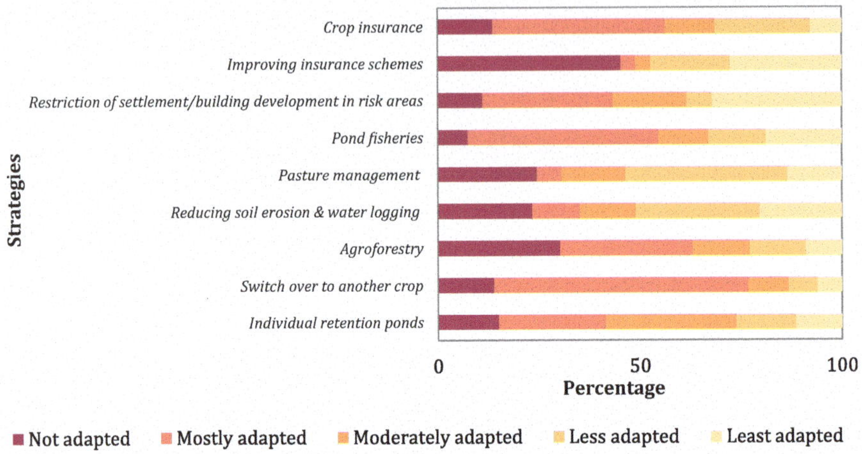

Fig. 15.3 Adaptation strategies adopted by local communities

Plate 15.1 Snaps captured during field survey (2019): (**a**) a big pond at Titha village, (**b**) fish pond in Jhara village, (**c**) a bird-protected fish pond in Gopalpur village and (**d**) remote location of village

Effectiveness of Early Warning Systems

Effectiveness of early warning systems was examined among the sampled households (Fig. 15.4). Among the early warning systems, own observation (74%) was

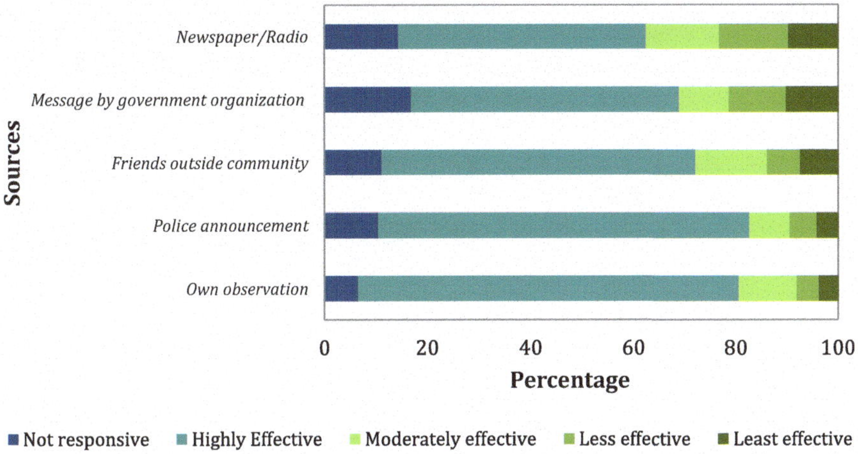

Fig. 15.4 Effectiveness of early warning system among the sampled households

found highly effective by the sampled households followed by police announcement (72%), friends outside the community (61%), information by government organization (52%) and newspaper and radio (48%). Own observation was considered highly effective by more than half (74%) of the sampled respondents followed by moderately effective (11%), not responsive (7%), less effective (4%) and least effective (4%). Announcements by the local policy were also identified as highly effective according to 72% of sampled respondents while nearly 10% stated that police were not responsive during floods followed by moderately effective (8%), less effective (5%) and least effective (4%). Most of the households (61%) also received disaster information from their friends living outside their communities. Nearly 52% of the sampled households disclosed that message from government was highly effective during flood events, while 17% stated that these were not so responsive followed by less effective (11%), least effective (10%) and moderately effective (10%). The main reason for this lack of information is remote locations of the houses and lack of communication. In case of hazard information from newspaper and radio, nearly 48% of sampled respondents identified it as highly effective followed by moderately effective (14%), not responsive (14%), less effective (13%) and least effective (10%).

Adaptation Strategies During Flood Hazard

Adaptation strategies during flood hazard involved perfect communication at community level, selection of appropriate location to stay, rehabilitation camps and

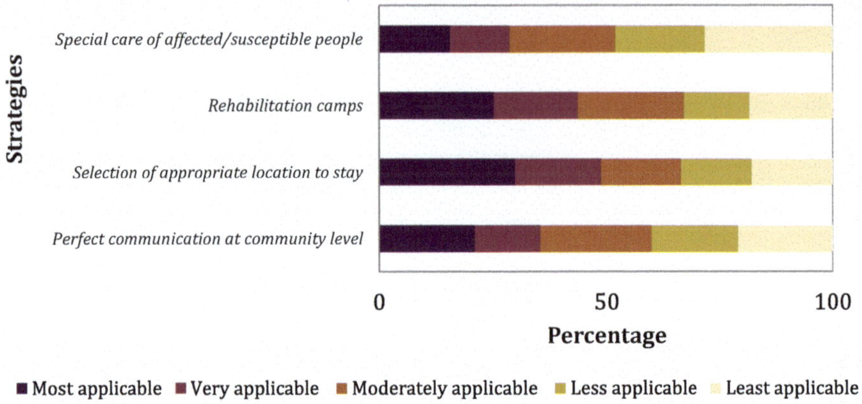

Fig. 15.5 Strategies adopted by sampled households during flood

special care of susceptible people. Selecting appropriate location to stay (30%) and staying in rehabilitation camps (25%) were mostly adapted by the respondents. Perfect communication at community level (24%) and special care of susceptible people (23%) were observed to be moderately adapted among the sampled households (Fig. 15.5).

Post Flood Adaptation Strategies

Post flood strategies consisted of exchanging help, preparing temporary settlement, coordinating with government and NGO, equal distribution of relief and first aid and probable health facilities. Post flood adaptation strategies were found moderate to less applicable among the sampled households (Fig. 15.6). Moderately adapted strategies among the sampled households were first aid and probable health facilities (30%) followed by equal distribution of relief and resources (30%) and preparing temporary settlement (23%). Coordinating with government and NGO (36%) and exchange help (31%) were found least adapted by the sampled households.

Flood Protection Strategies

Flood protection strategies involved technical flood protection (e.g. raise dykes, enlarge reservoirs, upgrade drainage systems), natural retention of flood water (e.g. floodplain restoration, change of land use), upgrading drainage systems, heightening existing infrastructure, diversion of excess water to irrigation fields, disaster response enhancement and improving insurance schemes against flood damage (Fig. 15.7). Improving insurance schemes against flood damage (45%),

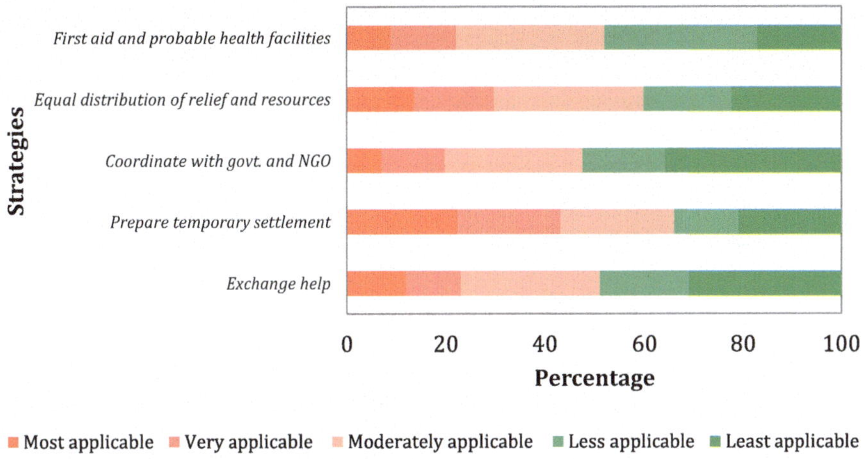

Fig. 15.6 Perception of sampled households on post flood strategies

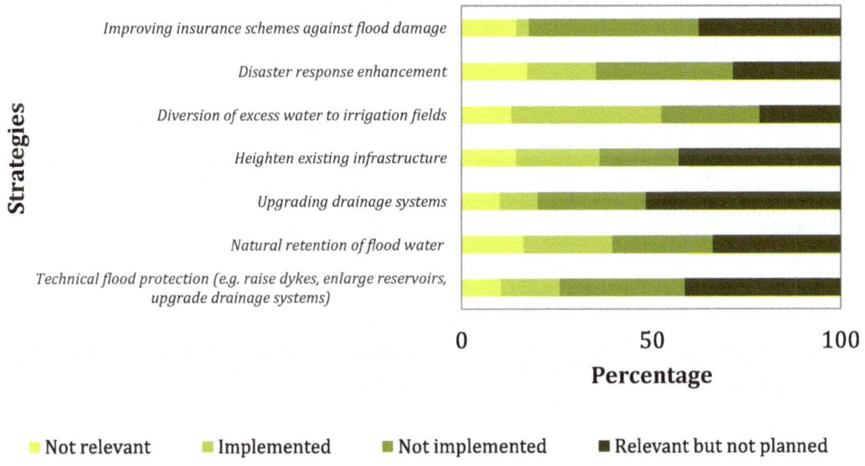

Fig. 15.7 Perception of sampled households on measures of flood protection

disaster response enhancement (36%) and diversion of excess water to irrigation fields (39%) were not implemented as disclosed by the sampled households. Heightening existing infrastructure (43%) was observed relevant but not planned followed by upgradation of drainage system (52%), technical flood protection (41%) and natural retention of flood water (34%).

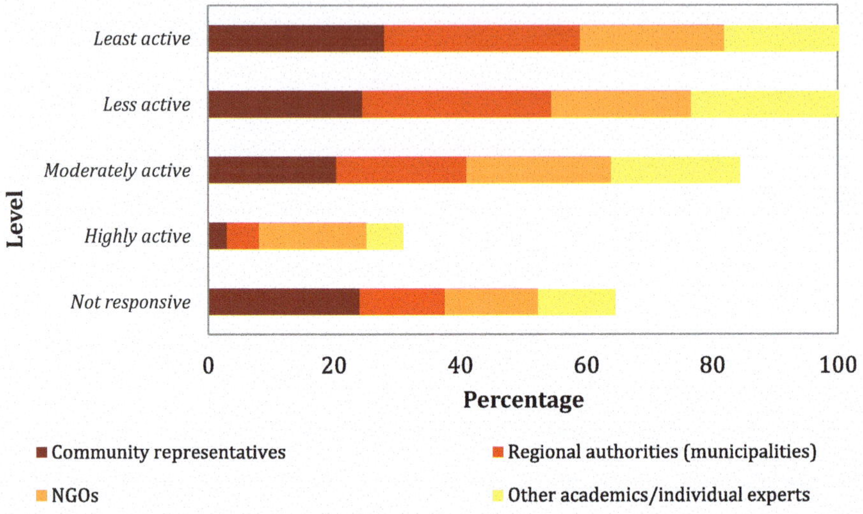

Fig. 15.8 Perception of sampled households on role of stakeholders

Role of Stakeholders

The role of community representatives, NGOs, regional authorities, academicians and researchers in lessening the flood risk was analysed among the sampled households (Fig. 15.8). The role of stakeholder analysis revealed that regional authorities (31%) followed by academicians and individual experts (31%) and community representatives (28%) were least involved in mitigating the flood hazard. Nearly 24% of the sampled households disclosed that local NGOs were moderately active (24%). Household characteristics revealed significant insights about the people living in the sub-basin. Most of the households were found having low socio-economic status. However, some villages were found connected by roads and found to be less impacted by flood.

Preferred Adaptation Strategies

Individual retention ponds, switch to other crops, pond fisheries, perfect communication at community level, selection of appropriate location to stay, rehabilitation camps, prepare temporary settlement, equal distribution of relief and resources, first aid and probable health facilities were preferred by all caste and gender groups (Table 15.2). The least preferred strategies among respondents identified were pasture management and improving crop insurance schemes. In case of village-wise analysis, Mohanpur, Minakhan, Uchildaha, Gopalpur, Amravati and Bakkhali villages performed well in all adaptation strategies except improving crop insurance

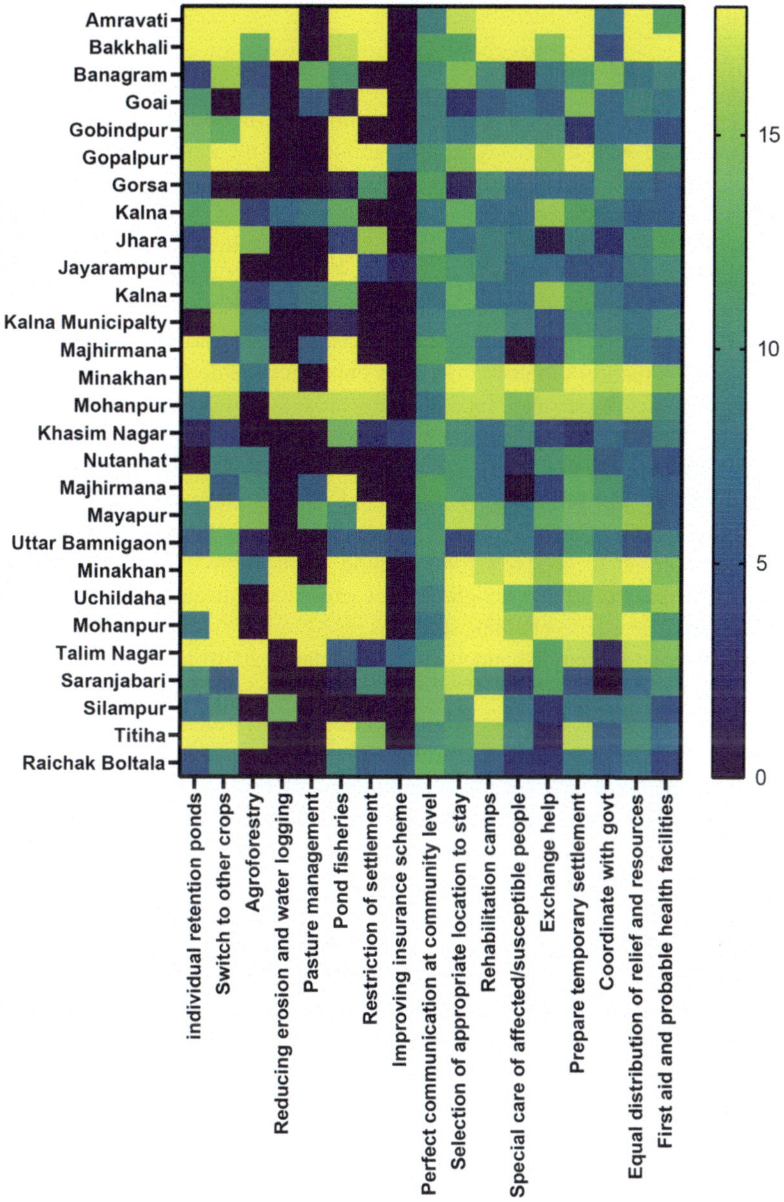

Fig. 15.9 Preferred adaptation strategies in the surveyed villages

scheme. The reason behind the highest preferability is the frequent flood events during the monsoon season (Fig. 15.9). Moreover, Amravati and Bakkhali villages are located in the southern coastal stretch where a series of multi-hazard events are prevalent and people tend to adapt more to the environmental casualties.

Discussion

The adaptation strategies discussed in this study are entirely based on the casualties induced by flood, traditional and natural methods. Earlier adaptation studies were mainly concerned with increasing understanding over the climate change implications. However, the local communities are more aware about the changes driven by climate and response accordingly. Recent climate change studies are providing better knowledge about the response of the communities in changing climatic conditions. The fifth assessment report of IPCC provided a detailed discussion on economics of adaptation citing various studies (Fankhauser, 2016). In the present analysis, most of the villages were found experiencing severe implications induced by floods, and sometimes their frequency is unpredictable due to changing climatic conditions. Studies have also indicated towards the increasing magnitude and frequency of the extreme weather events induced by climate change. Geographical locations, population growth, limited natural resources and assets and maladaptation increase the severity of these events (Ullah et al., 2018). The expensive implications of flood attracted the stakeholders and scholars to identify the suitable measures at global, regional and local levels for lessening implications. The flood adaptation includes various actions that strengthen the socio-economic and ecological systems against the adverse effects (Pathak, 2021).

In case of West Bengal, Bhattacharjee and Behera (2018) identified that shifting towards non-agricultural activities, elevating the houses and using flood-resistant crops are the prominent flood adaptation measures by farmers. Execution of good agronomic activities and climate information services may be effective adaptation strategies for the flood-affected rural communities (Koppa & Amarnath, 2021). On the other hand, in case of the northern mountainous region, Ghosh and Ghosal (2021) suggested that migration for better income opportunities, education, better transport connectivity, flood-resilient infrastructure and women self-help groups may be ideal adaptation measures for reducing flood vulnerability. The northeast parts of Bihar and Jharkhand, which lie in the *Ganga* Plain, have experienced severe incidents of flood where appropriate adaptation design is essential at village and household level (Jha & Gundimeda, 2019; Raghu et al., 2021). In the Bhagirathi sub-basin, most of the respondents engaged in agriculture have less land holdings, and they only grow crops for their sustenance. Flood-induced asset losses, health impacts, damages to agricultural land and changing climate conditions are the major complications of flood occurrences. Households with low income were found indebted at high-interest rates to support their needs. All these implications may only be reduced through effective adaptation strategies.

Climate change has also widened the flood implications that demand sustainable adaptation strategies (NITI Aayog, 2021). For strengthening this vision, the Government of India has initiated the estimation of sustainable development goals (SDGs) b NITI Aayog at state and union territory levels to reinforce the sustainable development of the states. NITI Aayog has classified SDG index score into aspirant (0–49), performer (50–64), front runner (65–99) and achiever (100). If we analyse

Fig. 15.10 Level of SDG goals in case of Jharkhand, Bihar and West bengal. (*Source*: NITI Aayog, 2021)

the level of overall SDG attainment of states lying in the sub-basin, all the states were found performer. All these states were found the aspirants of SDG2 (zero hunger) and SDG13 (climate action), while they found performer in SDG8 (decent work and economic growth). They were also identified as the front runner in case of SDG3 (good health and well-being), SDG6 (clean water and sanitation), SDG7 (affordable and clean energy) and SDG16 (peace and justice strong institutions). However, none of these states were recognized as achiever of any SDGs (Fig. 15.10). During fieldwork, it was observed that people have started constructing toilets in their houses with the help of government's welfare schemes such as *Swachh Bharat Mission* and WaSH (Water, Sanitation and Hygiene) programmes. However, the majority of the houses are still using dung cake as the main source of cooking fuel (43.1%). Here it is worth mentioning that recently, the households are increasingly using liquified petroleum gas (LPG), which accounts for 40% of the total cooking fuel. This increasing use of clean cooking fuel (LPG) may be attributed to welfare schemes such as *Pradhan Mantri Ujjwala Yojana* aiming to achieve energy-related SDGs (6 and 7), i.e. clean water and sanitation for all and affordable and clean energy in India (Plate 15.2).

In case of composite score, Jharkhand and Bihar were identified the performer while West Bengal is found the front runner. Moreover, there is a need for improvement especially in case of climate action (SDG13) goal of SDG

The present analysis has revealed that the adaptation strategies preferred by the households were largely pointing towards alternative liveihood, community participation during flood and provision of better healthcare systems. The low

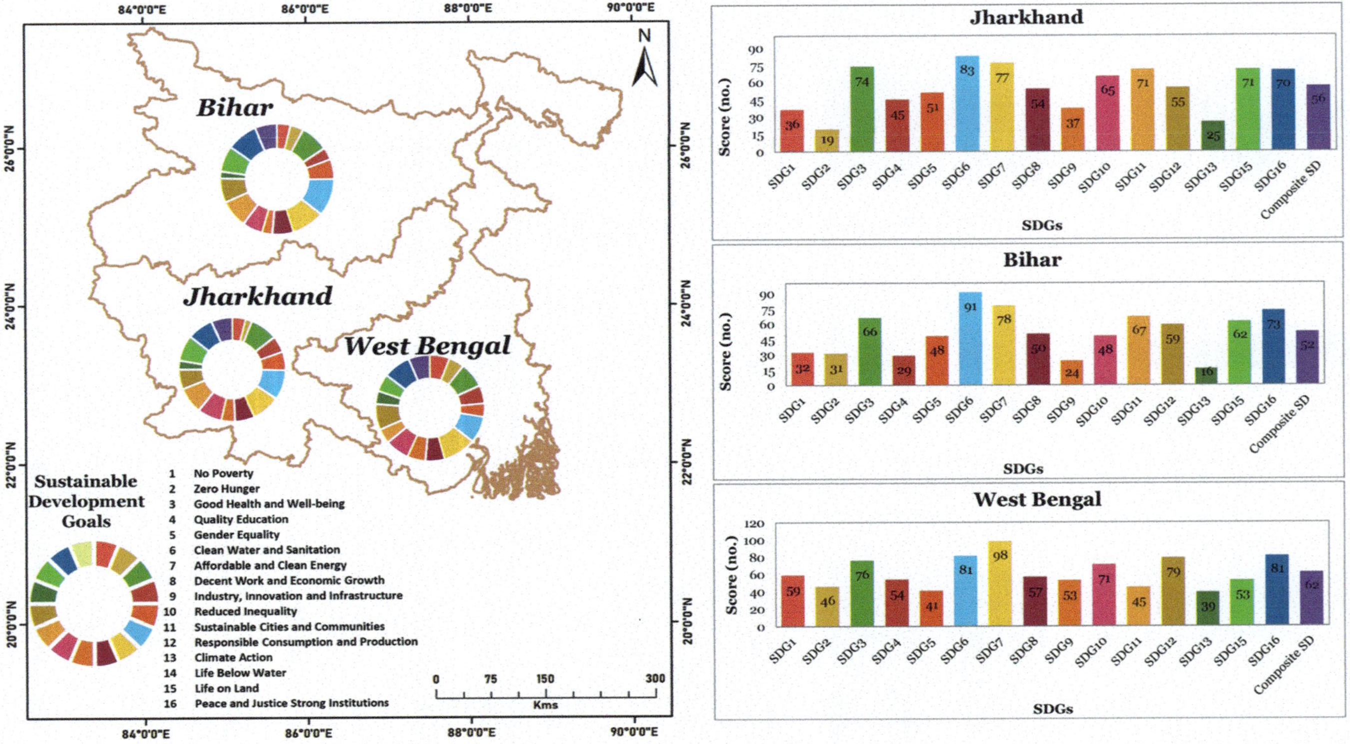

Plate 15.2 Typical view of kitchen in the surveyed households

performance of Raichak Boltala, Silampur, Khasim Nagar, Nutanhat and Kalna villages also indicates the high degree of flood implications and less adaptive capacity. The image of implications is much complicated at district or state level, which requires a unifying adaptation measure to enhance the adaptive capacity of the communities. However, this analysis is limited in terms of inclusion of only household-level adaptation but provides a concise interpretation of adaptation in surveyed villages. Furthermore, such an assessment may not only be helpful in identifying the adaptation strategies but to discriminate the level of adaptation in different social groups.

Conclusion

The present study has explored the adaptation strategies in one of the important hydrological units of India, i.e. Bhagirathi sub-basin. The analysis of household-level adaptation assessment is based on 432 households, and their perception of the effect of early warning system, adaptation by community, during and post flood strategies, protection strategies and role of stakeholder in selected villages was examined. Findings revealed that individual retention ponds, switch to other crops, pond fisheries, perfect communication at community level, selection of appropriate location to stay, rehabilitation camps, preparing temporary settlement, equal distribution of relief and resources, first aid and probable health facilities were preferred by all caste and gender groups. Pasture management and improving crop insurance schemes were found the least preferable. Village-wise analysis indicated towards the low performance of Raichak Boltala, Silampur, Khasim Nagar, Nutanhat and Kalna villages, whereas Mohanpur, Minakhan, Uchildaha, Gopalpur, Amravati and Bakkhali villages performed well in all adaptation strategies due to their better connectivity to urban areas.

SDG analysis has also raised the need for improvement in SDG1 (no poverty), SDG2 (zero hunger) and SDG13 (climate action) for achieving sustainable adaptation among communities while uplifting their socio-economic status. The study suggests that the policymakers should provide better economic opportunities, education levels and healthcare facilities, and encourage community-based adaptation strategies to mitigate the flood implications. Unbiased gender and community involvement, adequate support from government, crop insurance for individual support, strengthening the role of community-based organizations and modernizing indigenous flood control methods may effectively further enhance the response and adaptive capacity of the local communities.

References

Bhattacharjee, K., & Behera, B. (2018). Determinants of household vulnerability and adaptation to floods: Empirical evidence from the Indian State of West Bengal. *International Journal of Disaster Risk Reduction, 31*, 758–769. https://doi.org/10.1016/j.ijdrr.2018.07.017

Buchori, I., Pramitasari, A., Sugiri, A., Maryono, M., Basuki, Y., & Sejati, A. W. (2018). Adaptation to coastal flooding and inundation: Mitigations and migration pattern in Semarang City, Indonesia. *Ocean & Coastal Management, 163*, 445–455.

Census of India. (2011). Population enumeration data. Office of the Registrar General & Census Commissioner, India. https://censusindia.gov.in/2011census/population_enumeration.html. Accessed 6 Nov 2021.

Chersich, M. F., & Wright, C. Y. (2019). Climate change adaptation in South Africa: A case study on the role of the health sector. *Globalization and Health, 15*(1), 1–16.

Crane, T. A., Roncoli, C., & Hoogenboom, G. (2011). Adaptation to climate change and climate variability: The importance of understanding agriculture as performance. *NJAS-Wageningen Journal of Life Sciences, 57*(3–4), 179–185.

De Silva, M. M. G. T., & Kawasaki, A. (2018). Socioeconomic vulnerability to disaster risk: A case study of flood and drought impact in a rural Sri Lankan community. *Ecological Economics, 152*, 131–140.

Eckstein, D., Künzel, V., Schäfer, L., & Winges, M. (2019). *Global climate risk index 2020*. Germanwatch. www.germanwatch.org/en/cri. Accessed 2 Nov 2021

Fankhauser, S. (2016). *Adaptation to climate change, annual reviews of resource economics*. https://doi.org/10.1146/annurev-resource-100516-033554. Available at SSRN: https://doi.org/10.2139/ssrn.2869292.

Ghosh, M., & Ghosal, S. (2021). Climate change vulnerability of rural households in flood-prone areas of Himalayan foothills, West Bengal, India. *Environment, Development and Sustainability, 23*(2), 2570–2595. https://doi.org/10.1007/s10668-020-00687-0

Ghosh, A., & Kar, S. K. (2018). Application of analytical hierarchy process (AHP) for flood risk assessment: A case study in Malda district of West Bengal, India. *Natural Hazards, 94*(1), 349–368.

Hasan, M. S. U., & Rai, A. K. (2020). Groundwater quality assessment in the Lower Ganga Basin using entropy information theory and GIS. *Journal of Cleaner Production, 274*, 123077.

IPCC. (2012). Managing the risks of extreme events and disasters to advance climate change adaptation. In C. B. Field, V. Barros, T. F. Stocker, D. Qin, D. J. Dokken, K. L. Ebi, M. D. Mastrandrea, K. J. Mach, G.-K. Plattner, S. K. Allen, M. Tignor, & P. M. Midgley (Eds.), *A special report of Working Groups I and II of the Intergovernmental Panel on Climate Change* (582 pp). Cambridge University Press.

IPCC. (2021). Summary for policymakers. In V. Masson-Delmotte, P. Zhai, A. Pirani, S. L. Connors, C. Péan, S. Berger, N. Caud, Y. Chen, L. Goldfarb, M. I. Gomis, M. Huang, K. Leitzell, E. Lonnoy, J. B. R. Matthews, T. K. Maycock, T. Waterfield, O. Yelekçi, R. Yu, & B. Zhou (Eds.), *Climate change 2021: The physical science basis. Contribution of Working Group I to the sixth assessment report of the Intergovernmental Panel on Climate Change*. Cambridge University Press. (in press).

Jha, R. K., & Gundimeda, H. (2019). An integrated assessment of vulnerability to floods using composite index – A district level analysis for Bihar, India. *International Journal of Disaster Risk Reduction, 101074*. https://doi.org/10.1016/j.ijdrr.2019.101074

Koppa, N., & Amarnath, G. (2021). Geospatial assessment of flood-tolerant rice varieties to guide climate adaptation strategies in India. *Climate, 9*(10), 151. https://doi.org/10.3390/cli9100151

NDRCC. (2021). *2020 global natural disaster assessment report*. International Federation of Red Cross and Red Crescent Societies National Disaster Reduction Centre of China, Academy of Disaster Reduction and Emergency Management. https://reliefweb.int/report/China/2020-global-natural-disaster-assessment-report. Accessed 2 Nov 2021.

NITI Aayog. (2021). *SDG India Index & Dashboard 2020–21 – Partnerships in the Decade of Action*. NITI Aayog, Government of India, Sansad Marg, New Delhi-110001, India. sdgindiaindex.niti.gov.in. Accessed 1 Dec 2021.

Pathak, S. (2021). Determinants of flood adaptation: Parametric and semiparametric assessment. *Journal of Flood Risk Management, 14*(2), e12699. https://doi.org/10.1111/jfr3.12699

Raghu, P. T., Das, S., & Veettil, P. C. (2021). *Smallholder adaptation to flood risks: Adoption and impact of Swarna-Sub1 in Eastern India*.

Rehman, S., Sahana, M., Hong, H., Sajjad, H., & Ahmed, B. B. (2019). A systematic review on approaches and methods used for flood vulnerability assessment: Framework for future research. *Natural Hazards, 96*(2), 975–998. https://doi.org/10.1007/s11069-018-03567-z

Rehman, S., Hasan, M. S. U., Rai, A. K., Avtar, R., & Sajjad, H. (2021). Assessing flood-induced ecological vulnerability and risk using GIS-based in situ measurements in Bhagirathi sub-basin, India. *Arabian Journal of Geosciences, 14*(15), 1–17.

Sahana, M., Rehman, S., Sajjad, H., & Hong, H. (2020). Exploring effectiveness of frequency ratio and support vector machine models in storm surge flood susceptibility assessment: A study of Sundarban Biosphere Reserve, India. *Catena, 189*, 104450.

Sahana, M., Rehman, S., Ahmed, R., & Sajjad, H. (2021). Analyzing climate variability and its effects in Sundarban Biosphere Reserve, India: Reaffirmation from local communities. *Environment, Development and Sustainability, 23*(2), 2465–2492.

Smit, B., & Pilifosova, O. (2003). Adaptation to climate change in the context of sustainable development and equity. *Sustainable Development, 8*(9), 9.

Swart, R., Biesbroek, R., & Lourenço, T. C. (2014). Science of adaptation to climate change and science for adaptation. *Frontiers in Environmental Science, 2*, 29. https://doi.org/10.3389/fenvs.2014.00029

Tripathi, G., Pandey, A. C., Parida, B. R., & Kumar, A. (2020). Flood inundation mapping and impact assessment using multi-temporal optical and SAR satellite data: A case study of 2017 flood in Darbhanga district, Bihar, India. *Water Resources Management, 34*(6), 1871–1892.

Ullah, W., Nihei, T., Nafees, M., Zaman, R., & Ali, M. (2018). Understanding climate change vulnerability, adaptation and risk perceptions at household level in Khyber Pakhtunkhwa, Pakistan. *International Journal of Climate Change Strategies and Management*.

UNEP. (2021). *How climate change is making record-breaking floods the new normal*. https://www.unep.org/news-and-stories/story/how-climate-change-making-record-breaking-floods-new-normal. Accessed 2 Nov 2021.

Chapter 16
Farmers in the Environment of Climate Change: A Study of Adaptation and Coping Strategies in West Bengal

A. K. M. Anwaruzzaman and Samsul Hoque

Abbreviations

APY	Area Production and Yield
DBT	Direct Benefit Transfer
DAC&FW	Department of Agriculture, Cooperation & Farmers Welfare
DST and MOST	Department of Science and Technology and Ministry of Science & Technology
GoWB	Government of West Bengal
HYV	High Yielding Variety
ISFR	India State of Forest Report
IT	Information Technology
IPCC	Intergovernmental Panel on Climate Change
JDJB	JalDharo-JalBharo
NICRA	National Initiative on Climate Resilient Agriculture
NMSA	National Mission for Sustainable Agriculture
P&RD	Panchayat and Rural Development
PMFBY	PradhanMantriFasalBimaYojna
PMKSNS	Pradhan Mantri Kisan Samman Nidhi Scheme
SPSS	Statistical Packages for the Social Sciences
U.N.O	United Nations Organization

A. K. M. Anwaruzzaman (✉)
Department of Geography, Aliah University, Kolkata, West Bengal, India
e-mail: anwaruzzaman.geog@aliah.ac.in

S. Hoque
Department of Geography, Krishnagar Government College, Krishnagar, West Bengal, India

A. L. Singh et al. (eds.), *Climate Change, Vulnerabilities and Adaptation*,
https://doi.org/10.1007/978-3-031-49642-4_16

Introduction

A background note prepared by OECD (2015) to outline the impact of climate change on agriculture and to explore the possibilities to achieve a sustainable, productive and environment-friendly agricultural system opined that climate change has posed a serious challenge and is expected to impact negatively with increased temperature, increasing intensity and frequency of extreme weather events and degradation of soil as well as creating a scarcity of water.

Climate change is quintessentially affecting agriculture through abnormal temperatures, precipitation and environmental hazard limits (e.g. heat waves), changes in vermin and infections, changes in climatic carbon dioxide and ground-level ozone focuses, changes in the healthful nature of certain food varieties and changes in ocean level and observed through negative impact on crop production, particularly in the low-latitude countries (Folnovic, n.d.).

There are apprehensions by a number of climate experts in their studies that climate change is going to have ramifications for food security, both in the USA and also internationally, through changes in crop yields as well as food costs and impacts on food handling, stockpiling, transportation and retailing (USGCRP, 2014).

Climate-smart agriculture (CSA) may be an answer to the challenges—which is an integrated way (WB, 2021) to deal with good use of cropland, animal husbandry, agro-forestry and fisheries—that provides solutions to the interlinked difficulties of food security and accelerated climate change.

Under such a situation, farmers in different countries adopt different ways. Several coping strategies and adaptation in US agriculture are observed to overshadow the adverse impact on agriculture, incorporate changes what is delivered, altering the data sources utilized for creation, embracing new innovations, and changing administration methodologies (USGCRP, 2018).

The impact of climate change on agriculture and food production has been found in several studies. When change in long-term mean climate will have importance for global food production and may require adaptation, more serious dangers to food security might be presented by changes in year-to-year variation and more extreme weather events. Historically, it has been observed that the largest fall in the food production has been due to low rainfall events (Kumar et al., 2004; Sivakumar et al., 2005).

The impact of climate change on the growth of typical crops is studied in China where authors (Ma et al., 2019) found global rising of temperature being responsible for altered climatic patterns causing the effect on crop cycles, shortening crop cycle and also greatly impacting crop growth. Thus, the present study aims to fulfil the following objectives: to examine farmers' perception of climate change, to prepare an overview of farmers' adaptation in the climate change environment, to enumerate coping strategies and innovations practiced by farmers of West Bengal and to assess regional differences in climate change impacts on agriculture.

Methods and Materials

The data used for the present study are primary data collected through field surveys, supplemented by secondary data collected from published sources. As many as 209 practicing farmers, taking an equal number from each district (WB had 19 districts before the bifurcation of West Midnapore, Burdwan, Jalpaiguri and Darjeeling districts) have been surveyed with a structured survey schedule designed in Bengali for ease of understanding of the farmers. Before finalization of the survey schedule, a pilot survey was conducted in Malda, Murshidabad and Nadia districts for the convenience of the authors. Sample respondent farmers were selected from each district and were selected from clusters of three villages as per convenience of the authors. The sample respondent farmers were brought to a common place to have a 'Focus Group Discussion' to get in-depth information on different aspects of the impacts of climate change on agriculture, adaptations and coping strategies of farmers in the region. The secondary data related to climatic elements have been collected from NASA Climate Research and Services, India Meteorological Department and Ministry of Earth Sciences, Pune (India Metrological Department, 2008). The data related to APY of crops were collected from the Ministry of Agriculture & Farmers Welfare, Government of India.

The collected data have been analysed using different quantitative techniques where correlation coefficient and multiple linear regression etc. and descriptive statistics such as mean, standard deviation, coefficient of variance etc. were calculated using SPSS. A one-way analysis of variance (ANOVA) has been conducted using SPSS. Cartographic and diagrammatic representations for better display of relations have been made using QGIS.

Results and Discussion

This segment of the research begins by tracing farmers' perception towards climate change. Thereafter, the paper narrates the evidence of climate change observed by the farmers. The discussion then progresses to identify the problems of climate change faced by the farmers. Coping strategies adopted by the farmers, farmers' demands from the government to mitigate climate change-induced difficulties, adoption policies to overcome the causes of climate change and for the reduction of the severity of climate change in general and on the agricultural sector in particular are also key points of this section. Correlating climatic elements such as temperature and precipitation with leading crops such as rice and jute has been done, which is followed by the conclusion of this research paper in the last section.

Farmers' Perception on Climate Change

Though the farmers have different perceptions, the majority of the respondents opined that the climate is changing.

Figure 16.1 reveals that 96.65% respondents are of the opinion that climate is changing in the study area. The majority of the respondents (45.93%) perceived that the amount of rainfall has decreased in the last 20 years, while 29.19% reported that the amount of rainfall has become highly uncertain due to climate change (Fig. 16.2). The decreasing trends of rainfall is common in both summer and winter monsoon seasons in large parts of northern and central India, as reported by Annamalai et al.

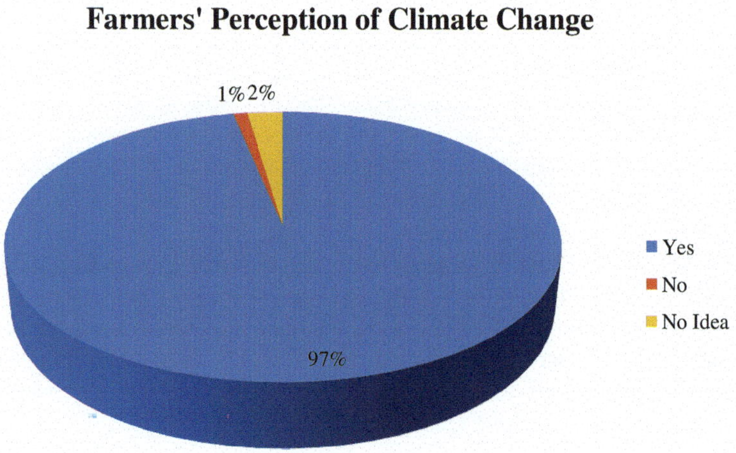

Fig. 16.1 Perception on climate change. (*Source*: Field Survey, 2021)

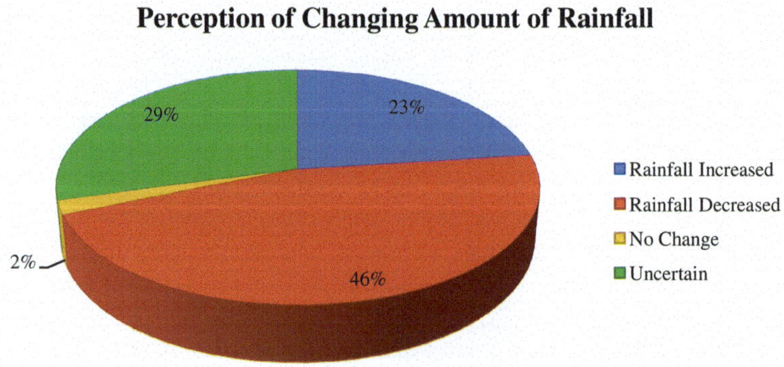

Fig. 16.2 Perception on changing amount of rainfall. (*Source*: Field Survey, 2021)

(as cited in Department of Science and Technology, Ministry of Science and Technology, 2016). The number of rainy days is also decreasing, while the temperature is gradually increasing, as observed by the respondent farmers in the study area (Field Survey, 2021). The perception of farmers is supported by various reports on climate change. The global total temperature increased by 0.78 °C over the period of 1850–2012 (IPCC Report, 2014). The Global mean annual temperatures likely to be increased by 1 °C and 3 °C by 2025 and 2100 respectively as per the indication of current estimates of change in climate (DST and MOST, 2016).

Evidences of Climate Change Observed by the Farmers

Worldwide, climate scientists have been working on climate change, and they are continuously trying to assemble evidences of climate change. These evidences of climate change are assembled from various authentic scientific sources such as innumerable weather stations across the globe, satellite images, weather satellites, aircrafts etc. for the last hundreds of years. Increasing temperatures over land and ocean surfaces, melting of glacial, polar and arctic sea ice, rising sea level, increasing extreme events are some of the major evidences at national and international level of Fifth Assessment Report (2014) by IPCC (IPCC, 2014).

The majority of the respondents perceived that the amount of rainfall and number of rainy days are gradually decreasing due to climate change, and they shared their observations and put forward their views on how environmental conditions are changing over time. The respondents, who believe that the amount of rainfall is decreasing, reported that the requirement of irrigation for *Aman* rice cultivation and less amount of water in the surrounding inland water bodies and rivers are some of the indicators of decreasing rainfall. As high as 74.00% respondents observed that the length of the rainy season is decreasing over time (Fig. 16.3).

On average, more than 35% respondents' observation is that the onset and retreat of monsoon is becoming highly uncertain. Reduced amount of moisture in the soil before the *Rabi* and *Kharif* cultivation; uncertainty of onset and withdrawal; length and variability of summer and winter seasons; and number of foggy days are some of the major indicators of climate change as perceived by the farmers of the study area. According to the perception of the majority of the respondents, in recent years, the frequency, length and severity of some climatic extreme events such as, thunderstorm, *Kalbaishakhi* (Nor'wester) etc. and its impact such as drought are getting more while the frequency of floods are getting less, but the length and severity of floods are increasing due to climate change over the periods (Field Survey, 2021). Department of Science and Technology, Ministry of Science and Technology of Government of India, on climate change programme (DST and MOST, 2016) also stated that the frequency and severity of the climatic extreme events are receiving more in major parts of the country because of climate change.

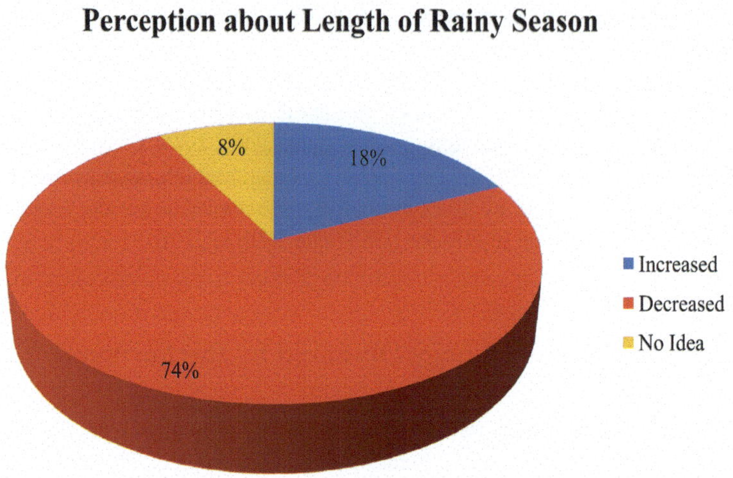

Fig. 16.3 Perception about length of rainy season. (*Source*: Field Survey, 2021)

Difficulties Faced by the Farmers

Almost the entire study area is composed of alluvial soils which offer a good foundation for agriculture. Like whole India, about two-third of the workforce in West Bengal directly and indirectly depends on farming, which has a multiplier effect on the economy of the study area. In spite of the crucial role of the economy, the farming sector of India in general and the study area in particular are facing difficulties due to climate change. The uncertainty, seasonal variability and decreasing amount of rainfall on the one hand and increasing temperature over the periods are the major challenges to the farmers. Besides these, some climate change-induced extreme events such as prolonged droughts, cyclonic storms and other natural hazards are the major concern for the farmers. About 25.84% respondent farmers reported that irrigation is becoming compulsory for cultivation throughout the years, while 10.53% said that they cannot cultivate *Aman* rice and 45.93% informed that farming is becoming expensive due to decreasing amount of rainfall. The majority of them (91.87%) also perceived that crop diseases are increasing due to climate change, which leads to low productivity and high expenditure in farming (Field Survey, 2021). Climate-induced disasters, especially droughts and cyclones, have become more disastrous in the recent years. These extreme events and disasters badly damage the crops. Even sometimes, they are not able to harvest their ripped crops due to the sudden occurrence of extremely severe climatic events. Deterioration of soil fertility, land degradation, desertification of agricultural land, increasing demand for irrigation round the year, increased crop diseases, increasing demand of pesticides and insecticides for each crop are on as a result the law of diminishing returns in agriculture in work and crop damage due to extreme climatic events are the major problems which the farmers of the study area are facing due to climate change

(Field Survey, 2021). The development reflected in the official reports also. Climate change and its adverse effects are the major challenges for Indian agriculture (DST and MoST, 2016). Hence, the consequences of climate changes on agriculture have devastated the lives and livelihood of poor farmers of the study area.

There is a regional variation of negative impacts on agricultural crops in terms of magnitude and variety due to climate change. In the Northern Plain, *Aman* and *Boro* rice, wheat, jute, potato, pulses, sugarcane, betel leaf, pineapple, tea plantation, pointed gourd; in Rarh region, *Aman* and *Boro* rice, wheat, potato, oil seeds, betelvine, pointed gourd, flowers have been impacted, while in Bengal Delta, *Aman* rice, *Boro* rice, potato, oil seeds, jute, sugarcane, pulses, banana and pointed gourd have been negatively impacted.

Sector-Wise Impacts Witnessed

The Indian agricultural sector is the most vulnerable sector to the risks due to climate change as two-third of Indian agricultural land is rain fed and even the irrigation system also depends on the monsoon rain. Other climatic extreme events such as frequent occurrences of floods, cyclones and thunderstorms are also major problems, which severely damage the Indian agriculture culture. In the study area, the Northern Plain and Gangetic Delta are highly flood prones which badly damage the agricultural system and crop production. Almost every year, a large amount of agricultural land with standing crops, mainly rice, pulses, jute and different types of vegetables is submerged due to the regular occurrence of floods in one part or another of these regions. The mighty Tista, Torsa, Sonkosh, Jaldhaka and Mahananda are the main rivers of the Northern Plain, which originate from the Himalaya and flow southward through the Northern Plain, flooding this region during the rainy season and affecting the lives and livelihoods of the farmers. But the Gangetic Delta, which comprises Murshidabad, Nadia, North 24 parganas, South 24 Parganas, Howrah, Hooghly, Kolkata, used to be flooded by mainly Ganga, Bhagirathi Jalangi, Hooghly, Churni rivers etc. The southern part of the Gangetic Delta is highly prone to cyclones along with floods. The very recent severe tropical cyclones 'Aila' (8 May in 2009), 'Amphan' (20 May 2020) and relatively weaker tropical cyclone 'Jawad' in (2021) with heavy rain severely damaged the standing crops and resultant crop production due to inundation of crops in the southern part of West Bengal. There has been widespread damage in the southern part of the Gangetic Delta, particularly North 24 Paraganas, Howrah, Hooghly, Kolkata, Nadia and parts of Murshidabad and the coastal part of West Bengal, specifically East and West Midnapurs and South 24 Paraganas, due to very severe tropical cyclone 'Amphan'. In the northern 24 Paraganas, north-eastern Nadia, southern part of Murshidabad and middle region of Hooghli district about 12.5% of the land under jute, i.e. 38,119 ha, was affected by severe tropical cyclone '*Amphan*' (Chakraborty et al., 2021). The Rarh Bengal, fringe region of Western Plateau proper, which consists of Birbhum, Bardhaman, West Midnapur, East Midnapore districts, used to get affected by cyclones and also drought prone. In this agro-climatic region, the occurrence of severe tropical cyclones and cyclones induced rain and resultant floods also affect the standing crops and crop productivity. The Western Plateau proper region, which consists of Purulia and Bankura, is highly drought prone. As per the opinions of the respondent farmers mainly from Gangetic Delta, Northern Plain and Western Plateau region, climate change negatively affected the cultivation of those crops that require either huge amounts of water or are not drought resistant such as rice, jute etc. As per their information, 29.41% farmers have stopped the cultivation of jute, while 15.31% abandoned both *Boro* and *Aman* rice cultivation due to scarcity of water and resultant high cost of crop production. But at the same time, they have increased the cultivation of crops which are less water thirsty and resistant to extreme heat. They have shifted to cultivate maize and vegetables to mitigate the problems of water scarcity and extreme heat. Fishing and animal husbandry are also negatively affected by contemporary climate change. About 46.41% respondents perceived that the fishing sector is damaged, while 22.97% perceived that animal husbandry is also negatively affected due to climate change (Field Survey, 2021). The perception and claim of the farmers have been supported by the crop data, which are provided by Agriculture Informatics Division, National Informatics Centre, Ministry of Communication and IT, Government of India. As per the record of Ministry of Agriculture and Farmers Welfare, the total rice and jute production of West Bengal significantly declined over the period of 1997–1998 to 2019–2020, while the maize production significantly increased in the state, particularly in Malda and Murshidabad during the same period of time.

Coping Strategies Adopted by the Farmers

Potential adaptation strategies to deal with the impact of climate change are developing the cultivation of crops tolerant to heat and salinity stresses and resistant to flood and drought, modifying crop management practices, improving water management, adopting new farm techniques such as resource-conserving technologies (RCTs) and crop diversification, improving pest management, better weather forecasts and crop insurance and harnessing the indigenous technical knowledge of farmers (DST and MOST, 2016, p. 9).

In the study area, farmers' perception is that climate change affected their agricultural sector negatively. Hence, they have adapted various measures to sustain agriculture and avoid losses of agricultural productivity due to current changing climatic conditions, especially temperature and precipitation.

To mitigate the problems of water scarcity, farmers are cultivating water-saving crops, more heat-resilient crops, high-yielding varieties of crops, cultivating profitable cash crops instead of food grains, applying more chemical fertilizers, insecticides and pesticides, installing tube wells at their own cost for irrigation using groundwater etc. Application of pesticides on each crop, changing the application pesticides, changing planting dates (standard meteorological week), crop diversification, changing crop variety are some of the coping strategies to mitigate the problems of enhanced crop diseases due to climate change. According to the respondents, 22.49% farmers emphasized on cash crops including vegetables instead of food crops, 15.31% have stopped *Aman* and putting emphasis on *Boro* rice cultivation, 17.22% emphasized on maize instead of rice cultivation, 17.22% emphasized on plantation farming instead of food crops, 20.57% started HYV rice cultivation, while another 7.18% still cultivate rice, which requires either less amount of water or have a short period of time (Field Survey, 2021). A large number of farmers in the study area, especially of Rarh and Western Plateau region, are harvesting rainwater, adapted micro-irrigation systems and cultivating a variety of crops which demand less water and are resistant to drought to sustain yield stability. A large section of farmers of the Northern Plain and Gangetic Delta informed that they have changed the land use pattern to avoid water scarcity. The huge amount of land, which was sown for food crops, especially rice and wheat, has been converted into plantation farming and non-agricultural purposes in these agro-climatic regions to maintain their livelihood. Yet another change in land use in the form of land being kept fallow for many cropping seasons where land degradation also has been witnessed (Field Survey, 2021).

Climate Change and Cropping Pattern

Change in cropping pattern is one of the major expressions of farmers' adaptation to the climate change environment. It is true that changes in cropping patterns may also be in response to the market price, particularly in the case of commercial crops. For the present study, one food crop, i.e. rice/wet paddy which is a leading food crop cultivated in West Bengal and jute which is a leading cash crop, has been considered for the analysis. Trends of cultivated land devoted to these crops vis-à-vis climatic elements such as precipitation and temperature have been analysed for the selected districts representing three broad agro-climatic meso regions of North Bengal Plain (Cooch Behar), Western Plateau and the plateau fringe (Birbhum) and Deltaic West Bengal (Nadia).

It has been observed from Figs. 16.4 and 16.5 that the overall trend in jute cultivation is declining over the period 1997–1998 to 2019–1920. Whereas there is clear declining trend in case of Cooch Behar and Nadia districts representing North Bengal Plain and Deltaic West Bengal, the Western Plateau (Birbhum) has recorded an overall declining trend but has witnessed many fluctuations. This has been possible due to the fact that the Western Plateau region under discussion has a poor record of jute cultivation and hence little bit of increase or decrease in the area under jute is exhibited through fluctuation in the graph. The entire crop weather calendar of jute cultivation requires a hot and humid climate with bounty rainfall. Even the post-harvest processing of jute retting requires a sufficient depth of flowing or stagnant water. The absence of water in the nearby water bodies may be a sufficient cause of a loss of the entire yield. The reduction in area under jute

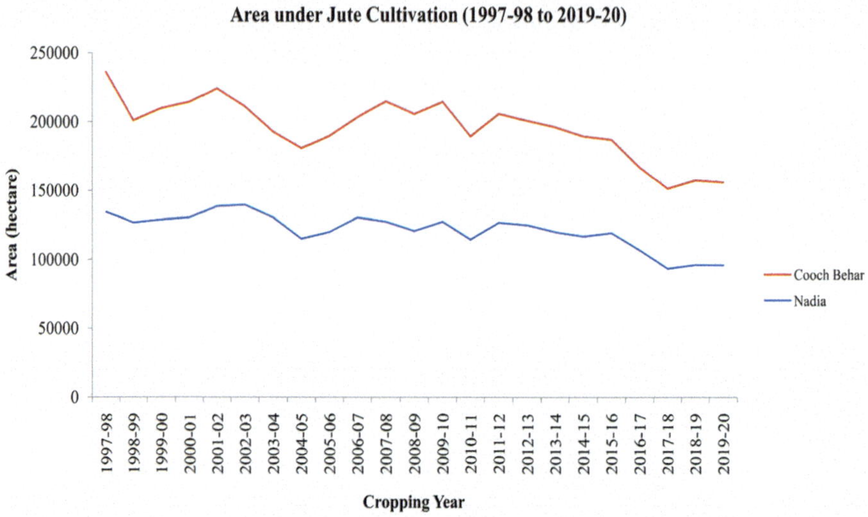

Fig. 16.4 Area under jute cultivation of Cooch Behar and Nadia districts during 1997–1998 to 2019–2020. (*Source*: Ministry of Agriculture and Farmers Welfare, Retrieved, November 20, 2021)

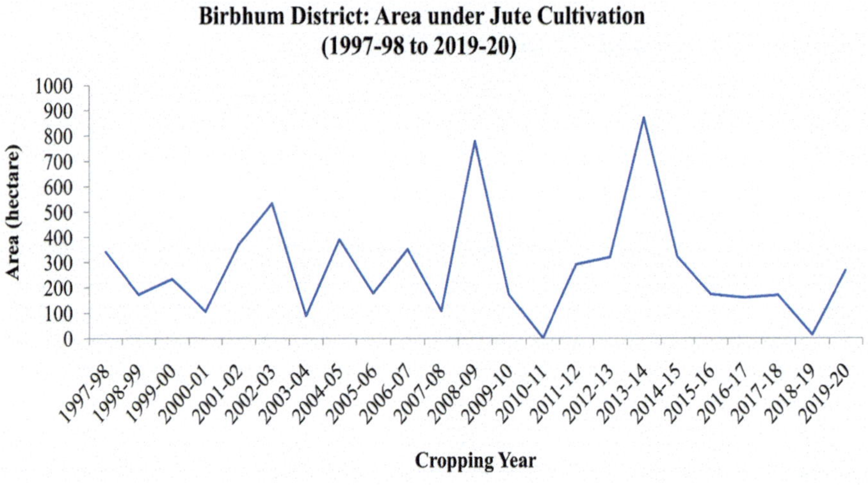

Fig. 16.5 Area under jute cultivation of Birbhum district during 1997–1998 to 2019–2020. (*Source*: Ministry of Agriculture and Farmers Welfare, Retrieved, November 20, 2021)

cultivation may, therefore, be related to a reduction in rainfall during the period observed.

Figure 16.6 on rice cultivation during the same period on the other hand shows a stagnant trend with a slight decline in the representative districts. Whereas Birbhum (Western Plateau) and Cooch Behar (North Bengal Plain) recorded overall stagnation during the entire period, Nadia (Deltaic West Bengal) recorded a decline.

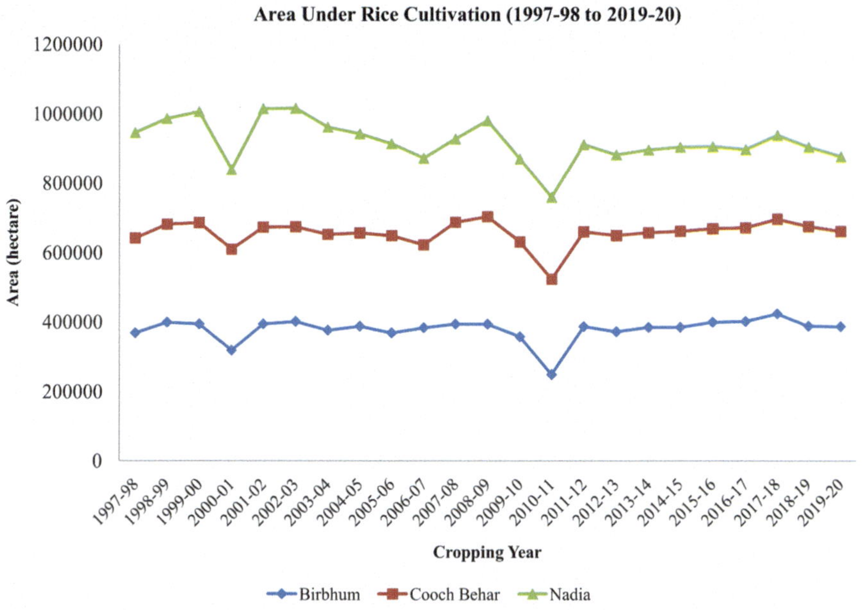

Fig. 16.6 Area under rice cultivation of Birbhum. Cooch Behar and Nadia district during 1997–1998 to 2019–2020. (*Source*: Ministry of Agriculture and Farmers Welfare, Retrieved, Nov 20, 2021)

As the entire paddy cultivation practiced today in West Bengal is wet paddy, it requires either sufficient rainwater or irrigation mostly surface or groundwater. In both cases, rainfed and surface irrigation, enough precipitation is the requirement of the process. The irrigation method used both in the *Kharif* (monsoon) and Rabi (winter-spring) is a flooding method which requires a huge quantity of water. Moreover, the rate of evapotranspiration is the highest during the Boro rice cultivation during winter–spring season due to very low vapour pressure in the atmosphere. The stagnation and decline of area under paddy cultivation again maybe thus related to insufficient rain and increased cost for irrigation based on groundwater.

Though there are many weather elements that change with climate change, for the convenience of the present study, precipitation and temperature (2 m above the surface) have been studied. Figure 16.7 depicts the trends of precipitation in the three representative districts. The data show a very slight increasing trends in the Birbhum and Nadia districts, whereas Cooch Behar has recorded a considerable increase during the period. In spite of an increasing trend, farmers have turned their face as revealed by them towards water-intensive paddy and jute cultivation due to loss inflicted during the prolonged dry spell prevailed during the latter half of the 1990s and first half of the 2010s.

The farmers opined that during the last decade or so, unreliable rainfall along with high cost of fuel used for irrigation has forced farmers to show less interest in the cultivation of jute and rice. The data show that Maize, particularly spring maize, is

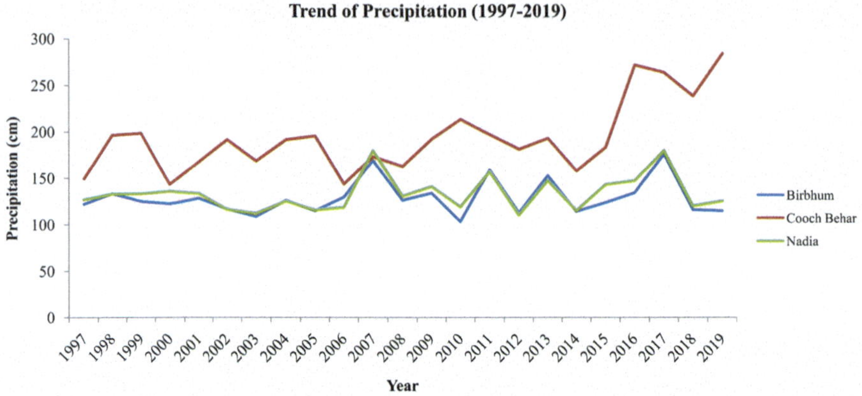

Fig. 16.7 Trends of precipitation of Birbhum. Cooch Behar and Nadia district. (*Source*: Authors' Development based on Power Data Access Viewer, NASA)

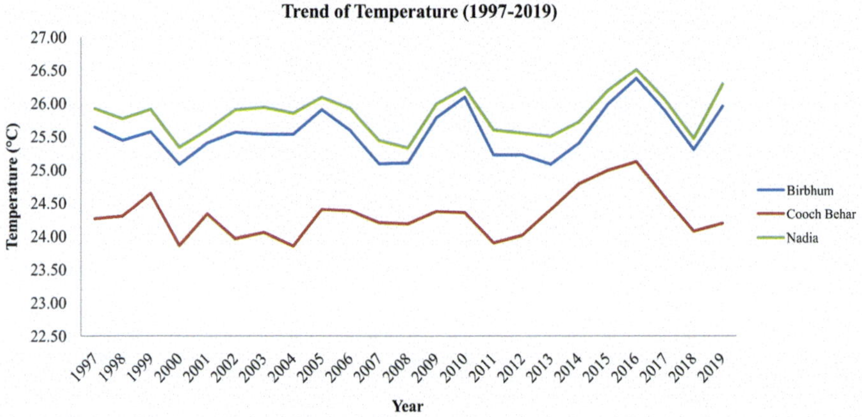

Fig. 16.8 Trend of temperature of Birbhum. Cooch Behar and Nadia district. (*Source*: Authors' Development based on Power Data Access Viewer, NASA)

slowly replacing *Boro* rice in the entire state which can be seen a jump of more than seven times during the period of 1997 to 2020.

Another important weather element, i.e. temperature in the study area, shows a stagnant to slight declining trends (Fig. 16.8). However, unusual peak temperature in 2015–2016 and the persistence of very high temperatures during 1980s and initial years of 1990s made farmers more apprehensive of relying on more water-intensive and less heat-resistant crops. It has been reported by the farmers during the field survey that they have experienced and apprehended increasing temperatures detrimental to their crops. Thus more heat-resilient and water-thirsty crops are the new choice of the farmers in the region.

Trend of Climatic Elements and Yield of Crops

After having observed the trends of cultivation of different crops in terms of area devoted to crops vis-à-vis trends of weather elements, it is pertinent to examine the trend of yield of different crops with respect to trends of climatic elements. For the convenience of this work, only temperature and precipitation have been observed during 1997–2020 in the representative districts of Birbhum (western Plateau), Nadia (Deltaic West Bengal) and Cooch Behar (North Bengal Plain). A multiple linear regression (MLR) model has been run taking yield of crops as a dependent variable and temperature and precipitation as independent variables. The MLR has been worked out for both rice and jute, the leading crops, first being the leading cereal and also staple food grain consumed by the people of the state, and jute, the leading cash crop cultivated in the major parts of the state.

The MLR analysis to examine the effect of changing temperature and precipitation on the yield of rice has been done with the help of Eq. (16.1).

$$y = b_0 + b_1x_1 + b_2x_2 + e \tag{16.1}$$

y = dependent variable (yield of rice).
x_1 = explanatory variable (temperature).
x_2 = explanatory variable (precipitation).
b_0 = intercept (constant term).
b_1 = slope coefficient of first explanatory variable (temp).
b_2 = slope coefficient of second explanatory variable (precipitation).
e = the model's error term (also known as the residuals).

Similarly, in the second Eq. (16.2) the effect of changing temperature and precipitation on the yield of jute has been done with the help of equation.

$$y = b_0 + b_1x_1 + b_2x_2 + e \tag{16.2}$$

where the independent variable (y) is the yield of jute. The remaining notation in the second equation remains as in the first equation.

It has been found from the MLR that there is a negative relation of the yield of rice and both temperature and precipitation. However, temperature is having a stronger impact on the yield of rice than precipitation in Birbhum district (Table 16.1). But both are found to be statistically insignificant as *P*-value is more than 0.05 in both cases.

In the case of Cooch Behar, it is found that both temperature and precipitation are found to have negative relations with the yield of rice, but temperature is having stronger relations as compared to precipitation. From the inspection of *P*-value, it is found that temperature does not have a statistically significant relation with yield of rice in Cooch Behar (*P*-value is more than 0.05) and precipitation has a statistically significant relation with yield of rice as the P-value in this case is less than 0.05 (Table 16.2).

Table 16.1 Impact of temperature (X_1) and precipitation (X_2) on rice production (Y)

Summary output of ANOVA statistics					
Representative district		Coefficients	Standard error	*t* Stat	*P*-value
Birbhum	Intercept	−3.376	3.730	−0.905	0.376
	Temperature (X_1)	0.234	0.143	1.643	0.116
	Precipitation (X_2)	0.003	0.003	1.168	0.257
Cooch Behar	Intercept	−5.539	6.397	−0.866	0.397
	Temperature (X_1)	0.251	0.267	0.938	0.359
	Precipitation (X_2)	0.008	0.002	3.385	0.003
Nadia	Intercept	−0.817	4.356	−0.188	0.853
	Temperature (X_1)	0.126	0.167	0.752	0.461
	Precipitation (X_2)	0.004	0.003	1.307	0.206

Source: Computed by Authors based on APY data, Ministry of Agriculture and Farmers Welfare, Government of India and Power Data Access, NASA

Table 16.2 Impact of temperature (X_1) and precipitation (X_2) on jute production (Y)

Summary output of ANOVA statistics					
Representative districts		Coefficients	Standard error	*t* Stat	*P*-value
Birbhum	Intercept	−47.467	59.147	−0.803	0.432
	Temperature (X_1)	2.470	2.260	1.093	0.287
	Precipitation (X_2)	0.018	0.043	0.424	0.676
Cooch Behar	Intercept	−10.381	24.187	−0.429	0.672
	Temperature (X_1)	0.726	1.011	0.718	0.481
	Precipitation (X_2)	0.021	0.009	2.313	0.032
Nadia	Intercept	1.303	20.354	0.064	0.950
	Temperature (X_1)	0.473	0.780	0.606	0.551
	Precipitation (X_2)	0.009	0.013	0.659	0.517

Source: Computed by Authors based on APY data, Ministry of Agriculture and Farmers Welfare, Government of India and Power Data Access, NASA

On the other hand, the yield of jute is found to be related negatively to temperature and precipitation in Birbhum and Cooch Behar and positively in Nadia during the period studied. It is however observed that yield of jute is much more strongly related to temperature than precipitation. However, both independent variables have a statistically insignificant impact on the productivity of jute as it is exhibited from the *P*-values that are more than 0.05. The relations between the explained and explanatory variables are found to be negatively related. Again it is found that temperature is much more strongly related than precipitation but negatively related. The *P*-value for temperature is more than 0.05 and hence has an insignificant relation with the yield of jute, but precipitation has a statistically significant impact as exhibited by the *P*-value less than 0.05, indicating a statistically significant relation. In the Nadia district, it is found that the yield of jute is positively related to both temperature and precipitation; however, precipitation has having weaker relation than temperature as exhibited by the coefficients. But the *P*-values for both cases are found to be more than 0.05, and hence, the independent variables are found to be statistically insignificant (Table 16.3).

Though the analysis has been done only for three selected districts, representing three agro-climatic regions but glancing through the data shows a similar trends for other districts too.

Government Initiatives Demanded by the Farmers

The Government of India has initiated timely action to address the issues of climate change. These initiatives have dealt with valuable inputs to offset the impact of climate change on major food crops along with horticulture and livestock productions. India has launched the National Mission for Sustainable Agriculture (NMSA) to come up with suitable adaptation strategies. The NMSA has identified various dimensions for adaptation to climate change such as improved seeds, farm practices and nutrient management; pest management; water use efficiency; markets; credit support; agricultural insurance; access to information and livelihood diversification etc. (DST and MOST, 2016).

The Government of India has initiated several welfare programmes to cope with climate change-induced farming problems and for the welfare of the farmers for sustainable agriculture and the continuation of better farming practices to maintain the stability of farm production. The Prime Minister of India launched *PMFBY* on 18 February 2016 (Department of Financial Service, 2016). The farmers who are suffering from crop loss or damage due to unforeseen events are getting financial support through PMFBY (Department of Agriculture, Cooperation and Farmers Welfare, 2016). The Government of India has implemented *PMKSNS*, popularly known as *PM-Kisan Scheme*, by which the farmers' families having cultivable land are getting an amount of Rs. 6000 per annum directly into the bank accounts under Direct Benefit Transfer mode by the Central Government (DAC & FW, 2020). The farmers of West Bengal are getting financial support up to a maximum amount of Rs. 10,000 per year and a minimum amount of 4000 per year from the Government of West Bengal through *Krishak Bandhu* scheme which is implemented in January 2019 for the assistance to the farmers for accelerating the of farming activities in West Bengal (Department of Agriculture, Government of West Bengal, 2019). The Government of West Bengal (n.d.) offers new pump sets with power connectivity to the farmers of West Bengal through *The Sech Bandhu Prakalpa*. The Water Resources Investigation and Development Department (WRIDD), GoWB (2011–2012) is harvesting rainwater in all kinds of water bodies, viz. reservoirs, ponds, tanks, canals etc., by launching a programme named *JDJB*. The agricultural labourers and marginal farmers along with other labourers are also getting jobs through this programme as it is engaged in re-excavation of ponds, tanks and other water bodies under MGNREGA in combination with P&RD Department, GoWB. As per the government record of Department of Statistics and Programme Implementation of GoWB (2015), a total of 145,743 water body structures have been

Table 16.3 Summary output of regression statistics for rice summary output of regression statistics for jute

Districts	Multiple *R*	R^2	Observations	Districts	Multiple *R*	R^2	Observations
Birbhum	0.382	0.1457	23	Birbhum	0.241169	0.058163	23
Cooch Behar	0.656	0.4301	23	Cooch Behar	0.516055	0.266313	23
Nadia	0.311	0.0966	23	Nadia	0.189783	0.036018	23

Source: Computed by Authors based on APY data, Ministry of Agriculture and Farmers Welfare, Government of India and Power Data Access, NASA

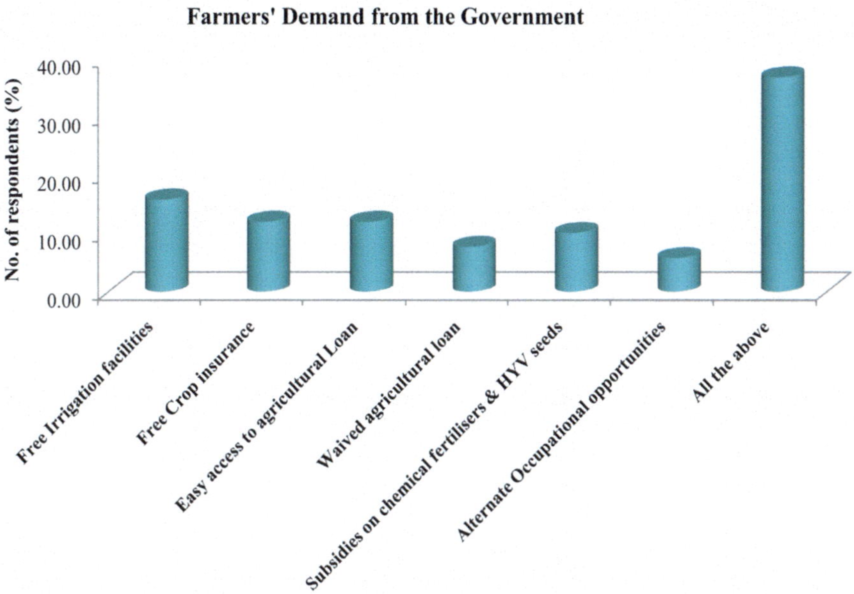

Fig. 16.9 Farmers demand from the government to offset the impact of climate change. (*Source*: Field Survey, 2021)

created or renovated as of December 2015 under JDJB programme, which meet the growing needs for water throughout the year. Under *West Bengal Accelerated Development of Minor Irrigation Project* (*WBADMIP*), the WRIDD of GoWB has constructed various minor irrigation systems such as check dams, river lift schemes, water retention structures, tube wells, pump dug wells etc., which lead to better water management practices and resultant growth of total cropped area, crop diversity, cropping intensity and ultimately growth of total production.

In spite of several welfare measures implemented by the Indian Government and Government of West Bengal, the farmers throughout the country could not take the benefits of these programmes due to some difficulties and terms and conditions attached to the said programmes. In these circumstances, the farmers of the study area have demanded some mitigating measures from the governments such as free irrigation facilities, free crop insurance, easy access to agricultural credits, subsidies on chemical fertilizers and HYV seeds, guarantee of MSP etc. for all farmers for all agricultural seasons to cope with climate change induced problems. Besides these, they have demanded to waive agricultural loans if crops are damaged by natural calamities (Fig. 16.9).

Strategies to Reduce the Severity of Climate Change

The Ministry of Agriculture, Government of India has launched the National Initiative on Climate Resilient Agriculture (NICRA) for long-term research to assess the impact and to formulate appropriate adaptive strategies to reduce the impact of climate change on Indian agriculture including horticulture, livestock and fisheries. The following strategies should be taken to reduce the severity of the impact of climate change on the agricultural crops including fisheries and livestock keeping:

I. New varieties of crops should be developed with higher yield potential and highly resistant to multiple stresses such as drought, flood, salinity. "Genetic engineering could play a pivotal role for 'gene pyramiding' to pool all desirable traits in a plant to get the 'ideal plant type' which may also be 'adverse climate tolerant' genotype". (DST and MOST, 2016, p. 9).

II. Emphasis on improved water conservation, water harvesting and improvement of irrigation accessibility, micro-irrigation systems should be developed and selection of suitable crop-based irrigation and water availability has to be promoted to sustain yield stability.

III. Improved technologies and investment are needed for Integrated Water Management Systems. Micro-watershed management and rainwater harvesting will be very effective alternatives for poor farmers in the rainfed areas especially in the Rarh and Western Plateau region of the study area. Micro-watershed management systems and proper rainwater harvesting will also be very fruitful in the Northern Plain because this region has numerous rivers and rivulets which originate from the Himalaya. Besides this, the agro-climatic region receives copious amounts of rainfall during the monsoon season. Improved irrigation methods like sprinkler irrigation, deep pumping, drip irrigation may be the better alternatives to sustain the tea production in the foothill areas of Darjeeling Himalaya round the year. Different types of permanent structures like water harvesting tanks, check dams etc. are needed for micro-watershed management and rainwater harvesting.

IV. The planting date of the crops could be adjusted by adopting a new crop weather calendar and it should be as per local changed climatic conditions so that the flowering and harvesting period do not coincide with any extreme events. The cropping pattern and crops should be changed in the areas where floods, cyclones, heat waves etc. occur frequently. For example, *Aman* rice–pulses and *Boro* rice, Maize–*Aman* rice–wheat, jute–*Aman* rice–Potato and other vegetables are good crop combinations in the Gangetic Delta region. Mushroom and Pisciculture are appropriate in the low lying and persistently water-logged areas.

V. Clearing drainage system is essential in the river bank erosion and flood-prone areas particularly in Malda and Murshidabad districts to avoid the severity of rain-induced flood during the rainy season. It would be wise to river plays in the floodplain region so that the flood severity will decrease and the revival of the soil fertility with deposition of new alluvium will benefit the farmers in the long run.

VI. Forecasting of weather and prediction of extreme climatic events should be done well in advance so that farmers get opportunities for crop management options which will minimize damage to crops.

VII. The Agriculture and Farmers Welfare Departments of both central and state governments should provide proper training in terms of integrated pest management, which will minimize crop diseases and cost of production. This department may help the farmer to identify proper crops for a location to minimize crop diseases and resultant high production.

VIII. Free crop insurance should be arranged for all the farmers, including sharecroppers and leaseholders, which will reduce the risk of crop failure due to climate-induced extreme events and increased attacks of pests and insects etc.

Adaptive Measures to Overcome the Causes of Climate Change

Climate change as a global problem is an urgent threat to millions of lives on the planet Earth. Though climate change is happening globally and it is a worldwide problem, the cause and effect of climate change vary from one region to another. Hence, people can minimize and reduce the pace of the causes of climate change by the rational utilization of natural resources and changing their consumer behaviour. The people of India in general and of West Bengal in particular can overcome or minimize the causes of climate change by adapting the following:

(a) *Limited use of fossil fuels*: Excessive and wasteful use of fossil fuels like coal, petroleum and natural gas is one of the reasons behind climate change. India needs to move from the use of fossil fuel-based conventional energy sources to renewable resources such as solar energy, hydrothermal power, biomass energy, geothermal power, wind energy etc. as soon as possible. As the study area comes under the hot and humid tropical region and has a long coastal area, it has huge potential for solar, hydro, wind, wave and tidal energy. Improved technology and adequate investment are needed to tap and harness these potentials.

(b) *Curtailing deforestation*: Deforestation is a very pertinent issue that India is facing today. Though, the India State Forest Report (2019) shows an increase in total forest cover of 0.65% as compared to the previous assessment, i.e. ISFR (2017), actual forest cover has not increased. Deforestation can accelerate climate change which leads to climate extremes, soil erosion, desertification, land degradation, increased crop diseases and cost of crop production, increased greenhouse gases in the atmosphere etc. One of the main causes of deforestation in India is expansion of agricultural land to meet the growing needs of the huge population of the country. The rapid urbanization and industrialization, mining and quarrying activities especially in the north-eastern states, construction of roads and railway tracks, irrigation canals, growing demands for wood-made furniture etc. are also responsible for deforestation. Improved agricultural practices, recycling paper, sustainable industrial development, judicious urban development, limited use of wooden furniture, proper forest management can minimize deforestation.

(c) *Start reforestation*: As per the India State Forest Report (2019), total forest cover of the country is 24.56% of the country's total geographical area, but it should have been at least 33% of the total geographical area for a healthy environment. There is a huge scope of reforestation in the coastal area, plateau region, degraded land, deforested land, governments' fellow land etc. in West Bengal and India. So, the governments at both the province and state level should take proper steps for reforestation in suitable areas.

(d) *Switch to sustainable and eco-friendly transport*: Being a densely populated country, India has a huge number of petrol and diesel vehicles which consume large amounts of fossil fuels every day. Battery-driven vehicles, electric vehicles and other eco-friendly vehicles shared traveling instead of use of personal cars etc. and mass transportation systems have to be promoted which can largely arrest pollution and resultant climate change.

(e) *Banned use of plastic*: Plastic has enveloped our lives, but it is a dangerous pollutant. It does not break down in nature for a long time. We can't burn it, we can't bury it. Extensive use of plastic has affected the ecosystem. So, the Governments should ban the production and use of harmful plastic by implementing stringent rules and people of the country must have to follow those. Besides this, governments should promote production and use of good alternatives to plastic, so that people can easily adapt to the new products spontaneously. Jute and other natural fibre-made alternatives are viable alternatives.

(f) *Lifestyle should be changed*: A large section of the people of the country have habituated with show-off culture and fashionable lifestyle, which leads production of greenhouse gas etc. causing climate change. For example, people used to purchase some unnecessary furniture, unnecessary foods and other unnecessary articles which exert pressure on the natural resources and resultant excess pressure on the planet Earth. People can make their house design as per the climatic condition to avoid the use of room heater and air cooler to maintain a comfortable room temperature. People also can minimize the use of refrigerators, which release greenhouse gases that lead to climate change. Architecture can always be so that air and light is used to minimize wasteful use of fossil fuels.

(g) *Move closer to the workplace*: In India, transportation is one of the leading sources of greenhouse gas emissions. If people used to stay close to their workplace, then use for vehicular movement will reduce.

People of the country should recall Principle-1, proclaimed in the Rio-Declaration on Environment and Development by U.N.O. (1992), i.e. '*Human beings are at the centre of concerns for sustainable development. They are entitled to a healthy and productive life in harmony with nature*'.

Conclusion

From the above discussion it may be concluded that along with the global concern, the climate of the study area is changing and it has been perceived by the farmers. With a changing climatic pattern, they are facing a number of challenges to continue their agricultural activities and to sustain the crop yield. Under the changed climatic conditions and resultant difficulties, they are exposed to climate vulnerability. The people have resorted to changing cropping patterns and adopting various measures to neutralize the negative impacts on the crop yield as well as their lives and livelihood. The appropriate initiatives and incentives to the farmers for adaptation of mitigative measures by the governments to withstand and cope with the problems of climate change and face the challenge due to exposure to vulnerability may solve the problem to a great extent. The sustainable development approach of the farmers to adopt climate-suitable crops and practices, along with the sustainable behaviour of the people, may reduce the severity and overcome the causes of climate change.

Acknowledgements At the outset, we express our deep sense of gratitude and sincere thanks to the editors of the book titled *Climate Change, Vulnerability and Adaptation in India* for giving us an opportunity to be an integral part of it. We would also like to convey our sincere thankfulness and gratitude to the surveyors who have given their valuable time for the collection of data across the study area in the ambiance of Covid-19 pandemic. It would be failing in our duties if we do not acknowledge the fact that the surveyors have visited the farmers' houses and interviewed the respondents to avoid the risk of infection of Covid-19. We were not in a position to complete this work without their heartfelt cooperation. Most importantly, we are indebted to the farmers who have cooperated by sharing their views and information without which this work would be incomplete. We are also highly grateful to the publisher for the publication of this valuable book, which is composed of very pertinent research work regarding climate change and adaptation in India.

Declaration on Plagiarism The authors would like to express that the paper is free from substantive plagiarism beyond permissible limits. The limited similarity may occur due to references and other unavoidable parts used in this paper. No part of this paper has been published or submitted anywhere else for publication. It will not be communicated anywhere else till the final discussion is taken by the present editor(s).

Declaration on Conflict of Interest The authors declare that there is no conflict of interest.

Financial Assistance/Sponsorship No financial assistance in any form is received for this work.

References

Chakraborty, A., Srikanth, P., Murthy, C. S., Rao, P. V. N. & Chowdhury, S. (2021). Assessing lodging damage of jute crop due to super cyclone Amphan using multi-temporal Sentinel-1 and Sentinel-2 data over parts of West Bengal, India. *Environmental Monitoring and Assessment*, 193:464. Springer. pp.1–18 Retrieved July 15, 2021 from https://doi.org/10.1007/s10661-021-09220-w

Department of Agriculture. (2019). *Krishak Bandhu*. Government of West Bengal. Retrieved December 5, 2021, from https://krishakbandhu.net/about_kb

Department of Agriculture, Cooperation and Farmers Welfare. (2016). *Operation guidelines of Pradhan Mantri Fasal Bima Yojana*. Ministry of Agriculture & Farmers Welfare KrishiBhawan. https://agricoop.nic.in/sites/default/files/PMFBY_Guidelines.pdf

Department of Agriculture, Cooperation and Farmers Welfare. (2020). *Operational guidelines of Pradhan Mantri Kisan Samman Nidhi Scheme*. Ministry of Agriculture & Farmers Welfare Krishi Bhawan. Retrieved December 5, 2021, from https://pmkisan.gov.in/Documents/RevisedPMKISANOperationalGuidelines(English).pdf

Department of Financial Service. (2016). *Pradhan Mantri Fasal Bima Yojana (PMFBY)*. Ministry of Finance, Government of India. Retrieved January 4, 2022, https://financialservices.gov.in/insurance-divisions/Government-Sponsored-Socially-Oriented-Insurance-Schemes/Pradhan-MantriFasal-BimaYojana(PMFBY)#:~:text=The%20Pradhan%20Mantri%20Fasal%20Bima,scheme%20in%20Rabi%202016%2D17

Department of Science & Technology, Ministry of Science &Technology. (2016). Climate change and agriculture in India:A Thematic Report of National Mission on Strategic Knowledge for Climate Change (NMSKCC) under National Action Plan on Climate Change (NAPCC). In A. Gupta & H. Pathak (Eds.), *Climate Change Programme Strategic Programmes Large Initiatives and Coordinated Action Enabler (SPLICE) Division* (pp. 2–68). Department of Science & Technology, Ministry of Science & Technology Government of India. Retrieved January 4, 2022, from https://dst.gov.in/sites/default/files/Report_DST_CC_Agriculture.pdf

Department of Statistics and Programme Implementation. (2015). State Statistical Handbook. Bureau of Applied Economics and Statistics. Government of West Bengal. Retrieved July 30, 2020 from http://www.wbpspm.gov.in/publications/Statistical%20Hand%20Book

Folnovic, T. (n.d.). *Climate change impacts on agriculture*. Retrieved October 10, 2021, from https://blog.agrivi.com. https://blog.agrivi.com/post/climate-change-impacts-on-agriculture

Forest Survey of India. (2019). *India state of forest report 2019*. Ministry of Environment, Forest & Climate Change Government of India. Dehradun, Uttarakhand. Retrieved January 2, 2022, form http://www.indiaenvironmentportal.org.in/files/file/isfr-fsi-vol1.pdf

Government of West Bengal. (n.d.). *Sech Bandhu*. https://wb.gov.in/government-schemes-details-sechbandhu.aspx

India Metrological Department. (2008). Climate of West Bengal. Controller of Publications, Government of India. Civil Lines, New Delhi-110054. 3–43, Retrieved July, 15, 2022, from https://imdpune.gov.in/library/public/Climate%20of%20WestBengal.pdf

IPCC. (2014). Climate change 2014: Synthesis report. In Core Writing Team, R. K. Pachauri, & L. A. Meyer (Eds.), *Contribution of Working Groups I, II and III to the fifth assessment report of the Intergovernmental Panel on Climate Change* (p. 151). IPCC. Retrieved November, 5, 2021 from https://www.ipcc.ch/site/assets/uploads/2018/05/SYR_AR5_FINAL_full_wcover.pdf

Kumar, K. K., Kumar, K. R., Ashrit, R. G., Deshpande, N. R., & Hansen, J. W. (2004). Climate impacts on Indian agriculture. *International Journal of Climatology, 24*, 1375–1393. https://doi.org/10.1002/joc.1081

Ma, J., Weng, B., Bi, W., Xu, D., Xu, T., & Yan, D. (2019). Impact of climate change on the growth of typical crops in karst areas: A case study of Guizhou Province. *Advances in Meteorology, 2019*, 1–16. https://doi.org/10.1155/2019/1401402

Ministry of Agriculture & Farmers Welfare. (n.d.). *Special Data Dissemination Standard Division, Directorate of Economics & Statistics*. Government of India. Retrieved November 20, 2021, from https://aps.dac.gov.in/APY/Public_Report1.aspx

NASA. (n.d.). *Power data access viewer*. Retrieved December 16, 2021, from https://power.larc.nasa.gov/data-access-viewer/

OECD. (2015, April). *Agriculture and climate change: Towards sustainable, productive and climate-friendly agricultural systems*. Retrieved October 28, 2021, from oecd.org: https://www.oecd.org/agriculture/ministerial/background/notes/4_background_note.pdf

Sivakumar, M. V. K., Das, H. P., & Brunini, O. (2005). Impacts of present and future climate variability and change on agriculture and forestry in the arid and semi-arid tropics. *Climate Change, 70*, 31–72. https://doi.org/10.1007/s10584-005-5937-9

UNO. (1992). *Report of the United Nations conference on environment and development*. General Assembly, United Nations. Retrieved January 2, 2022, from https://www.iau-hesd.net/sites/default/files/documents/rio_e.pdf

U. S. Global Change Research Program. (2014). *Highlights of climate change impacts in the United States: The third national climate assessment*. Retrieved October 20, 2021, from https://www.globalchange.gov/. https://www.globalchange.gov/browse/reports/highlights-climate-change-impacts-united-states-third-national-climate-assessment

U. S. Global Change Research Program. (2018). *Volume II: Impacts, risks, and adaptation in the United States: Fourth national climate*. U.S. Global Change Research Program. Retrieved October 20, 2021, https://nca2018.globalchange.gov/downloads/NCA4_2018_FullReport.pdf

Water Resources Investigation and Development Department. (2015). *Jal Dharo-Jal Bharo*. Government of West Bengal. Retrieved January 20, 2022, from http://www.wbwridd.gov.in/index.php/wridd/jal_dharo_jal_bharo

Water Resources Investigation and Development Department. (n.d.). *West Bengal Accelerated Development of Minor Irrigation Project (WBADMIP)*. Government of West Bengal. Retrieved January 20, 2022, from https://www.wbadmip.org/documents/Irrigation_system_Development.pdf

World Bank. (2021, April 5). *Climate-smart agriculture: Overview*. Retrieved November 2, 2021, from https://www.worldbank.org/. https://www.worldbank.org/en/topic/climate-smart-agriculture

Chapter 17
Modelling the Impact of Climatic Parameters on Dynamics of Malaria Transmission in Chhattisgarh State of India

Shambhavi Krishna and Shailendra Rai

Introduction

Malaria has always emerged as a problem for India for centuries. Currently, one-fourth of the world population is infected by malaria disease (Source: World Malaria Report 2019). In the year 2019 and 2020 countries under all high burden to high impact (HBHI) only India reported increases in malaria cases and deaths, which are estimated to be 14 million more cases and 47,000 more deaths in 2020 than in 2019 (Source: World Malaria Report 2021). In India, mostly the backward, poor, forested and remote areas having rural and tribal populations are mainly affected. The states of India with the maximum burden of diseases are as follows: Orissa, Chhattisgarh, southern Madhya Pradesh, Jharkhand, Rajasthan, Gujarat, Maharashtra, Karnataka, Goa, Meghalaya and Mizoram. The states of Orissa, Chhattisgarh, Jharkhand, Meghalaya and Madhya Pradesh contribute the maximum to the country in terms of malaria cases, accounting for nearly 45.47% and 63.64% deaths, with Orissa ranking first, followed by Chhattisgarh in the second position in reference to the malaria disease burden of the country (Source: World Malaria Report 2020). The malaria disease involves three main pillars, i.e. vector, parasite and host. There are basically nine Anopheles vectors present in India due to its diverse geoecological patterns. The two main parasites involved in the disease are *Plasmodium falciparum* (more in remote, forested and tribal areas) and *Plasmodium vivax* (more in plains). As mentioned above, among nine different vectors of Anopheles, *Anopheles*

S. Krishna (✉)
K Banerjee Centre of Atmospheric and Ocean Studies, University of Allahabad, Prayagraj, UP, India

S. Rai
K Banerjee Centre of Atmospheric and Ocean Studies, University of Allahabad, Prayagraj, UP, India

M. N. Saha Centre of Space Studies, University of Allahabad, Prayagraj, UP, India

A. L. Singh et al. (eds.), *Climate Change, Vulnerabilities and Adaptation*,
https://doi.org/10.1007/978-3-031-49642-4_17

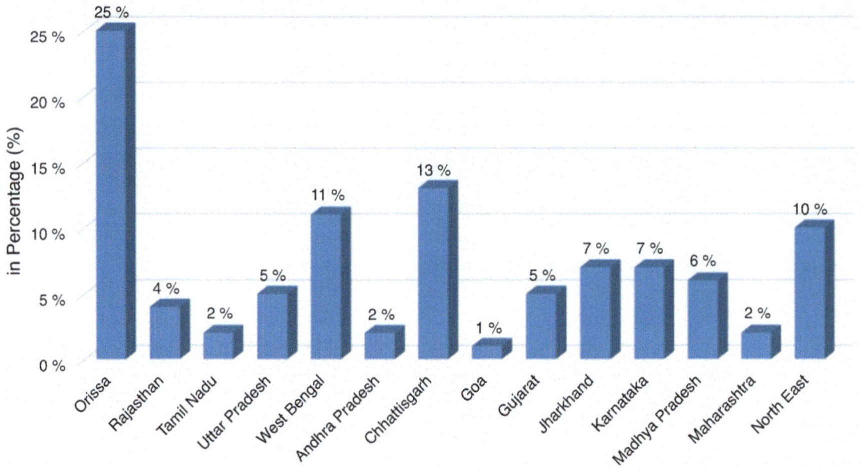

Fig. 17.1 Percentage distribution of malaria incidences in different states of India

culicifacies contributes to 70% of malaria cases in the states with undulating forest cover, deep valleys, hills and hillocks such as Orissa, Chhattisgarh and other central and eastern regions of India (Subbarao et al., 2019).

For a long period of time, Chhattisgarh is the state that has contributed the maximum to the malaria cases and deaths in India after Orissa. It is observed from Fig. 17.1 that Orissa in the year 2002 contributed 25% of malaria cases followed by Chhattisgarh with 13% of the contribution among total annual malaria incidences in the country (Source: NVBDCP Report 2002, India), which rose to 18% in the year 2019 as compared to other states contributing 82 percent of total malaria positive cases having contribution of only 2.29% to the total population of India (Source: Malaria Mukhta Baster report 2019). From Fig. 17.2, it is seen that annual malaria incidence in Chhattisgarh shows maximum variation during the time period of 2012–2019, which is a matter of deep research. The reason might be the launch of a program by the WHO named High Burden to High Impact (HBHI) in four states of India in July 2019, and Chhattisgarh is among those states that reported a decline of 23.20% of the malaria API in 2019 than in 2018 (Source: World Malaria report 2020). But the other factor involved could be the impact of climate change over the variations seen in weather variables, namely rainfall and temperature, in the region of Chhattisgarh in reference to the human health.

There have been studies and research conducted till now in various parts of the world through various models in data science for the better understanding of identifying endemic regions of malaria and its transmission in many new areas of the world. For the past many years, several mathematical models have provided an explicit framework for the understanding of malaria dynamics with the help of various weather variables such as temperature and precipitation including population dynamics. The main mathematical models famous across the globe used for studying climate suitability for transmission of malaria disease are LMM_RO (Ermert et al.,

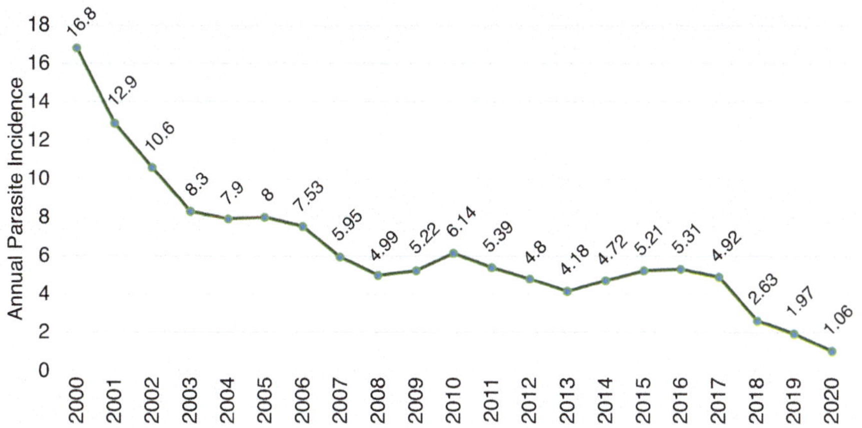

Fig. 17.2 Year-wise annual parasite incidence (API) in Chhattisgarh during 2000–2020. (Source: Malaria Mukhta Baster Report 2019)

2011), MIASMA (Lieshout et al., 2004), VECTRI (Tompkins & Ermert, 2013) and MARA (Kleinschmidt et al., 2001). Among these, the VECTRI model is not only responsible for studying the impact of weather parameters basically mean rainfall and temperature on malaria transmission but can also be used for the analysis in terms of surface hydrology and population dynamics of a particular area.

Therefore, the focus of this study involves the investigation of malaria transmission in the state of Chhattisgarh by the use of the VECTRI model through temporal variability including both climatic and non-climatic factors. VECTRI model aims at the study of the impact on the larvae, parasite and vector populations due to rainfall and temperature variations. The outmost important feature of the VECTRI model is that it can be resolved to a spatial resolution of 10 km or less (Tompkins & Ermert, 2013). The basic parameterization structure of the VECTRI model involves the characteristics of *Anopheles gambiae* vector and *Plasmodium falciparum* parasite over the African region (Tompkins & Ermert, 2013). But due to absence of *Anopheles gambiae* in the study area, the species present is *Anopheles culicifacies and thus the modifications are done accordingly*. Thus, it involves some modifications in the parameters of the model over the Chhattisgarh for the study of the transmission of malaria for a period of 20 years (2000–2019) as well as its dynamics, and other factors provided include population, rainfall and temperature of the study area for the smooth running of the VECTRI model.

Data and Methods

Study Area

The area of research is Chhattisgarh, which is a forested state situated in east-central India, surrounded by Odisha, Telangana, Maharashtra, Madhya Pradesh and states of Uttar Pradesh and Jharkhand. The location of Chhattisgarh, India, is at a latitude of 21.29 and a longitude of 81.82, with the GPS coordinates of 21° 17′ 42.4752″ N and 81° 49′ 41.6352″ E (Bharos et al., 2019). The state basically experiences a tropical climate, which is hot and humid during summer and the temperature reaches up to 49 °C. The state receives monsoon from June to September, an average of 1292 millimetres of rain (Chakraborty et al., 2017). Winter starts in November and ends in January. But the matter of concern is that the state is prone to many extreme weather events and natural hazards such as cyclones, floods, heat waves and droughts, which are responsible for affecting the economy, agriculture, industrial sector and health of the population present in the state. Moreover, in the present time, the other thing to worry about the state is the changes seen in the weather variables due to climate change giving rise to the vector-borne diseases, especially malaria. Figure 17.3 shows the location of Chhattisgarh in India, highlighting different districts as well as the number of malaria cases in the state from the year 2008–2020.

VECTRI Model Description

VECTRI, being a mathematical dynamic model, is used to research malarial transmission and demonstrate the effect of weather variability along with population. The model was officially introduced at the University of Addis Ababa in November 2011. It is used as a research tool to understand the reasons behind the occurrence and transmission of malaria on a regional scale with spatial resolutions of 10 km or less so as to increase the accuracy of the analysis (Tompkins & Ermert, 2013). It is a grid cell distributed dynamical model. It includes the study of malaria transmission through surface hydrology, population density, biting rates and chances of transmission. The model can be used to observe the difference in transmission of the disease at various levels such as in terms of topography, rural–urban areas and many others. The nexus with population in the model can be upgraded to include other factors such as migrations, urbanization, socio-economic status, immunization and others, thus allowing regional and continental-wide simulations (Tompkins et al., 2012).

The basic physics and parameters of this model are of *Anopheles gambiae* vector and *Plasmodium falciparum* parasite, but for the state of Chhattisgarh, the stimulations are made according to the *Anopheles culicifacies* vector and *Plasmodium falciparum* parasite. To further perform the experiment, data from the India Meteorological Department (IMD) of daily maximum and minimum temperature at 1° × 1° resolution with rainfall at 0.25° × 0.25° resolution and population data

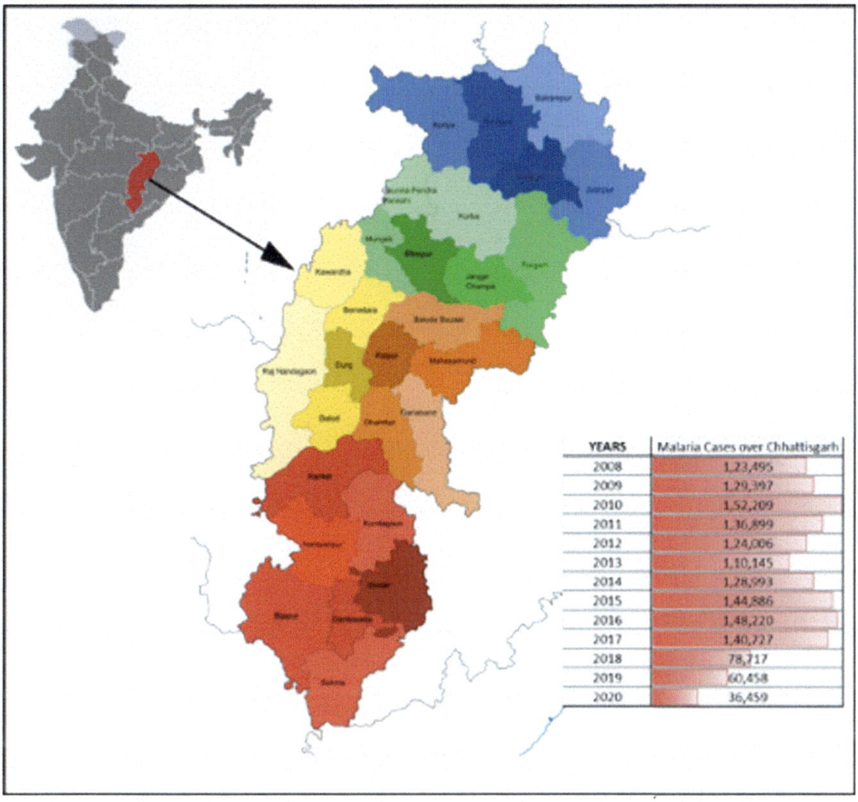

YEARS	Malaria Cases over Chhattisgarh
2008	1,23,495
2009	1,29,397
2010	1,52,209
2011	1,36,899
2012	1,24,006
2013	1,10,145
2014	1,28,993
2015	1,44,886
2016	1,48,220
2017	1,40,727
2018	78,717
2019	60,458
2020	36,459

Fig. 17.3 Malaria cases observed in Chhattisgarh during 2008–2020 with its location in the map of India

from worldpop.org for the state from 2000 to 2019 are used as input in the VECTRI model. The final analysis is carried out by using MATLAB software.

Result and Discussion

The following is the result of the experiment, which includes the discussion of the analysis being performed over the region of Chhattisgarh in terms of variations of the weather variables, namely temperature, rainfall, mosquito population and entomological inoculation rate (EIR) to study the pattern of malaria transmission over the state.

The need of the hour is to understand the changes in the climate and be prepared for the future. This study will help us to predict the rain for the next monsoon, as well as the rise in sea levels noticed due to warmer sea temperatures. The main purpose of studying the monthly variation is to observe and predict the malaria incidences and

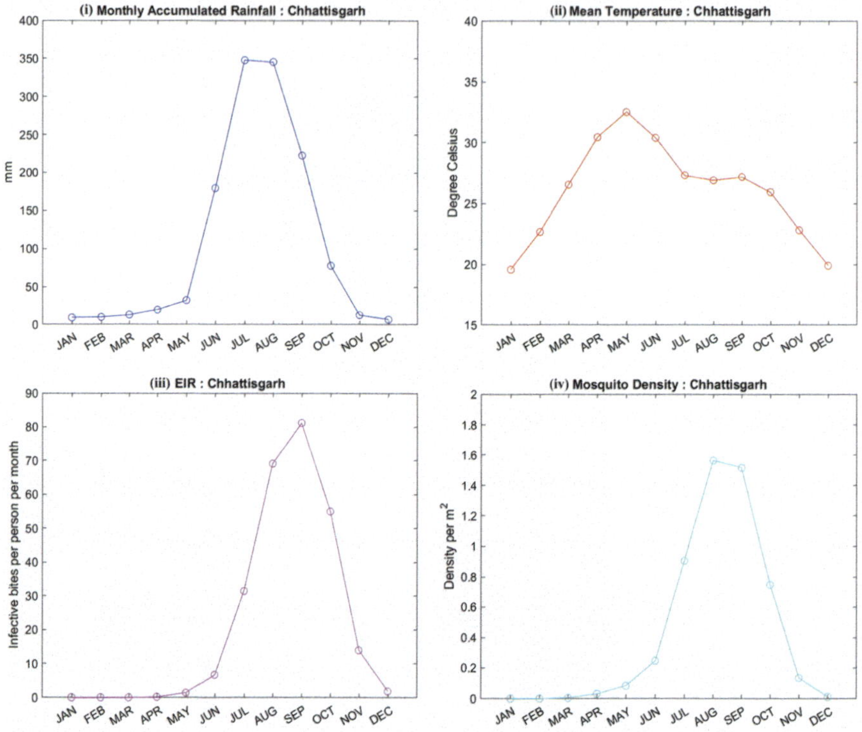

Fig. 17.4 Monthly variations seen over the state of Chhattisgarh during the time period of 2000–2019. (i) Accumulated rainfall, (ii) surface mean temperature, (iii) entomological inoculation rate (EIR), (iv) mosquito density

its transmission, respectively, in the state through the peaks seen in different months across the year. From Fig. 17.4, it is analysed that the range of rainfall observed during pre-monsoon season is 10–30 mm, while in monsoon, it rises to 185–350 mm and falls to the value of 90 mm and below during the post-monsoon season. On the other hand, the temperature during March–May is around 26–33 °C and during the months from June to September is 27–31 °C, while it drops to 18–24 °C during other months of the year. It is observed from the figure that the highest value of mosquito density is 0.3–1.6 per square metre during the months of monsoon and the highest value of EIR 9–82 infective bites/person/month also coincides with the highest value of the mosquito density and the maximum value of malaria transmission is seen during August and September because the value of mosquito density and EIR, respectively, are at its peak. The result analysed shows that the malaria transmission loses its peak during the season of winter till spring in Chhattisgarh, whereas the transmission is mainly seen during the months of monsoon (June–September) and at its best during July–September with a decrease seen after October.

Now, shifting the focus to Fig. 17.5, which shows the relationship between weather variables, namely rainfall and temperature, to the mosquito density and

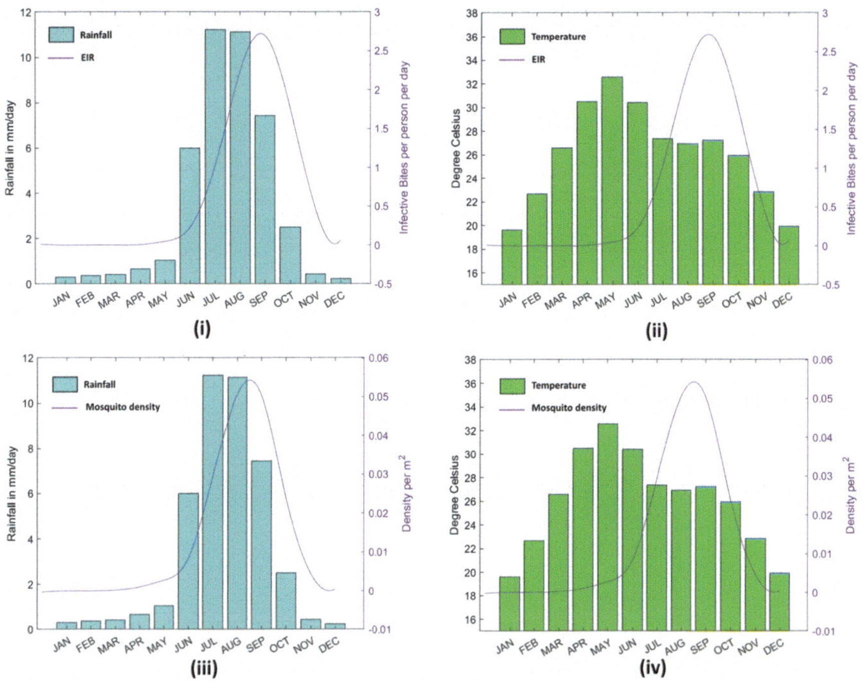

Fig. 17.5 The graph shows the relationship between VECTRI-simulated rainfall with (i) entomological inoculation rate (EIR), and (iii) mosquito density and surface mean temperature with, (ii) EIR, and (iv) mosquito density over Chhattisgarh during averaging period of 2000–2019

Entomological Inoculation Rate (EIR) through the monthly variations for the state of Chhattisgarh for the given period of study. Figure 17.5 shows a nonlinear relationship between the temperature, rainfall with the mosquito density and Entomological Inoculation Rate (EIR). Another observation includes the rise in the values of the vector density and EIR more with rainfall than with temperature, which establishes a strong correlation between rainfall with the mosquito density and Entomological Inoculation Rate (EIR) as compared to the temperature, because the value of mosquito density and Entomological Inoculation Rate (EIR) increases when the temperature ranges between 26 and 28 °C and the rainfall observed is 11 mm per day. On the other hand, this trend declines after October when the threshold crosses a certain limit. The further analysis also includes the computation of the values of the correlation coefficient for the region of Chhattisgarh for a given time period to demonstrate the effect of temperature and rainfall on mosquito density and Entomological Inoculation Rate (EIR). The rainfall and mosquito density with a correlation coefficient of 0.84 has a strong relation but in between mosquito density and temperature it is 0.2. The correlation coefficient value of Entomological Inoculation Rate (EIR) with rainfall is 0.69, and the value of Entomological Inoculation Rate (EIR) with temperature is 0.12. This shows a strong relationship between rainfall and the two factors from the VECTRI model, namely mosquito density and EIR.

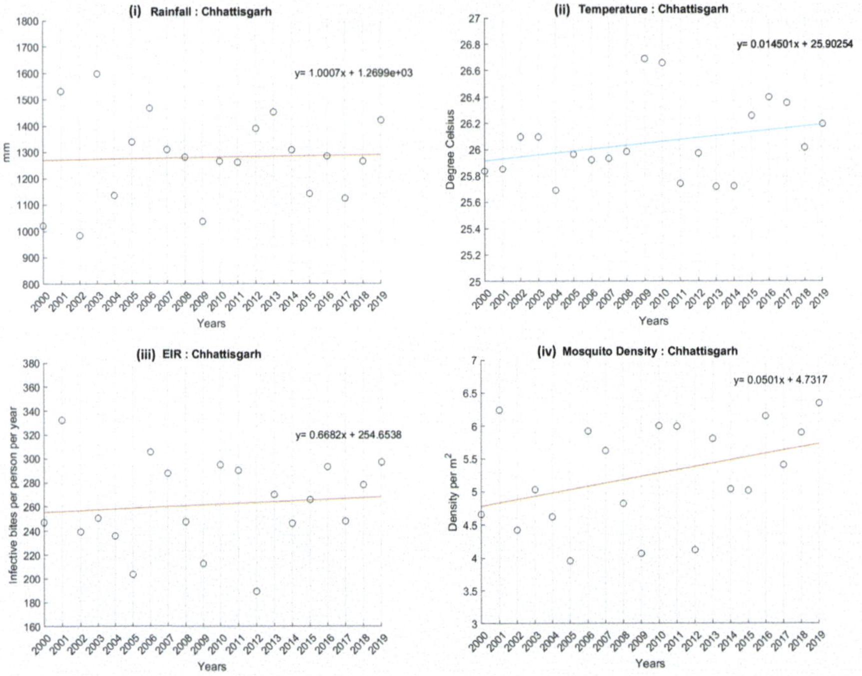

Fig. 17.6 Analysis of VECTRI model-generated output of annual mean. (i) Accumulated rainfall, (ii) surface mean temperature, (iii) entomological inoculation rate (EIR) and (iv) mosquito density during 2000–2019 over Chhattisgarh

Figure 17.6 shows the results of the two weather variables, temperature and rainfall, along with the VECTRI model variables, mosquito density and EIR. As seen in the figures, the rainfall and temperature show a linear rise of 1.0007 mm/year and 0.014 °C/year, respectively, from 2000 to 2019. The highest value of rainfall was 1600 mm recorded in 2003. The highest value of temperature was 26.7 °C observed in 2009 and 2010. The years 2001, 2003, 2016 and 2019 contribute towards minimum changes in mosquito density and Entomological Inoculation Rate (EIR). A variation between 40 and 63 per square metre and 190–330 number of infective bites per person per year in terms of mosquito density and Entomological Inoculation Rate (EIR), respectively, are observed.

Conclusion

It is observed that rainfall and temperature are the main factors responsible for the development of malaria vector, which helps in determining the intensity of malaria transmission in Chhattisgarh. The variations in the climatic parameters, namely temperature and rainfall, are also found to be associated with the species of mosquito

(*Plasmodium falciparum*). The results show a strong positive correlation of malaria vector density with rainfall and temperature. It is found that the increase in the cases of malaria incidence is seen more during the months of July, August and September, with observed precipitation of 6–11 mm/day and temperature of 26 °C–28 °C are seen more affirmative for the growth and development of mosquito as well as parasite involved. When the rainfall exceeds the limit of the threshold, there is an increase in the malaria cases. Furthermore, more attempts are to be made for the detailed study of malaria transmission in the state through spatial and seasonal variations for more better understanding of the VECTRI model, with more focus to be laid on the hydrological parameters being important for the biology of the malaria parasite.

Acknowledgements I would like to thank Dr. Adrian Tompkins, who helped me in providing details for setting up the VECTRI model developed by ICTP, Italy, for helping me with the calibration of the model and in handling data during the early stage of the research.

References

World Health Organisation. (2019). World Malaria Report 2019.

World Health Organisation. (2020). World Malaria Report 2020.

World Health Organisation. (2021). World Malaria Report 2021.

Subbarao, S. K., Nanda, N., Rahi, M., et al. (2019). Biology and bionomics of malaria vectors in India: Existing information and what more needs to be known for strategizing elimination of malaria. *Malaria Journal, 18*, 396. https://doi.org/10.1186/s12936-019-3011-8

Malaria Mukhta Baster Abhiyan- State Chhattisgarh. (2019). http://nhm.gov.in

NVBDCP. (2002). National Vector Borne Disease Control Programme, report available at, http://nvbdcp.gov.in/Doc/Annual-report-NVBDCP-2002.pdf

Ermert, V., Fink, A. H., Jones, A. E., et al. (2011). Development of a new version of the Liverpool malaria model. I. Refining the parameter settings and mathematical formulation of basic processes based on a literature review. *Malaria Journal, 10*, 35. https://doi.org/10.1186/1475-2875-10-35

Lieshout, V. M., Kovats, R. S., Livermore, M. T. J., & Martens, P. (2004). Climate change and malaria: Analysis of the SRES climate and socio-economic scenarios. *Global change*, 1487–1499.

Bharos, A., Mandavia, A., Naidu, R., & Badesha, A. (2019). *Distribution range extension of Grey headed lapwing (Vanellus cinereus) in Chhattisgarh and eastern Madhya Pradesh* (pp. 11–13). Jharkhand.

Tompkins, A. M., & Ermert, V. (2013). A regional-scale, high resolution dynamical malaria model that accounts for population density, climate and surface hydrology. *Malaria Journal, 12*, 65.

Kleinschmidt, L., et al. (2001). An empirical malaria distribution map for West Africa. *Tropical Medicine & International Health*, 779–786.

Chakraborty, A., Tigga, A., & Malini, B. (2017). Assessment of climates in Chhattisgarh plain – A moisture regime approach. *Transactions of the Institute of Indian Geographers, 39*, 25–33.

Tompkins, A., Ermert, V., & Lowe, R. (2012). *VECTRI: A new dynamical disease model for malaria transmission.*

Chapter 18
Statistical Analysis of Rainfall and Temperature Variability in Chotanagpur Plateau, India

Aastha Gulati and Suresh Chand Rai

Introduction

Even though our perception and knowledge of climate change/variability are still growing, the acceptance of alterations to the global climate has come into existence. Accurate assessment of meteorological parameters is of prime importance for development planning, agricultural progress and food security. Information about the temporal variability of precipitation and temperature is important for scientific and practical purposes. Trend detection in precipitation time series is especially crucial for the management and optimum utilization of available water resources. Regional scale investigations of precipitation trends reveal a declining precipitation tendency over several parts of the world (Giakoumakis & Baloutsos, 1997; Piervitali et al., 1998; Gan, 1998; Ventura et al., 2002; Basistha et al., 2008), while a growing trend was observed in North Carolina (Boyles & Raman, 2003) and Mainland Spain (Mosmann et al., 2004). Studies on temperature trend analysis have demonstrated that global surface temperature has augmented by 0.6 °C over the last 100 years, with 1998 being the warmest year (IPCC, 2007). It also projects a decrease in agricultural production for slight increases in regional temperature (1–2 °C), which would upsurge the danger of starvation, particularly in seasonally dry and tropic regions. It has been reported that in India mean annual temperature had been rising at a rate of 0.05 °C/decade over 1901–2003, generally contributed by the increase of maximum temperature (0.07 °C/decade) rather than by the increase of minimum temperature (0.02 °C/decade) (Kothawale & Rupa Kumar, 2005). As a result, the diurnal temperature range shows an increase of 0.05 °C/decade. The diurnal temperature range (DTR), i.e. the difference between maximum and minimum temperatures, is

A. Gulati
TERI University, New Delhi, India

S. C. Rai (✉)
Department of Geography, Delhi School of Economics, Delhi University, New Delhi, India

A. L. Singh et al. (eds.), *Climate Change, Vulnerabilities and Adaptation*,
https://doi.org/10.1007/978-3-031-49642-4_18

an important index of climate change and is receiving considerable attention in recent times. Several studies in the last few decades have focused on spatial-temporal trend detection of temperature (Karl et al., 1993; Srivastava et al., 1992; Rupa Kumar et al., 1994; Easterling et al., 1997; Rebetez & Beniston, 1998; Jones et al., 1999; Yadav et al., 2004; Lobell, 2007; Jhajharia & Singh, 2011; Jain et al., 2012). Despite such extensive studies on precipitation and temperature trends, very little information was available on the analysis of these important meteorological parameters for the Chhotanagpur Plateau of India. Therefore, the present study was designed to study the climatic variability in terms of precipitation and temperature in the Chhotanagpur plateau.

The Study Area

The state of Jharkhand, part of the Chhotanagpur Plateau, is the 13th most populated state in India and cares about 3% of the country's population (www.censusindia.gov.in). Around 43% of the population in the state is below the poverty line, much higher than the national average of 26% (Ghosh, 2009). The Jharkhand state has been mentioned to be worse than Zimbabwe and Haiti (Ghosh, 2009). Ranchi is the state capital and receives an average annual precipitation of 1300–1500 mm, which happens generally between June and September. Rainy season precipitation is considered by high intensity and short duration, which are accomplished by producing large volumes of runoff. The mean summer and winter temperatures are 23.8 °C and 8.6 °C, respectively. The population of the region is largely dependent on rain-fed subsistence agriculture, thus making precipitation and temperature highly important parameters for livelihood and food security.

Methodology and Data Used

The climatological data analysed in this chapter were collected from the agro-meteorological station of Birsa Agricultural University, Ranchi. Daily precipitation and temperature records spanning 30 years (1982–2011 for precipitation and 1982–2012 for temperature) were assessed, which is considered to be adequate for a valid mean statistic (Burn & Elnur, 2002). The data have been analysed on a monthly, annual and seasonal basis. The four seasons that the region is witnessing are winter (January–February); pre-monsoon (March–May); monsoon (June–September) and post-monsoon (October–December). The number of rainfall days (where rainfall, $P \geq 2$ mm) was also assessed. The whole data set was checked for missing values, which were equated to data from other instruments in the same station or from other stations close to it. Average monthly data were considered only if less than three consecutive data were missing (Ventura et al., 2002). The Kruskal–Wallis test was applied to annual data sets to verify the homogeneity (Sneyers,

1990). Mann–Kendall test is applied for trend identification (Mann, 1945; Kendall, 1975), while Sen's estimator of the slope was used to ascertain the magnitude of the trend (Sen, 1968).

Results and Discussion

Trend Analysis of Precipitation

The observed rainfall pattern shows a high seasonal and annual variability. There was variability among the years, with a standard deviation of 278.9 mm. The mean annual precipitation received was 1501 mm (during 1982–2011). The rainy season contributed about 82.6% of the annual rainfall, with a maximum (30%) contribution in July, followed by August, September and June. The input of pre- and post-monsoon rainfall to the annual rainfall was 7.7% and 7.0%, respectively. The winter precipitation (January and February) contributed the least (2.7%), demonstrating that irrigation amenities are required for rabi crops (the second crop in the region). The mean annual rainfall has revealed a declining trend (Fig. 18.1) over 30 years (1982–2011). This shows a drier climate in the future. The number of annual precipitation days (days with $P \geq 2$ mm) also displays a decline in trend (Fig. 18.1).

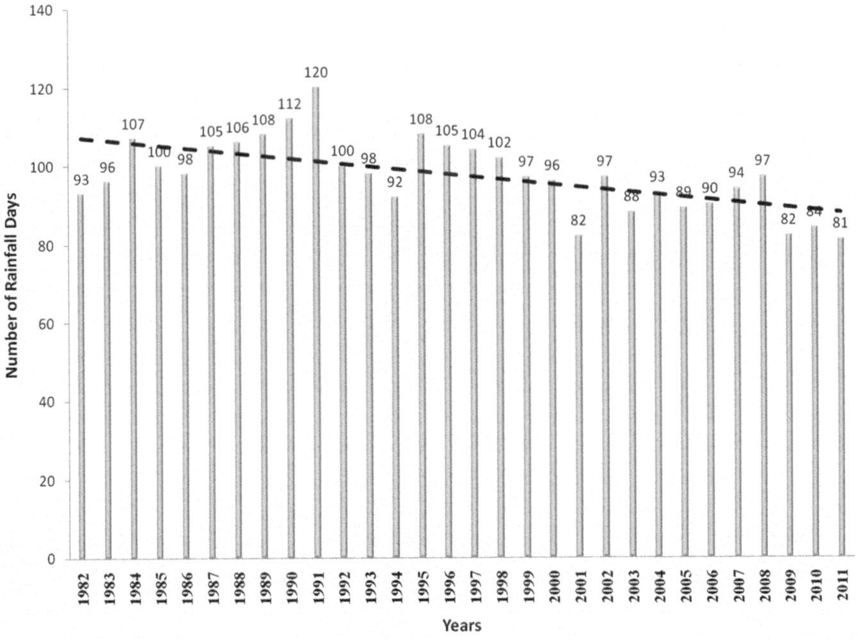

Fig. 18.1 Number of rainfall days (1982–2011)

Table 18.1 Sen's Estimator of Slope for monthly rainfall (mm/year)

Year	Jan	Feb	Mar	Apr	May	Jun	Jul	Aug	Sep	Oct	Nov	Dec
1982-1991	-0.89	0.60	0.61	-2.02	-0.04	0.20	-1.50	2.50	1.04	-1.74	1.60	0.86
1992-2001	-1.62	-1.46	-0.94	-1.76	0.92	**-0.94**	1.94	-0.72	0.77	-0.90	-0.49	-0.70
2002-2011	2.14	1.90	0.67	*-1.60*	-1.67	1.07	0.29	-1.90	-1.05	0.01	*-2.76*	-0.19
1982-2011	0.20	0.57	0.53	-0.58	*-0.45*	-2.65	0.46	*-0.65*	**-1.23**	0.49	**-0.62**	-0.71

Bold values and red italicized values indicate statistical significance at 99% and 95% confidence levels as per the Mann–Kendall test, respectively (+ for increasing and – for decreasing)

The rainy season sets in around the 24th standard meteorological week (11–17 June) and withdraws itself around the 44th standard meteorological week (29 October–4 November), staying in for 20 weeks or 140 days and the actual rainy season lasts for 13–29 weeks or 91–203 days. Information on the onset and withdrawal of the rainy season is very important for the proper planning of cropping activities dependent on rainfall.

The decadal analysis of monthly rainfall data revealed that of 48 slope values, 27 were negative, indicating a declining trend. Though the values of Sen's slope estimator are relatively larger, very few values were significant. For the whole data series (1982–2011), 7 months exhibited negative values, of which four were significant (Table 18.1).

Trend investigation was also completed on the seasonal scale to examine whether there are trends in the data at this scale. Figure 18.2 shows the plots of seasonal and annual rainfall for the region for the period 1982–2011, along with the trend lines.

The precipitation pattern of the monsoon season shows the least scatter among all the seasons. The yearly precipitation trend also does not show much scatter, but precipitation for pre- and post-monsoon periods has a large variation from 1 year to the other. The winter period displays a relatively higher scatter in the latter half of the data series. Figure 18.3 illustrates the box–whisker plots of the slopes of the seasonal and annual rainfall time series. The box extends from the 25th percentile to the 75th percentile, with the line representing the median and the dot representing the mean.

The whisker plots display the minimum and the maximum values. Except for the annual series and the winter season, the distance between the median to minimum is less than the distance between the median to maximum, indicative of high variability of the slopes till the median during pre-monsoon, monsoon and post-monsoon seasons. High-intensity precipitation in the region is an identified reason for soil erosion. Overland flow from agricultural fields is also recognized to carry the much-desired nutrients from the fields. The variation in different levels of erosive rainfall (25–50 mm, 50–75 mm, 75–100 mm and more than 100 mm) events in 24 h has been analysed for three decades. A growing trend in the number of events over the decades, i.e. from 17.3 (1982–1991) to 20.9 (2002–2011), has been detected. An upsurge in several erosive events may degrade the situation by leaving the top fertile soil barren and unproductive. Monthly analysis of the erosive events indicates that

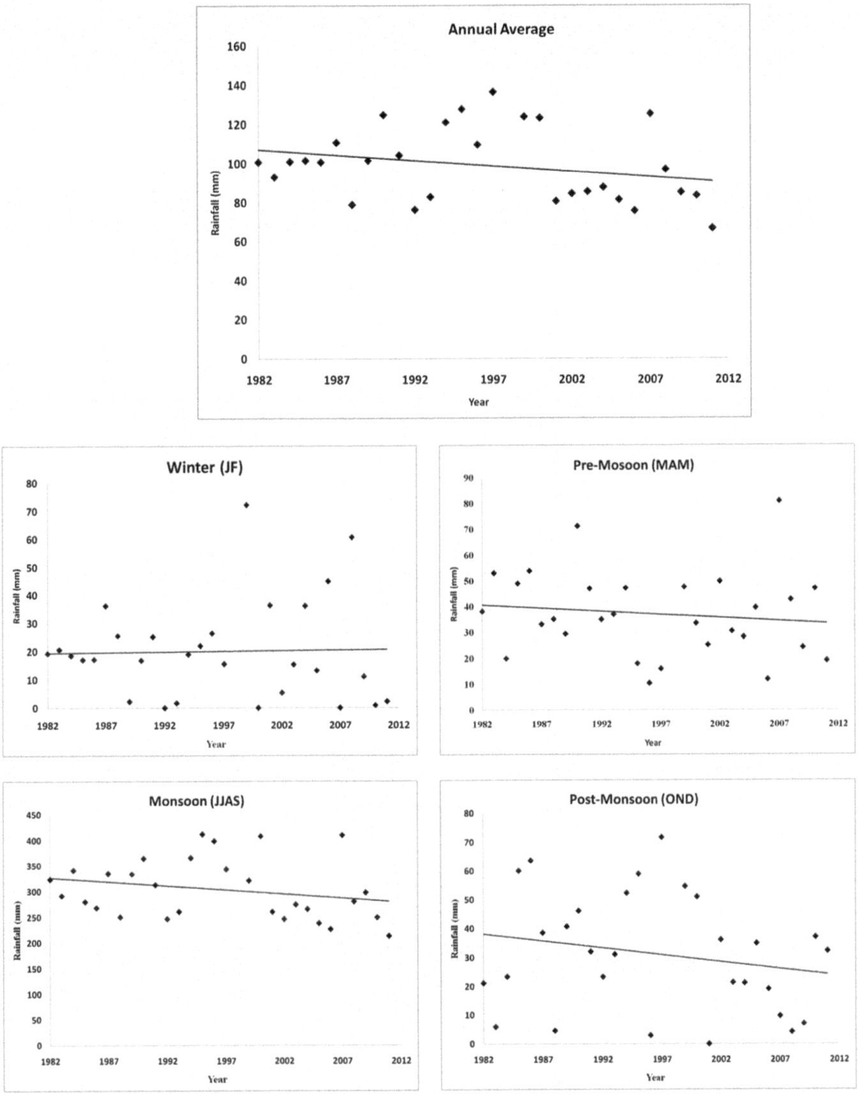

Fig. 18.2 Plots of seasonal and annual rainfall for the region for the period 1982–2011, along with trend lines

the erosive rainfall in the range of 25–50 mm was the highest in July (Fig. 18.4). This range is highly conducive towards splash, sheet and rill erosions.

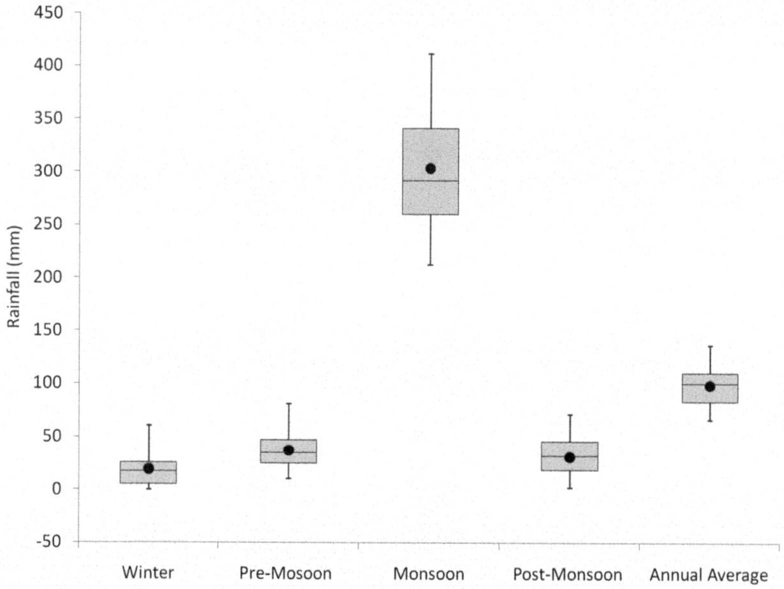

Fig. 18.3 Box–whisker plot of slopes of seasonal and annual rainfalls

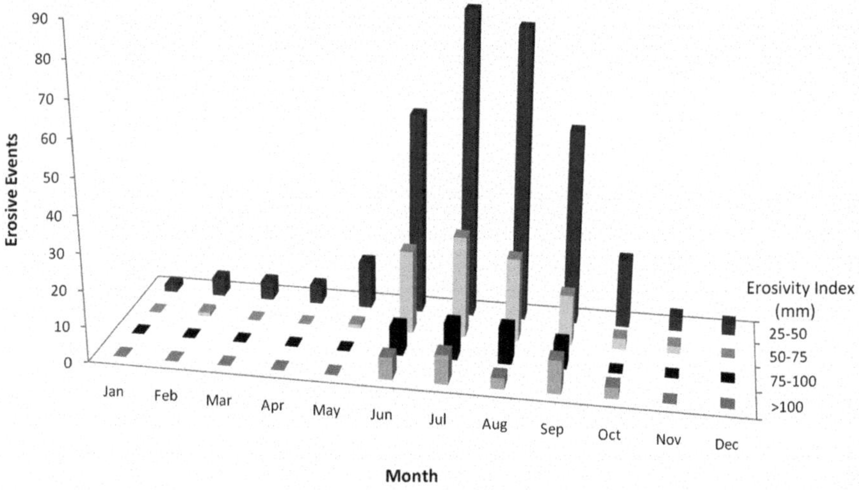

Fig. 18.4 Monthly occurrence of erosive events in different erosivity ranges

Trend Analysis of Temperature

The investigation of monthly maximum temperature for the period 1982–2012 displays a growing trend for 8 months and a statistically significant declining trend for June. At the same time, the minimum temperature data had a falling trend for

Table 18.2 Sen's Estimator of Slope for monthly temperature from 1982 to 2012 (°C/year)

1982-2012	Variable	Jan	Feb	Mar	Apr	May	Jun	Jul	Aug	Sep	Oct	Nov	Dec
1	Max Temp	0.006	0.040	-0.034	0.003	-0.026	*-0.045*	0.034	0.011	0.006	0.011	0.007	-0.024
2	Min Temp	*-0.088*	**-0.090**	-0.032	-0.023	-0.015	-0.019	**0.126**	-0.029	*-0.067*	-0.033	*-0.052*	**-0.091**
3	Temp Range	**0.088**	**0.110**	-0.013	0.018	-0.015	-0.023	**-0.092**	0.036	*0.064*	*0.046*	*0.058*	*0.069*

Bold values and red italicized values indicate statistical significance at 99% and 95% confidence levels as per the Mann–Kendall test, respectively (+ for increasing and − for decreasing)

11 months, of which five were significant. The month of July exhibited a significant increasing trend. The diurnal temperature range also had a rising trend for 8 months, and it was statistically significant for all but 2 months, which implies that the range of temperatures is increasing. However, a significant decreasing trend was seen for July (Table 18.2).

Figures 18.5 and 18.6 show the plots of seasonal and annual time series (1982–2012) and linear trend lines for Tmax, Tmin and DTR for the region. Among the seasons, the pre-monsoon and post-monsoon seasons show the most scatter, while the winter season shows the least scatter for Tmax. For Tmin, pre-monsoon shows the maximum scatter, while post-monsoon shows the minimum.

However, pre-monsoon showed the least scatter for DTR, while most scatter was observed for winter. An increasing trend (+0.36 °C/decade) in DTR was obtained through the Mann–Kendall test. The DTR increase observed at the site is consistent with the findings for India (Rupa Kumar et al., 1994) but is in total contrast with the global trends in DTR (Easterling et al., 1997; Vose et al., 2005). The minimum to maximum box–whisker plots of Sen's slopes of the seasonal and annual temperature time series show almost a normal behaviour for Tmax, with a slightly skewed distribution for pre-monsoon, monsoon and post-monsoon seasons (Fig. 18.7a).

The plots of Tmin (Fig. 18.7b) show a highly skewed distribution for monsoon, followed by post-monsoon season. Decadal analysis (1982–1991, 1992–2001 and 2002–2012) was done to ascertain the temporal trends. For the periods 1982–1991 and 1992–2001, all the temperature variables, maximum temperature, minimum temperature and temperature range, had a rising trend. The trend values for March, June and December were significant at a 95% confidence level. For the period 2002–2012, the temperature variables had a mixture of rising and falling trends, but the maximum and minimum temperatures had statistically significant positive and negative slopes for 5 and 7 months, respectively (Table 18.3).

Based on the investigation, it can be specified that the maximum and mean temperatures in the region display a rising trend and the increase is notably significant in statistical terms for the post-monsoon and winter seasons. This behaviour of the data indicates the presence of an element of the seasonal cycle in the temperature

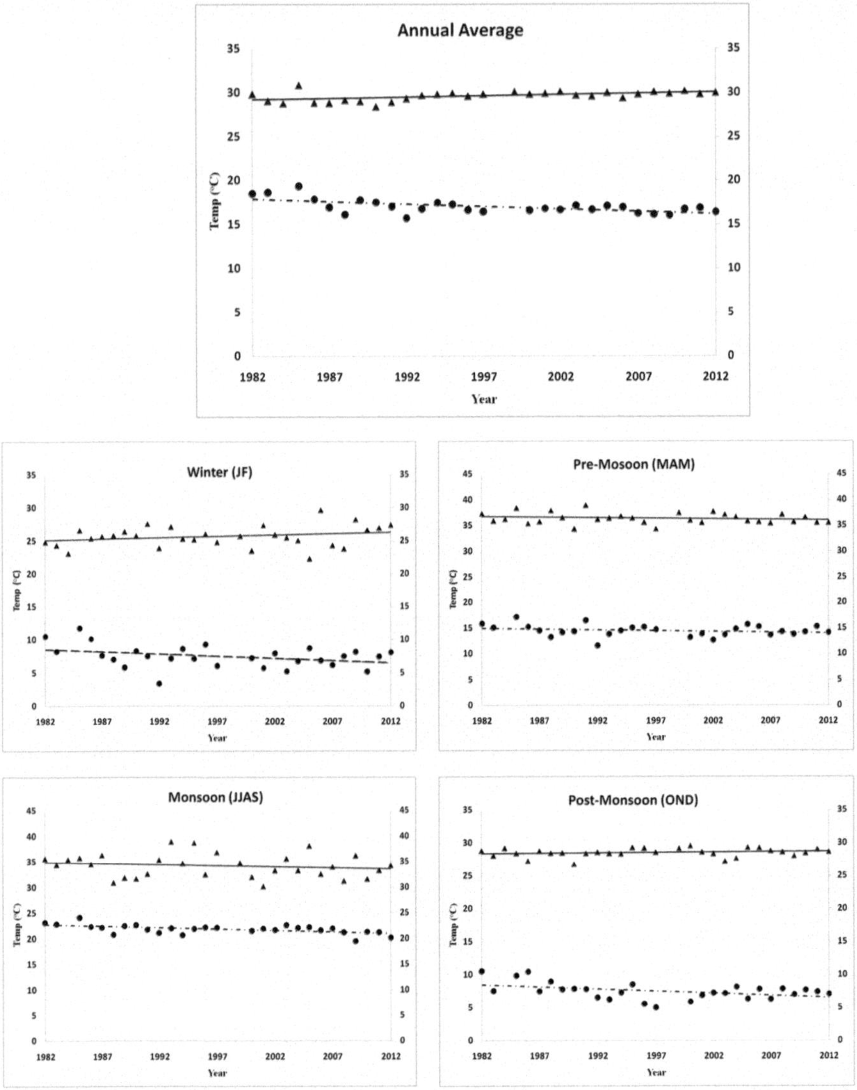

Fig. 18.5 Plots of seasonal and annual temperatures (maximum and minimum) for the region for the period 1982–2012 along with trend lines

trends over the plateau, which is likely to spell trouble for agriculture, water availability and human–livestock health. This damage is further carried forward to affect the livelihood and food security of the farmers who are directly dependent on this climate-sensitive sector for their subsistence. Anthropogenic activities like land use/cover change, changing lifestyle, urbanization and industrialization are affecting the climate significantly. The impact of these activities could be minimized by taking appropriate adaptation and mitigation measures.

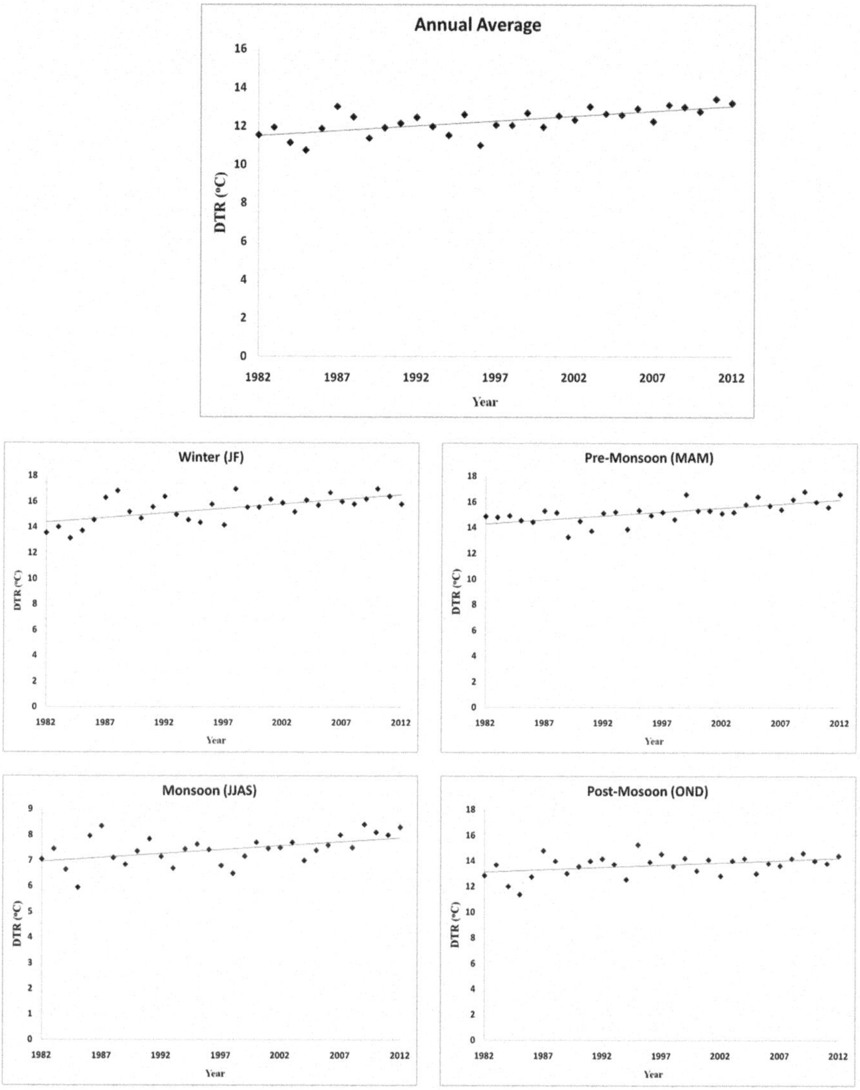

Fig. 18.6 Plots of the seasonal and annual temperature ranges for the region for the period 1982–2012, along with trend lines

Adaptive and Mitigation Measures

It is well observed that the climate is playing a significant role in farmers' livelihood and food security. However, the reasons identified as causes for this were those that have beyond farmers' control such as climate and terrain. Causes such as poor agricultural practices and ineffective soil and water conservation measures were

(a)

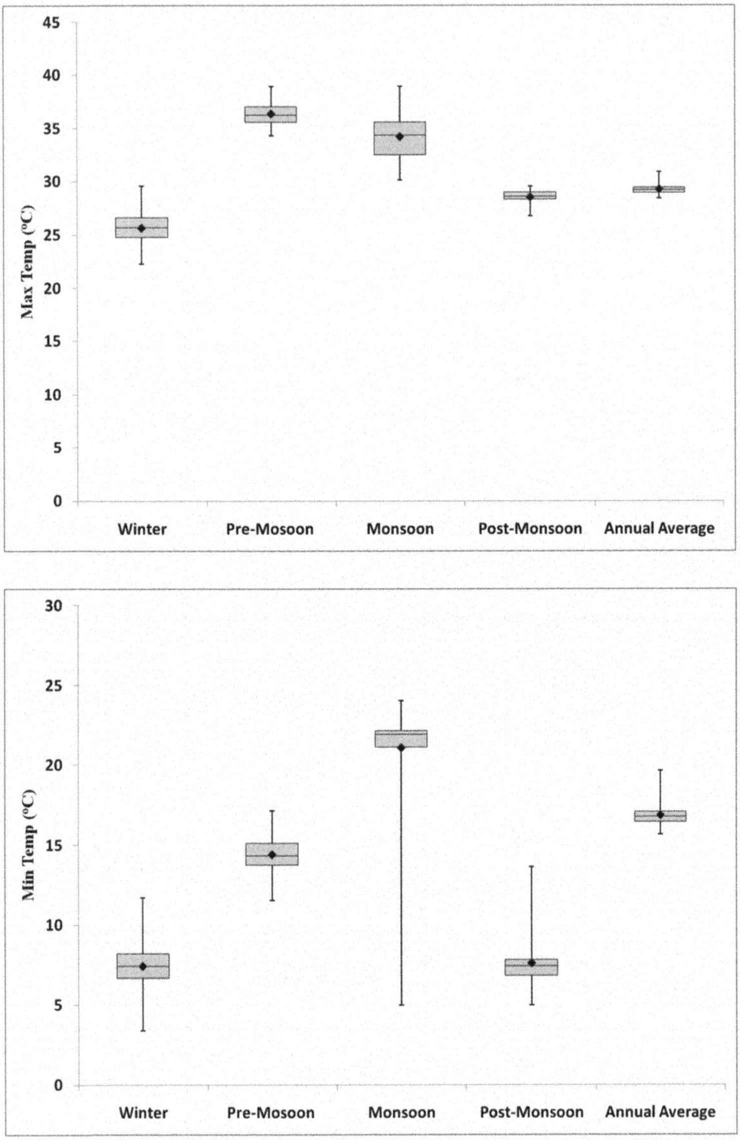

(b)

Fig. 18.7 Box–whisker plot of Sen's slopes of seasonal and annual maximum temperature (**a**) and minimum temperature (**b**)

not seen to have a linkage with climate change. However, in addition to increased crop yields, soil and water conservation measures were perceived to improve soil health as well as the market value of the produce. It was observed during the field survey that the farmers were aware of various adaptive measures, but the rate of adoption was low. Most farmers possessed the traditional skills of sustainable

Table 18.3 Sen's Estimator of Slope for monthly temperature for three decades (°C/year)

1982-1991	Variable	Jan	Feb	Mar	Apr	May	Jun	Jul	Aug	Sep	Oct	Nov	Dec
1	Max Temp	-0.16	0.29	*-0.31*	-0.08	0.01	*-0.46*	-0.03	-0.04	0.06	-0.09	-0.06	-0.35
2	Min Temp	-0.49	0.31	*-0.13*	-0.29	0.05	*-0.20*	0.55	-0.20	0.15	-0.24	-0.16	-0.39
3	Temp Range	0.26	-0.06	*-0.34*	0.11	-0.05	*-0.32*	-0.62	0.14	-0.12	0.17	0.08	0.01
1992-2001													
1	Max Temp	-0.04	0.04	*-0.11*	-0.12	0.03	*-0.76*	-0.16	0.14	0.02	0.08	0.19	*-0.04*
2	Min Temp	-0.16	0.50	*0.16*	-0.03	0.08	**-0.11**	0.01	-0.16	0.09	0.03	0.16	*-0.11*
3	Temp Range	0.09	-0.45	*-0.40*	-0.06	-0.12	**-0.77**	-0.32	0.27	-0.03	0.05	0.02	*0.08*
2002-2012													
1	Max Temp	0.19	0.22	*-0.02*	-0.01	-0.18	**-0.13**	*-0.24*	*0.07*	*0.06*	0.07	0.01	*-0.03*
2	Min Temp	0.04	**-0.12**	*0.06*	-0.04	**-0.08**	*-0.04*	**-0.06**	-0.01	**-0.09**	*-0.13*	-0.02	0.00
3	Temp Range	0.15	0.33	*-0.08*	0.04	-0.10	*-0.09*	-0.18	0.08	0.15	0.21	0.06	*-0.03*

Bold values and red italicized values indicate statistical significance at 99% and 95% confidence levels as per the Mann–Kendall test, respectively (+ for increasing and – for decreasing)

agricultural practices (such as soil conservation, water harvesting, inter-cropping, harvesting and post-harvesting practices.); however, only a few had adopted these measures. Of the total respondents interviewed, only 12% had adopted all these measures. There are still many farmers (46%) who are not receptive to change or the adoption of new skills.

The adoption of agroforestry practices to combat climate change is one of the best adaptive measures. There is plenty of scope and potential to promote leguminous crops and green manure in the area. The adoption of green manure and organic manure will be more beneficial to increase the agronomic yield. Following the on-field observations and the results of this study, it would be advisable to take up measures for the benefit of the community. Uplifting the small and marginal farmers of the state is likely to let the agricultural sector bloom and infringe upon the high rates of migration towards the towns. Knowledge and skill pieces of training for land and water management, soil conservation, inter and mixed cropping, use of appropriate water harvesting technology, etc., should be imparted to the local community. These skills will help them combat climate change.

Conclusion

Based on the results obtained, it can be specified that the maximum and mean temperatures in the region display a rising trend and the increase is notably significant in statistical terms for the post-monsoon and winter seasons, indicative of the presence of an element of the seasonal cycle in the temperature trends over the plateau. The effects of this can be crucial for the agricultural sector and the population dependent on it. An increase in temperature during the critical stages of plant growth can hamper crop yields drastically. The observed decreasing trend in precipitation along with an increasing trend in the number of erosive events is expected to be highly conducive towards sheet, splash and rill erosion, which can harm the fertile topsoil of the agricultural fields. The results obtained in this study should prove useful, providing relevant information for decision-makers for investment and conservation-related policies in the area.

Acknowledgements We would like to thank HSBC Bank and TERI University for awarding the Climate Change Scholarship to conduct the present research work. We would also like to duly acknowledge Birsa Agricultural University, Ranchi, and Jharkhand Forest Division for helping with data collection and fieldwork.

References

Basistha, A., Arya, D. S., & Goyal, K. (2008). Analysis of historical changes in the Indian Himalayas. *International Journal of Climatology*. https://doi.org/10.1002/joc.1706

Boyles, R. P., & Raman, S. (2003). Analysis of climate trends in North Carolina (1949–1998). *Environment International, 29*, 263–275.

Burn, D. H., & Elnur, M. A. H. (2002). Detection of hydrological trends and variability. *Journal of Hydrology, 255*, 107–255.

Easterling, D. R., Horton, B., Jones, P. D., Peterson, T. C., Karl, T. R., Parker, D. E., Salinger, M. J., Razuvayev, V., Plummer, N., Jamason, P., & Folland, C. K. (1997). Maximum and minimum temperature trends for the globe. *Science, 277*, 364–367.

Gan, T. Y. (1998). Hydroclimatic trends and possible climatic warming in the Canadian Prairies. *Water Resources Research, 34*(11), 3009–3015.

Ghosh, J. (2009). Food for all. *Frontline, 26*(17), 131.

Giakoumakis, S. G., & Baloutsos, G. (1997). Investigation of trend in hydrological time series of the Evinos River Basin. *Hydrological Sciences Journal, 42*(1), 81–88.

IPCC. (2007). Climate change 2007: The physical science basis. In S. Solomon, D. Qin, M. Manning, Z. Chen, M. Marquis, K. B. Averyt, M. Tignor, & H. L. Miller (Eds.), *Contribution of Working Group I to the fourth assessment report of the Intergovernmental Panel on Climate Change*. Cambridge University Press.

Jain, S. K., Kumar, V., & Saharia, M. (2012). Analysis of rainfall and temperature trends in Northeast India. *International Journal of Climatology*. https://doi.org/10.1002/joc.3483

Jhajharia, D., & Singh, V. P. (2011). Trends in temperature, diurnal temperature range and sunshine duration in Northeast India. *International Journal of Climatology, 31*, 1353–1367. https://doi.org/10.1002/joc.2164

Jones, P. D., New, M., Parker, D. E., Martin, S., & Rigor, I. G. (1999). Surface air temperature and its changes over the past 150 years. *Reviews of Geophysics, 37*, 173–199.

Karl, T. R., Jones, P. D., Knight, R. W., Kukla, G., Plummer, N., Razuvayev, V., Gallo, K. P., Lindsay, J., Charlson, R. J., & Peterson, T. C. (1993). A new perspective on recent global warming: Asymmetric trends of daily maximum and minimum temperature. *Bulletin of the American Meteorological Society, 74*, 1007–1023.

Kendall, M. G. (1975). *Rank correlation methods*. Charles Griffin.

Kothawale, D. R., & Rupa Kumar, K. (2005). On the recent changes in surface temperature trends over India. *Geophysical Research Letters, 32*, L18714. https://doi.org/10.1029/2005GL023528

Lobell, D. B. (2007). Changes in diurnal temperature range and national cereal yields. *Agricultural and Forest Meteorology, 145*, 229–238.

Mann, H. B. (1945). Nonparametric tests against trend. *Econometrica, 13*, 245–259.

Mosmann, V., Castro, A., Fraile, R., Dessens, J., & Sanchez, J. L. (2004). Detection of statistically significant trends in the Summer precipitation of Mainland Spain. *Atmospheric Research, 70*, 43–53.

Piervitali, E., Colacino, M., & Conte, M. (1998). Rainfall over Central–Western Mediteranean Basin in the period 1951–1995. Part 1: Precipitation trends. *Nuovo Cimento, 21C*(3), 331–334.

Rebetez, M., & Beniston, M. (1998). Changes in sunshine duration are correlated with changes in daily temperature range this century: An analysis of Swiss climatological data. *Geophysical Research Letters, 25*(19), 3611–3613.

Rupa Kumar, K., Krishna Kumar, K., & Pant, G. B. (1994). Diurnal asymmetry of surface temperature trends over India. *Geophysical Research Letters, 21*(8), 677–680.

Sen, P. K. (1968). Estimates of the regression coefficient based on Kendall's tau. *Journal of the American Statistical Association, 63*, 1379–1389.

Sneyers, R. (1990). *On the statistical analysis of series of observations* (WMO tech. note no. 143). Geneve.

Srivastava, H. N., Dewan, B. N., Dikshit, S. K., Rao, P. G. S., Singh, S. S., & Rao, K. R. (1992). Decadal trends in climate over India. *Mausam, 43*, 7–20.

Ventura, F., Pisa, P. R., & Ardizzoni, E. (2002). Temperature and precipitation trends in Bologna (Italy) from 1952 to 1999. *Atmospheric Research, 61*, 203–214.

Vose, R. S., Easterling, D. R., & Gleason, B. (2005). Maximum and minimum temperature trends for the globe: An update through 2004. *Geophysical Research Letters, 32*, L23822. https://doi.org/10.1029/2005GL024379

Yadav, R. R., Park, W. K., Singh, J., & Dubey, B. (2004). Do the western Himalayas defy global warming? *Geophysical Research Letters, 31*, L17201. https://doi.org/10.1029/2004GL020201

Chapter 19
Experience of Climate Change and Adaptation in Daily Living: Evidence from the Suru Valley of Ladakh Region

Kacho Amir Khan, Aparajita Chattopadhyay, and Cho Cho Zainab Huriya

Introduction

Mountains across the continents act as water reservoirs for billions of people and provide ecosystem services to the mountain communities. Climate change impact on the world's mountain system is visible where it is affecting drinking water supplies and water for livestock, hydropower generation, agricultural productivity and natural disaster. In the Himalayas too, climate change is seen as a major concern where its possible impacts are affecting the socio-economy, environment and culture of the communities (Liu & Rasul, 2007). It may continue to adversely affect the habitats of the mountains and their inhabitants. Study of the temperature trend in the Himalayas and surrounding regions shows that temperature increases in the uplands are greater when compared with the lowlands (Schickhoff et al., 2016). In the Himalayas, the changes in temperature and other precipitation patterns are largely affecting the volume of the glaciers, water supplies, agriculture etc. (Dimri & Dash, 2012). The change detection studies of the Kang Yatse and Lungser Ranges show rising losses in the glaciated area, while the examples of the Central Ladakh Range show significantly lower decreasing rates (Schmidt & Nüsser, 2017). The expected changes in the severity and number of extreme events of precipitation are seen as significant results of the many predicted consequences (Shekhar et al., 2010).

K. A. Khan (✉)
Department of Geography, University of Ladakh, Ladakh, India

Department of Population and Development, International Institute for Population Sciences, Mumbai, India

A. Chattopadhyay
Department of Population and Development, International Institute for Population Sciences, Mumbai, India

C. C. Z. Huriya
Department of Sociology, University of Jammu, Jammu, India

A. L. Singh et al. (eds.), *Climate Change, Vulnerabilities and Adaptation*,
https://doi.org/10.1007/978-3-031-49642-4_19

Secondary impacts of extreme events like flash flooding, landslides, glacier lake outburst floods etc. cause significant damage across the area in severe events such as a cloudburst (Thayyen et al., 2013). The origin of the glacial lake level rise is unknown; however, it could be related to a tunnel blockage or an increase in volume of meltwater due to a huge heatwave and excessive precipitation observed by the people (Schmidt et al., 2020). The scientific knowledge and data on human adaptation to climate change are less; however, people feel the effect of climate change as per their experiences on their well-being, livelihoods and the availability of natural resources (Sharma et al., 2009). The sense of climate change by people can be recognized as an important contribution to environmental issues, agricultural development and potential solutions (Weber, 2010). Different forms of ice reservoirs, generally referred to as 'artificial glaciers', developed in Ladakh and promoted as effective adaptation solutions in order to deal with recurring water scarcity (Nusser et al., 2019). Therefore, local people's awareness and understanding of climate change is a crucial necessity to improve coping strategies leading to better climate change adaptation (Mehta et al., 2010).

Ladakh region has miniscule climate studies (Le Masson & Nair, 2012). The impacts of climate change, such as glacier loss, depletion of groundwater supplies and other environmental degradation, have contributed to a significant challenge to the traditional living style and the livelihoods of the people of Ladakh (Jorgyes, 2010). In the Himalayan region, local communities have been dealing with climatic changes for ages (Berkes & Jolly, 2002; Byg & Salick, 2009; Chaudhary et al., 2011). Communities where their livelihood is largely dependent on ecosystem services are better aware of the changes and apply indigenous knowledge of adaptation (Chaudhary & Bawa, 2011; Chaudhary et al., 2011; Vedwan & Rhoades, 2001). Based on recorded data and models, the perceived awareness of local people about climate change and its impacts will vary from that of actual climate change data. To deal with ongoing climate change, indigenous knowledge is of utmost importance in a particular space (Ford et al., 2012; Becker et al., 2012). The knowledge and awareness of climate change from the local communities may support scientific research and may also help in precise findings and feedback in policy formulation for climate change adaptation approaches (Reidlinger & Berkes, 2001; Chaudhary & Bawa, 2011).

The present study of Suru Valley in the Ladakh region is the first of its kind where information is collected on climate change perception and its associated impacts using different qualitative tools. The key purpose of the present study is to know how men and women in the region perceive climate change, its potential impacts on their day-to-day lives and adaptive measures to these climatic changes. The study also explores the seasonal calendar and its changes, and communities' perceptions on hazards. The research can be helpful in framing effective strategies for climate change adaptation in cold deserts and high-altitude mountain valleys of the world that are equally ecologically vulnerable.

Methods and Materials

Study Area

The study was performed in the 12 villages of the Suru Valley (six villages each from the upper and lower Suru Valley) in the Union Territory of Ladakh, India, and the valley is one of the most inhabited valleys in the Ladakh region. The Suru Valley lies in the trans-Himalaya and is a vast mountainous area between the southwestern Great Himalaya Range and the north-eastern Indus Valley (Fig. 19.1). It occupies the south-east portion of the Ladakh Himalaya. The valley is populated by the river Suru (one of the tributaries of the river Indus) flowing from the glaciers of Rangdum and Nun Kun Peaks. It is an arid and desert-like landscape with vegetation only along the water stream at elevations ranging from 2500 to 5000 m. Inside a rain shadow, the valley falls and gets limited rainfall. Agriculture and livestock rearing are the predominant livelihoods for most people in this area, and the people's livelihoods are supported by limited natural resources. However, changing climate patterns are not well understood here due to the harsh, cold weather, minimal monitoring provisions and weather stations operating in this region.

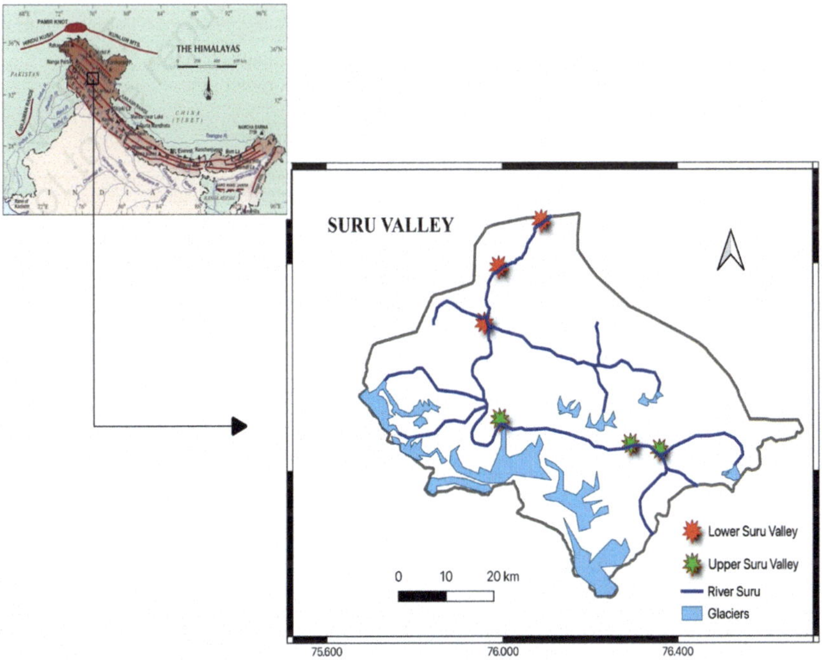

Fig. 19.1 Map of the study area, depicting three villages of lower Suru Valley (altitude <3500 m) and three villages of Upper Suru Valley (altitude >3500 m) considered in the study

Methods of Data Collection

The research integrates 120 semi-structured interviews of adults aged 18+ and 12 focus group discussions (FGDs) in various villages in the Suru Valley of the Ladakh region with a number of other participatory rural assessment (PRA) instruments. Information on various aspects of changing climate, water, agriculture, livestock and livelihood opportunities, awareness, responses to climate change and its associated impacts was collected using PRA tools, i.e. seasonal calendars, group discussions, community ranking of hazard, previous experiences through in-depth interviews etc. The information was collected during the months of June 2019 to November 2019 in different villages of Suru Valley of Ladakh region. In order to develop general perceptions of changing climate and its effects, in-depth interviews, group discussions and meetings were held with the villagers (Table 19.1). The participants for the FGDs were among the participants of in-depth interviews who were also willing to give FGDs. The study was conducted in the lower and higher altitudes of Suru Valley where 60 interviews for men and women each in the higher (above 3500 m) and lower altitudes (below 3500 m) were interviewed in equal ratios. A total of 12 FGDs (six each for men and women) were conducted (Table 19.2). The

Table 19.1 Overview of the assessment tools

Assessment tools	Purpose	Source of information
Semi-structured interviews (a) In-depth interviews (b) Life histories	Information on the perceived impacts of climate change in people's daily life and their adaptive responses to these changes Temporal changes in climate and also other socio-economic changes in the region	Men Women Village head
Focus group discussion (FGDs) (a) Seasonal calendar (b) Communities ranking of hazard	To understand the perceived gender differential impacts, their capacity to adapt to these impacts, and their needs To describe and classify the risks that have the greatest effect on lives and livelihoods. The outcome is demonstrated using a radar map where 1 stands for 'has a marginal effect' and 5 stands for 'has a significant effect' on the community's lives and livelihoods	Men Women

Table 19.2 Sampling distribution of respondents for the in-depth interview

	Suru Valley			
	Lower Suru Valley		Upper Suru Valley	
Villages	Men	Women	Men	Women
Minjee	10	10	–	–
Saliskote	10	10	–	–
Sanko	10	10	–	–
Parkachik	–	–	10	10
Shaqma Karpo	–	–	10	10
Rangdum	–	–	10	10
Total	120			

semi-structured questionnaires were both written in English and Urdu. Later, the translations were done from the local language (Purki, Ladakhi) and Urdu to English. Some of the verbatims are also included in the study. Proper consent from participants was also taken before the interview where only those participants who gave consent were interviewed.

Results

The respondents were in equal numbers for men and women. The basic demographic characteristics of the surveyed respondents are given in Table 19.3. Fifty-five percent of the surveyed respondents were in the age group 25–54 years and 28.33% in the age group 18–24 years. Only 16.67% of respondents were in the age group 55–64 at the time of the survey. The age groups were divided as 15–24 years (early working age), 25–54 years (prime working age) and 55–64 years (mature working age) where only adults aged 18+ were included. Elderly adults aged 65 and above were also excluded from the survey. So the majority of the respondents belong to the prime working-age group. Education plays a very important role in climate change awareness. The majority of the respondents (33.33%) and (31.67%) had primary education and no education, respectively. 23.33% of the respondents had attained secondary education, and

Table 19.3 Percentage distribution of respondents by selected background characteristics

Sex	Numbers	Percentage
Male	60	50
Female	60	50
Age group		
15–24	34	28.33
25–54	66	55
55–64	20	16.67
Highest education qualification		
No education	38	31.67
Primary	40	33.33
Secondary	28	23.33
Higher	14	11.67
Religion		
Muslim	80	66.67
Buddhist	40	33.33
Occupation		
Agricultural and livestock activity	46	38.33
Daily wage worker	34	28.33
Business	8	6.67
Government employee	18	15
Not working	14	11.67

only 11.67% of the respondents completed their higher education. Muslim religion (66.67%) constituted the majority, while the rest followed the Buddhist religion (33.33%). It is interesting to see that the majority (38.33%) of the respondents were either engaged in agriculture or livestock for their livelihood. This shows how people's livelihood heavily depends on the agriculture sector, which is highly sensitive to climate change. 28.33% of the respondents reported getting their source of income as a daily worker where they work as labourers, porters (for tourists and armed forces) and other construction workers in the nearby towns. This also included some working as daily wage workers under different central government schemes like the Mahatma Gandhi National Rural Employment Guarantee Act program (MGNREGA). Fifteen percent reported to have a government job when compared with 6.67% who said they had their own business. 11.67% of the respondents reported that they were currently not engaged in any kind of economic activity. This included mainly the respondents who were studying.

Perceived Climate Change, Its Impacts and Adaptive Measures

The results indicate that about less than two-thirds of the respondents have ever heard about the word climate change, while there are still more than one-third who have not heard about climate change. About 56.67% of the respondents said that they heard about climate change, while 43.33% of the respondents have not heard about climate change at all (Fig. 19.2). Looking further at the gender differential of awareness of climate, the study found that among the respondents who have ever heard about climate change, 61.76% of them were men, while only 38.24% of them were women. Thus, a good percentage of people still do not know about climate change, especially women in the region.

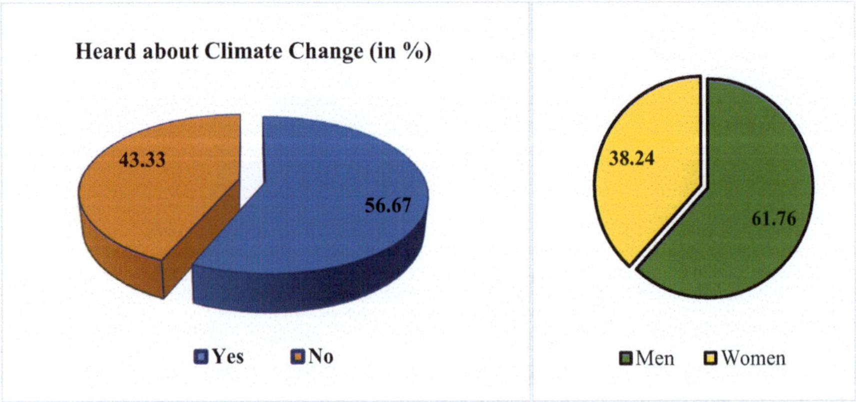

Fig. 19.2 Percent of respondents who have ever heard about "climate change"

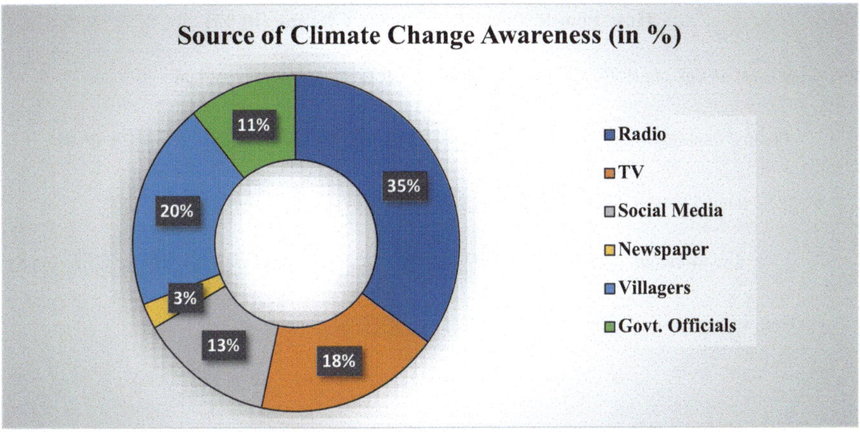

Fig. 19.3 Percentage of sources of climate change information (multiple responses)

Information about climate change was heard from various sources when called upon from where the respondent heard about climate change (Fig. 19.3). Radio and television seem to be very effective in disseminating information on climate change in the region where more than 50% of the respondents heard about climate change information through either radio or television. Twenty percent of the respondents said that they heard and shared information among the villagers and community about climate change. Social media is an effective tool, but due to less and poor internet connectivity, only 13% of the respondents heard it through social media. About 11% also said that they heard about climate change through some government awareness campaigns on climate change. People in the region have no access to newspapers, so there is hardly anyone who had heard about climate change through newspapers.

Five major changes that people perceive as climate change in the region are warm temperatures, less snowfall, early flowering season, increase in number of extreme events like flash floods and more irregular rainfall and snowfall (Fig. 19.4). Less snowfall and floods/flash floods are seen as a major concern where it is perceived by 93.33% and 85% of the respondents as climate change, respectively. Sixty-five percent of the respondents perceive irregular rainfall/snowfall as changing climate. The increase in temperature is also felt by the people where 55% of the respondents perceive it as climate change. Early flowering season (36.67%) is perceived as the least where only 36.6% of the respondents see it due to climate change.

The people in the region reported changes in temperature, rainfall, snowfall and flowering season (Table 19.4). People stated that temperature and extreme events have increased; rainfall and snowfall have decreased in the region over the period of time. Talking about climate change, a 42-year-old man replied:

> Climate has changed a lot during the last 10–20 years, you will hardly see any snow on the mountains, the glaciers are melting fast and you will also find very less snowfall as compared

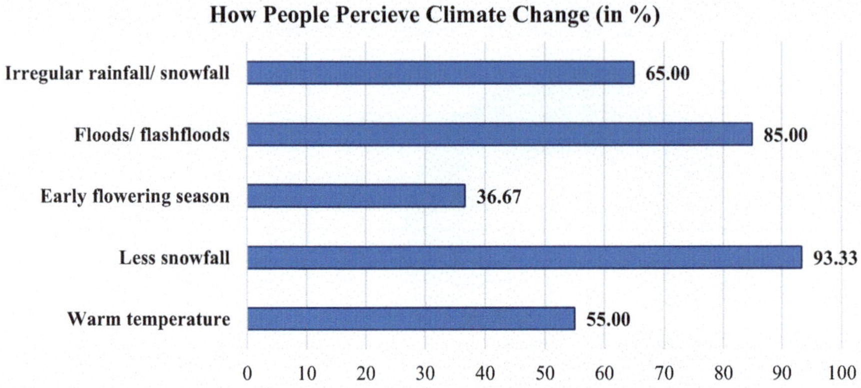

Fig. 19.4 Percentage of responses on perception of climate change (multiple responses)

Table 19.4 Perceived climate change, its impact and adaptive measures of the people in the Suru Valley of Ladakh region

Communities' perception of change	Experienced impacts on livelihood systems	Coping and adaptation
Slightly higher temperature with less rainfall	Lack of fodder and pastures in the mountains for animals Land becoming less productive Highly temperature-sensitive crops like barley, buckwheat etc. are less productive	Decreasing the number of domestic animals in a household. Barter fodder for money or manure Less land area under cultivation; buying food from markets More fruits and vegetables are grown replacing cereal crops
Less snowfall during winters	Spring drying; lower flow rate in springs and streams Less water for domestic, agricultural and livestock purposes	Construction of *zrgings* (ponds) for water storage Traditional water-sharing system in almost all of the villages The native residents get more water sharing than the latter settled residents
Onset of early flowering season and warmer summers	Early ripening of fruits and early fall of leaves Shift of one crop a year to two crops in a year Increased incidence of pests and insects to crops and trees	Drying the fruits and vegetables then storing it in dry form Increase labour, time and resources Increase use of insecticides and pesticides
Increase in the number of extreme events like flashfloods	Damage to houses, agricultural fields, trees and livestock	Houses are built away from the flood-prone areas; more tree plantation is done Mud roof is replaced by steel roof (though very less seen)
Unusual snowfall in spring season (flowering season)	Damage to orchid plants and other flowering trees like apricot trees, apple trees etc.	Supporting the branches of such trees with different tools from breaking and falling, well in advance

> to some 20 years ago, the water is also less and some of the streams are dried up because of the less snowfall.

Less snowfall and rainfall have led to a lack of fodder and pastures in the mountains for animals; land has become less productive, and the staple foods of people in the region like barley and buckwheat have become less lucrative economically. It has also led to drying of springs and decreased water flow in the streams. A 45-year-old man replied by saying:

> If there is less snowfall in winter that means you have less water in summer for the agriculture and livestock. What should we do then?

These impacts have led to some coping and adaptive measures like decrease in the number of domestic animals and barter fodder for money and manure to support the feeding of animals. People started shifting from agriculture to more stable income-based jobs (as people think in the region) like government services, daily wage workers, porters (for tourists and armed forces) or planting more income-based crops like fruits and vegetables, replacing the traditional crops like barley, wheat and buckwheat. Alfa-alfa crop is mainly grown as a fodder for the livestock.

The impacts of less snowfall and rainfall were seen higher in women than in men. A woman in her fifties said:

> Whenever there is less snowfall in winter, you have less water during summer and you will find women getting more worried than men because women do most of the work in the field from ploughing, watering and harvesting.

Women have demonstrated strong adaptative capacity in dealing with climate change and have been seen to be closer to nature. Many women have water management capabilities, awareness and skills, and sustainable farming practices that are relevant in mitigating climate change impacts. To overcome the scarcity of water, people are using the traditional sharing of water in the villages on a rational basis.

In the lower part of the valley, warmer climate has led to production of more diverse fruits and vegetables, which were earlier not possible due to comparatively cold climate. People responded that there is an early flowering season and an early onset of spring; people start ploughing and preparing their agriculture fields during the early week of March, though earlier it used to be in mid-April. Many of the respondents likewise felt that climate change has some positive consequences. A 38-year-old woman said:

> Earlier we used to grow a single crop in a year due to cold and harsh climate but the situation is now different as the climate is warmer. There is a shift in ploughing and harvesting season, we seed early in the month of March-April and July-August respectively and during rest of the months we grow another crop.

At the same time, there was early ripening of fruits, falling of leaves from trees and increase in the incidence of insects damaging crops and other fruit trees that had led people to be more dependent on insecticides and pesticides for controlling damages caused by insects. The practice of using high chemical fertilizer is replacing the old manure system in order to produce more crops.

Erratic rainfall and flash floods have increased during the last two to three decades, resulting in damage to crops and livestock in the region. A 50-year-old man replied when asked about flash floods and the damage caused by it:

> Whenever there is rainfall, it is sudden and you will get flash flood that damages the trees, crops and houses as well. Roads and highways were washed away from the erratic rainfall causing landslide and flash floods leading to cut off of many villages from the town. The loss of bridges and roads affects the livelihoods and prospects of people for mobility, connectivity and employment.

Seasonal Calendar of the Region

The start of a calendar year may differ from community to community. People in the region consider 21 March (*Navroz*) as a new year and also as the start of the sowing season. Table 19.5 reveals the seasonal calendar in the region that lists climate-induced events. The main events in the region during a calendar year were listed as rainfall, snowfall, water shortage, avalanches, flash floods, crop pests/diseases, livestock diseases and food shortage. It was observed that snowfall occurred mostly during the months of December, January, February and March, while rainfall was observed in March, April, August and September. But recent events of snowfall in May and June were also listed by the participants. This unusual snowfall led to damage to the flowering buds of fruit trees in the region. Explaining about the unusual snowfall in the seasonal calendar during the spring season, a 50-year-old man replied:

> Usually the snowfall occurs between the months of December to March, but for the past few years we saw snowfall in the month of April and May and even June. April and May are the months when there is flowering season of fruit trees in the region and this unusual events have caused damage to flowers that ultimately affects the production of fruits especially apricots and apples.

Table 19.5 An example of the seasonal calendar of overall Suru Valley [the number of stars (*) indicates intensity]

	Mar	Apr	May	Jun	Jul	Aug	Sep	Oct	Nov	Dec	Jan	Feb
Rainfall	*	**					*					
Snowfall	**	*	*	*						*	***	***
Water shortage				**	***	**						
Avalanches	***										*	**
Flashflood						***	**					
Crop pests/ diseases				**	***	***	**					
Livestock diseases												
Food shortage	**										**	***

Water and food shortage as reported by people were much higher and existed over a longer period. This resulted in women's suffering more as they were responsible for water and food management in the household and watering the field. Women were more aware and concerned about the period of months when there is shortage of water and food in the household and in the field. Flash foods were seen more in the region due to more erratic rainfall that usually happens in the months of August and September.

Crop pests and tree insects were the major problems in the area. Almost every participant reported about the pest and insect attack damaging their crops and trees. There was a surge in the pests and insects due to increase in summer temperature and that was mostly seen in the summer months of June, July and August (Table 19.5).

Community Ranking of Hazard

Figure 19.5 provides the communities' ranking of hazards and positioning of the hazards that had the greatest effect on their lives and livelihoods. For each identified danger or phenomenon, the participants received a stone and had to agree on where to position them on the radar map. As shown in Fig. 19.5a, men identified less snowfall and flash floods as the two risk factors with the greatest impact on their lives and livelihoods in the upper Suru Valley, while women identified less snowfall and water scarcity as the two major risks impacting their lives and livelihoods. Both men and women identified avalanches with the least impact. In the lower Suru Valley (Fig. 19.5b), we observed different effects of hazards. Here women identified water scarcity as a hazard with the strongest influence on their lives and livelihoods, while men reported less snowfall as the most severe hazard, followed by flash floods.

Discussion

The study highlights the importance of people's perceptions of climate change in a scarce climate data region. It also emphasizes the significance of local knowledge on climate change and the changes due to climate change, enabling how indigenous people interpret climate change and react to these changes in general. The study further investigates the gender differential impacts of changing climate and the adaptive measures to these climatic changes. The study is among the few research studies focusing on the Suru Valley of Ladakh Tans-Himalayas, to understand the climate change perception and its associated impacts using different qualitative tools. The study found that snow melt water from the mountains and glaciers plays a very important role in sustaining life in the region as they form the only source of water used for field irrigation, livestock and other domestic purpose. So, wherever there is settlement and vegetation, there is a river or a stream.

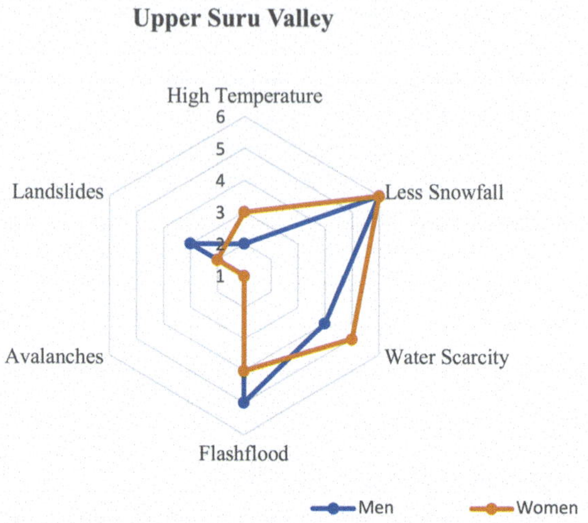

(a)

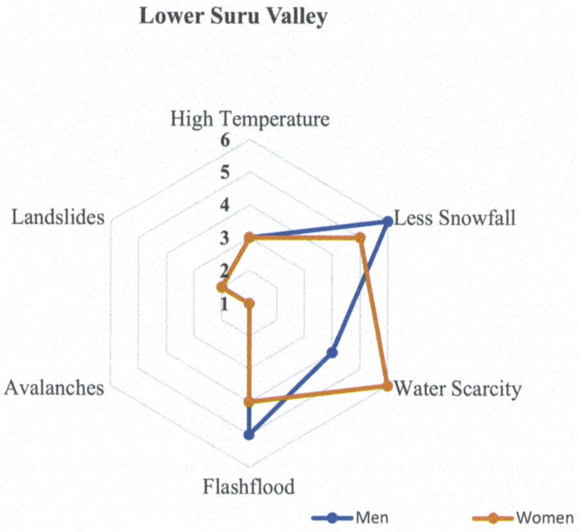

(b)

Fig. 19.5 Communities ranking of hazard severity for (**a**) the upper Suru Valley and (**b**) lower Suru Valley

Results indicate that still more than one-third of the people depend mostly on agriculture and livestock for their livelihood, which is highly sensitive to climate change. An increase in temperature and extreme events and a decrease in rainfall and snowfall over time will have a huge impact on people's livelihoods. Results show that temperature and extreme events have increased, while snowfall and rainfall have reduced in the region over time. Less snowfall and rainfall have led to a lack of fodder, land has become less productive and the staple foods of people in the region like barley and buckwheat have become less productive. Decreased snowfall and rainfall have also led to drying of water springs in the region. The changes like water scarcity, less pastures and less agricultural production have led to adaptative measures like decrease in the number of domestic animals and barter fodder for money and manure. The lower Suru Valley has a warmer climate and has led to the production of more diverse fruits and vegetables replacing cereal crops. At the same time, there is early ripening of fruits, falling of leaves and increase in the incidence of insects damaging crops. Unusual snowfall in the spring season has led to damage to the flowering buds of fruits and trees in the region. Similar observation was also found in the upland villages of West Bengal and Nepal (Chaudhary & Bawa, 2011). Declining snowfall in Nepal has contributed to a longer dry season over the past 6 years, reducing crop production and increasing hunger (Leduc, 2008). In Ladakh, people's staple foods like barley and buckwheat are replaced by more income-generating plants like vegetables and fruits. People prefer to consume more rice and wheat that are available from the market.

The study further highlights that water and food shortage from the agricultural fields have increased. Women recognized water scarcity as a hazard that has the toughest impact on their lives and livelihood, while men reported less snowfall as the most severe hazard. Gender differential impacts were visible in the perception of different hazards that affected the people in the region. The impact of water scarcity and less snowfall was more perceived by women when compared with men, as women have the role of watering the agricultural field and quenching water for households. Thus, the impacts were more visible on women than on men. For example, in Vietnam, similar findings were found in which, as each drought takes its toll, women are fetching water from the farthest water sources (Shaw et al., 2008). In Senegal, women were seriously affected by water shortages than men as women had to drive further in search of water, spending more time looking for different water source wells (Dankelman et al., 2008).

The findings of the research explored that in the Suru Valley, agriculture and livestock were heavily affected due to climate change. Crop pests and plant insects were major problems that caused severe damage to crops and other plants. These changes have led to reduction in agricultural production and forced community members to diversify their livelihood activities that are more market oriented. However, in some parts of the lower Suru Valley, higher temperatures brought opportunities, as people grew a diversity of fruits and vegetables, which earlier was not possible due to harsh climate. Crops like wheat, barley and buckwheat were grown from April to July, while crop like alfa-alfa was grown from August to October. Thus, two types of crops were grown in a single year due to warmer climates. A similar finding was also observed in Uttarakhand where farmers recorded that as a result of rising temperatures, fruit trees like apple and orange began to flower twice a year where they increased the quantity but decreased the fruit quality (Macchi et al., 2017).

Women did not have the same information, facility and access to knowledge on climate forecasts as men. Men used radio as an information channel, while women preferred seasonal predictions given by their husbands or some other male household member. Thus, men's better response to hazards like avalanches and flash flood was due to higher awareness through better access to communication facilities. As disaster warnings and alerts mostly come from TV, radio or cell phones, men use these gadgets more often (Ahmad & Fajber, 2009). Similarly, in South Africa, women in the communities did not have the same information and access to forecast climate information where men preferred radio as a means of transmission, while women favoured seasonal forecasts (Archer, 2003).

Conclusions

Climate change and its perceived impacts from the perspective of the local population are important as the region has limited weather monitoring facilities. Understanding the awareness of the population about climate change and their dealing with climate change will help to establish successful strategies and policy formation in the region. People in the villages of the Suru Valley depend significantly on nature, its resources and ecosystem services for their daily needs and are therefore more vulnerable to potential climate change impacts. The adaptive response of communities to changing climate and its impact is more of a short-term strategy in the region and therefore requires a more efficient option of long-term adaptation that focuses on the expertise of mountain communities. The adaptive responses should also include women's capabilities, as women are the key food producers in the region and are more involved in crop production. Incorporating the need and helping women in climate change actions will help in minimizing the uncertainty of climate change impacts and will increase the adaptive responses of the communities. This study could be useful for policymakers to establish successful community-led adaptation actions and strategies to mitigate the effects of climate change on vulnerable and often neglected communities in the region.

Funding This study did not receive any specific support grant from funding agencies in the public, commercial or not-for-profit sectors.

References

Ahmed, S., & Fajber, E. (2009). Engendering adaptation to climate variability in Gujarat, India. *Gender and Development, 17*(1), 33–50. https://doi.org/10.1080/13552070802696896

Archer, E. (2003). Identifying underserved end-user groups in the provision of climate information. *Bulletin of the American Meteorological Society, 84*(11), 1525–1532. https://doi.org/10.1175/BAMS-84-11-1525

Becker, A., Inoue, S., Fischer, M., & Schwegler, B. (2012). Climate change impacts on international seaports: Knowledge, perceptions, and planning efforts among port administrators. *Climatic Change, 110*, 5–29. https://doi.org/10.1007/s10584-011-0043-7

Berkes, F., & Jolly, D. (2002). Adapting to climate change: Social-ecological resilience in a Canadian western Arctic community. *Conservation Ecology, 5*(2), 18. https://doi.org/10.5751/ES-00342-050218

Byg, A., & Salick, J. (2009). Local perspectives on a global phenomenon-climate change in Eastern Tibetan villages. *Global Environmental Change, 19*, 156–166. https://doi.org/10.1016/j.gloenvcha.2009.01.010

Chaudhary, P., & Bawa, K. S. (2011). Local perceptions of climate change validated by scientific evidence in the Himalayas. *Biology Letters, 7*, 767–770. https://doi.org/10.1098/rsbl.2011.0269

Chaudhary, P., Rai, S., Wangdi, S., Mao, A., Rehman, N., Chettri, S., & Bawa, K. (2011). Consistency of local perceptions of climate change in the Kangchenjunga Himalaya landscape. *Current Science, 101*(4), 504–513. Retrieved March 27, 2021, from http://www.jstor.org/stable/24078981

Dankelman, I., et al. (2008). *Gender, climate change and human security: Lessons from Bangladesh, Ghana and Senegal.* http://www.gdnonline.org/resources/WEDO_Gender_CC_Human_Security.pdf

Dimri, A. P., & Dash, S. K. (2012). Wintertime climatic trends in the western Himalayas. *Climatic Change, 111*, 775–800. https://doi.org/10.1007/s10584-011-0201-y

Ford, J., Vanderbilt, W., & Berrang, F. L. (2012). Authorship in IPCC AR5 and its implications for content: Climate change and Indigenous populations in WGII. *Climatic Change, 113*(2), 201–213. https://doi.org/10.1007/s10584-011-0350-z

Jorgyes, S. (2010). Ladakh's cultural and traditional adaptation to its high latitude harsh climatic environment and challenges ahead due to its recent life style and climatic changes. *Himalayan Study Monographs, 11*, 196–200.

Le Masson, V., & Nair, K. (2012). Does climate modeling help when studying adaptation to environmental changes? The case of Ladakh, India. Climate change modeling for local adaptation in the Hindu Kush-Himalayan region. *Community, Environment and Disaster Risk Management, 11*, 75–94. https://doi.org/10.1108/S2040-7262(2012)0000011011

Leduc, B. (2008, June). *Case study: Gender and climate change in the Hindu Kush Himalayas of Nepal.* WEDO Gender and Climate Change Workshop in Dakar, Senegal. http://www.wedo.org/wp-content/uploads/nepalcasestudy.pdf

Liu, J., & Rasul, G. (2007). Climate change, the Himalayan mountains, and ICIMOD. *Sustainable Mountain Development, 53*, 11–14.

Macchi, M., Gurung, A. M., & Hoermann, B. (2017). Community perceptions and responses to climate variability and change in the Himalayas. *Climate and Development, 7*, 414–425. https://doi.org/10.1080/17565529.2014.966046

Mehta, P. S., Sharma, A. K., & Negi, K. S. (2010). Indigenous knowledge system and sustainable development with particular reference to folklores of Kumaun Himalaya, Uttarakhand. *Indian Journal of Traditional Knowledge, 9*(3), 547–550.

Nüsser, M., Dame, J., Kraus, B., Baghel, R., & Schmidt, S. (2019). Socio-hydrology of "artificial glaciers" in Ladakh, India: Assessing adaptive strategies in a changing cryosphere. *Regional Environmental Change, 19*, 1327–1337. https://doi.org/10.1007/s10113-018-1372-0

Reidlinger, D., & Berkes, F. (2001). Contributions of traditional knowledge to understanding climate change in the Canadian Arctic. *Polar Record, 37*, 315–328. https://doi.org/10.1017/S0032247400017058

Schickhoff, U., Singh, R. B., & Mal, S. (2016). Climate change and dynamics of glaciers and vegetation in the Himalaya: An overview. In R. B. Singh, U. Schickhoff, & S. Mal (Eds.), *Climate change, glacier response, and vegetation dynamics in the Himalaya* (pp. 1–26). Contributions Toward Future Earth Initiatives.

Schmidt, S., & Nüsser, M. (2017). Changes of high-altitude glaciers in the Trans-Himalaya of Ladakh over the past five decades (1969–2016). *Geosciences, 7*(2), 27. https://doi.org/10.3390/geosciences7020027

Schmidt, S., Nüsser, M., Dame, J., & Baghel, R. (2020). Cryosphere hazards in Ladakh: The 2014 Gya glacial lake outburst flood and its implications for risk assessment. *Natural Hazards, 104*(3), 2071–2095. https://doi.org/10.1007/s11069-020-04262-8

Sharma, E., Chettri, N., Tsering, K., et al. (2009). *Climate change impacts and vulnerability in the Eastern Himalayas*. ICIMOD.

Shaw, R., et al. (2008). *Drought-management considerations for climate-change adaptation: Focus on the Mekong region*. International Environment and Disaster Management (IEDM) and Oxfam, Vietnam. http://www.oxfam.org.uk/resources/policy/climate_change/downloads/vietnam_climate_change_report.pdf

Shekhar, M., Chand, H., Kumar, S., Srinivasan, K., & Ganju, A. (2010). Climate-change studies in the western Himalaya. *Annals of Glaciology, 51*(54), 105–112. https://doi.org/10.3189/172756410791386508

Thayyen, R. J., Dimri, A. P., Kumar, P., & Agnihotri, G. (2013). Study of cloudburst and flashfloods around Leh, India during August 4-6, 2010. *Natural Hazards, 65*(3), 2175–2204. https://doi.org/10.1007/s11069-012-0464-2

Vedwan, N., & Rhoades, R. (2001). Climate change in the western Himalayas of India: A study of local perception and response. *Climate Research, 19*, 109–117. https://doi.org/10.3354/cr019109

Weber, E. U. (2010). What shapes perceptions of climate change? *WIREs Climate Change, 1*, 332–342. https://doi.org/10.1002/wcc.377

Chapter 20
Unveiling Precipitation and Temperature Patterns in Kashmir Valley, India

Sana Rafi and Raghupathi Balasani

Introduction

The impact of climate change on the monthly, seasonal and annual precipitation and air temperature has received a great deal of attention by scholars worldwide (Jain & Kumar, 2012). One of the most important consequences of climate change would be an uncertainty of the precipitation distribution and temperature variation both spatially and temporally. However, the spatio-temporal quantitative estimation of precipitation and temperature is required for various purposes such as water resource management and climate change studies. It is more important in the Indian context as the country harbours an agrarian economy dependent heavily on the summer monsoon. According to the study conducted by MoEF 2010, increasing trend in temperature, precipitation as well as extreme temperature, precipitation, intensity of precipitation and number of rainy days were observed. The analysis was based on the climate change in the Indian Himalayan region assessment for 2030s with respect to 1970s. The study suggests that the number of rainy days in the Himalayan region in the 2030s may increase by 5–6 days on an average, and the intensity of rainfall is likely to increase by 1–2 mm/day. The annual rainfall is likely to increase from 5% to 13%, while some areas are showing an increase up to 50%. With respect to the annual temperature, a likely increase from 0.90 +/− 0.60 to 2.6 +/− 70 °C in 2030 was highlighted. Understanding the trends and variability of climatic factors is essential for appreciating the impact of climate change (Wani et al., 2017). They also mentioned that the magnitude of the variability of the factors changes according to the locations. Therefore, the need for continuous precipitation and temperature studies requires an emphasis, particularly in fragile locations such as Himalayas.

S. Rafi (✉)
Department of Geography, Faculty of Natural Sciences, Jamia Millia Islamia, New Delhi, India

R. Balasani
JNTU, Hyderabad, India

A. L. Singh et al. (eds.), *Climate Change, Vulnerabilities and Adaptation*,
https://doi.org/10.1007/978-3-031-49642-4_20

This study is therefore an attempt to detect trends and variability in precipitation and temperature for the period 1901–2018 in Kashmir Valley, a union territory situated in the northwestern Himalayas. This study is mainly focused on district-wise monthly, seasonal and annual trend and variability detection in precipitation and temperature. Non-parametric Mann–Kendall (MK) test and Sen's slope estimation method were applied for the objective. MK test is one of the most widely used non-parametric test for analysing trends for hydro-meteorological data and climate-based studies (Drápela & Drápelová, 2011; Hirsch et al., 1982; Kumar et al., 2005; Yang & Tian, 2009; Motiee & McBean, 2009; Karmeshu, 2012; Wani et al., 2017; Yadav et al., 2014; Yue & Wang, 2004). This test does not require data to be normally distributed, and it has low sensitivity to abrupt breaks due to inhomogeneous time series (Tabri et al., 2011; Karmeshu, 2012).

Literature Review

Changes in annual temperature and precipitation in selected states in northeastern USA were found by Karmeshu, (2012). While Salma, et al. (2012) found changes in rainfall in Pakistan from 1988 to 2006 and an upward-moving trend till 2030. Pant and Kumar, (1997) detected a significant warming trend of 0.57 °C per 100 years in annual and seasonal air temperature, with higher warming magnitude being in winter and post-monsoon seasons for the time series of 1881–1997 in India. Their study shows a significant negative trend over northwest India and no significant trend in any part of the country was observed. According to Bhutiyani, et al. (2007), an increasing trend was found in maximum, minimum, mean and diurnal temperature ranges over the northwestern Himalayas. The frequency of occurrence of hot days and hot nights during 1970–2005 over India was also analysed and an increasing trend in hot days and nights and a decreasing trend in cold days and nights were found for the country. A significant decreasing trend for cold days over the western Himalayas and western coast has also been observed. Basistha, et al. (2009), studied the rainfall data from 1901 to 1980 from 30 rain gauge stations in the Indian Himalayas. An increasing trend in precipitation has been found during (1901–1964) and a decreasing trend for the last two decades (1965–1980). Das et al. (2015) studied trends in 27 precipitation indices for 10 major rain gauge stations of the northeast region of India on a monthly, seasonal and annual basis during 1961–2010. The study revealed that the annual and monsoon rainfall have no significant trends. Significantly increasing trends during pre-monsoon season are found at Imphal and Tezpur, at Guwahati during the post-monsoon and at Cherrapunjee during winter. On a monthly basis, an increasing trend in rainfall is observed at Tezpur during April, at Imphal and Kailashahar during May and September and at Guwahati during October, while the rainfall of June at Imphal has a decreasing tendency during the period 1961–2010. The absence of any significant trend in rainfall amount on a monthly, seasonal and annual scale and in frequency of rainfall intensities of ≥2.5 mm/day, 1–3 cm/day, 4–6 cm/day ≥7 cm/

day and wet spells of length of 2 days, 3 days and 4 days at most of the stations confirmed that there is no clear-cut indication of any major changes in rainfall pattern over NER in terms of rainfall intensity and persistence. However, further research is needed for a better understanding of the temporal pattern of rainfall in the region. Jain, et al. (2012) analysed the trend for rainfall during 1901–1984 at 11 stations in Himachal Pradesh and found an increasing trend in annual rainfall at 8 stations and a significant decreasing trend in monsoon rainfall at eight stations. Likewise, Wani, et al., (2017) analysed the trends and variability in the annual temperature and rainfall for the district of Mandi, Himachal Pradesh (India), during 1981–2010 and found an increasing trend in annual maximum and minimum temperature and a decreasing trend in the monsoon's maximum and minimum temperatures although it is statistically not significant. The amount of annual rainfall does not show any significant trend, but the monsoonal rainfall has shown an increasing trend that is also statistically not significant. The analysis of rainfall series created using a network of 1476 rain gauge stations across India indicates a significant increasing trend in monsoon rainfall in Jammu and Kashmir (Rajeevan et al., 2006; Guhathakurta & Rajeevan, 2008). Similar results were provided for Kashmir Himalayas during 1910–2000 with the analysis conducted on 129 stations of India by Sen Roy & Balling, (2004). Moreover, Kumar & Jain, (2010) analysed the rainfall data during 1903–1982 at Srinagar, Kulgam, Handwara, Kokernag and Quazigunds stations. They have found a decreasing trend in annual rainfall at Srinagar, Kulgam and Handwara stations with the largest decrease of 20.16% of mean/100 years and the smallest for Srinagar. The decreasing trend in winter rainfall was statistically significant at Kulgam and Handwara and non-significant increasing trends in the pre-monsoon and post-monsoon seasons. Quazigund and Kokernag experienced decreasing annual rainfall, but Srinagar showed an increasing annual trend for the period 1962–2002. Annual, winter, pre-monsoon and post-monsoon rainfall increased, whereas monsoon rainfall decreased non-significantly at Srinagar during the last century. All stations experienced a decreasing trend in winter and monsoon rainy days, but Srinagar and Handwara also witnessed a decreasing trend in annual rainy days, and Kulgam showed the opposite trend. The most significant implication of climate change is an increase in the frequency and intensity of extreme weather events leading to hydro-meteorological disasters (Mavromatis & Stathis, 2011; Ramesh & Goswami, 2007). This is one of the anticipated effects of climate change. The Valley of Kashmir has experienced several disasters of recurrent nature that have resulted in the heavy loss of life, livelihoods and properties. There is a need for research and studies in various aspects including trend and variability detection for various climate parameters (Smith, 2010); for better planning, assessment capacity analysis and building resilience of communities to withstand pressure and shocks.

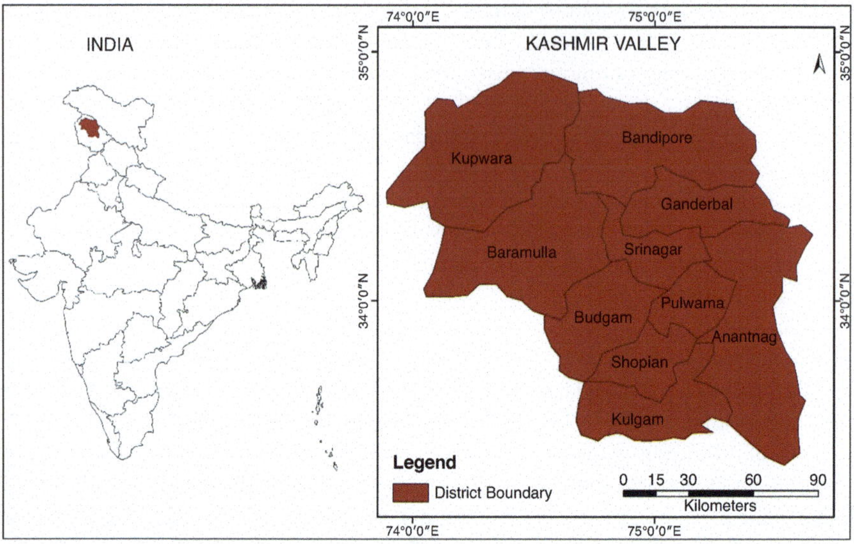

Fig. 20.1 Location map of the study area

Study Area

Kashmir Valley bounded on the southwest by the Pir-Panjal range and on the northeast by the greater Himalayas is one of the union territories of India and accounts for about 0.48% of the geographical area of the country. It lies in the extreme north of India between 33°22–34°22 N latitude and 73°59–75°30 E longitude (Fig. 20.1), having a dimension of 135 km long from northeast to southwest and 32 km wide covering an area of 15,941.95 sq. km with an average elevation of 1800 m above mean sea level while the surrounding mountains, which are always snow clad, rise from 3000 to 4000 m above mean sea level. It has a population density of 432 persons per sq. km having a population count of 6,888,475 (Census, 2011). Broadly, the valley consists of 10 districts, average of rainfall and temperature in each district has been presented in (Table 20.1). The climate of Kashmir Valley is temperate, and the year is divided generally into four seasons. The winter season (December–February) is followed by pre-monsoon or hot weather season (March–May). June–September constitutes the southwest monsoon season, and October and November is of post-monsoon period. This seasonal classification is given by the Indian Meteorological Department (IMD) and hence is generally adopted for such studies.

Table 20.1 District-wise geographical area, population, population density, mean temperature and rainfall of the Kashmir Valley

S. No.	District	Geographical area (Sq. km)	Population	Density (Sq. km)	Rain (mm)	Mean Temp. (°C)
1	Srinagar	1978.95	12,36,829	625	616.1	6.9
2	Kulgam	404	4,24,483	1051	1109.3	11.4
3	Shupian	312	2,66,215	853	1184.5	11.1
4	Ganderbal	259	2,97,446	1148	679.8	8.3
5	Budgam	1361	7,53,745	554	732.9	9.8
6	Bandipore	345	3,92,232	1137	552.8	4.9
7	Anantnag	3574	10,78,692	302	622.1	7.2
8	Pulwama	1086	5,60,440	516	1027.4	9.3
9	Baramulla	4243	10,08,039	238	747.7	10
10	Kupwara	2379	8,70,354	366	651.5	7.2
Total		15,941.95	68,88,475	432	791.51	8.61

Material and Method

Data Used

The monthly, seasonal and annual averages of precipitation and temperature are generally based on a gridded time-series dataset (CRU TS v. 4.02) at 0.5° resolution released on 18 November 2018 (Harris et al., 2014), covering the time period of 1901–2017 and sourced from http://www.cru.uea.ac.uk/cru/data/hrg/. The data have been extracted through Google Earth Pro software for each district.

Mann–Kendall Test

For detecting and analysing the trend in monthly, seasonal and annual precipitation and temperature, the advanced Mann–Kendall statistical test was applied using Addinsoft's XLSTAT 2018 software. To determine the magnitude of the trend, Sen's slope estimator was applied. The graphs were prepared in Microsoft Excel 2016, with a parametric trend line imposed on the graph.

Mann–Kendall test is a non-parametric test, often referred to as the MK test. It is used to detect monotonic trends in the data collected over time. It does not require data to be normally distributed, or linear. However, it does require that there is no autocorrelation. For the time series $xi, \ldots, x_n$, the MK test uses the following statistics Eq. (20.1).

$$S=\sum_{i=1}^{N-1}\sum_{j=i+1}^{N}\operatorname{sign}\left(x_j-x_i\right) \quad (20.1)$$

where N is the number of data points, $xj, \ldots, x_i = \theta$, which is computed as

$$\operatorname{sign}\theta=\begin{Bmatrix} 1 \text{ if } \theta>1 \\ 0 \text{ if } \theta=1 \\ -1 \text{ if } \theta<1 \end{Bmatrix}$$

The variance (S) for the S-statistic is defined by Eq. (20.2)

$$\operatorname{Var}(S)=\frac{N(N-1)(2N+5)-\sum_{i=1}^{N} t_i(t_i-1)(2t_i+5)}{18} \quad (20.2)$$

In which t_i denotes the number of ties to the extent i. The summation term is the numerator is used only if the data series contains tied values. The standard test statistics Zs is calculated using the Eq. (20.3).

$$Zs=\begin{Bmatrix} \frac{S-1}{\sqrt{\operatorname{Var}(S)}} & \text{if} \quad S>0 \\ 0 \quad \text{if} & S=0 \\ \frac{S+1}{\sqrt{\operatorname{Var}(S)}} & \text{if} \quad S<0 \end{Bmatrix} \quad (20.3)$$

The test statistic Zs is used as a measure of the significance of trend. In fact, this test statistic is used to test the null hypothesis, H_0. If $|Zs|$ is greater than $Z\alpha/2$, where α represents the chosen significance level (e.g. 5% with $Z_{0.025} = 1.96$), then the null hypothesis is invalid implying that the trend is significant (Karmeshu, 2012).

For detecting a true trend in a time series, a consideration of the autocorrelation is to be given. Hamed and Rao (1998) suggest a modified Mann–Kendall test, which calculates the autocorrelation between the ranks of the data after removing the apparent trend. The adjusted variance is given by Eq. (20.4):

$$\operatorname{Var}[S]=\frac{1}{18}[N(N-1)(2N+5)]\frac{N}{NS^*} \quad (20.4)$$

Where $\frac{N}{NS^*}=1+\frac{2}{N(N-1)(N-2)}\sum_{i=1}^{P}(N-i)(N-i-1)(N-i-2)pS(i)$.

N is the number of observations in the sample, NS^* is the effective number of observations to account for autocorrelation in the data, p_s (i) is the autocorrelation between ranks of the observations for lag i, and p is the maximum time lag under consideration (Sinha & Cherkauer, 2008).

On running the MK test, another statistic that is obtained is Kendall's tau. It is a measure of correlation carried out on the ranks of the data. It can take any value between −1 and +1, with a positive correlation indicating that the ranks of both variables increase together, while a negative correlation indicates that as the rank of one variable increases, the other decreases.

Theil–Sen's Estimator

The magnitude of the trend in a time series can be determined using a non-parametric method known as Sen's estimator (Sen, 1968). It is fairly resistant to outliers. This method assumes a linear trend in the time series. In this method, the slopes (T_i) of all data pairs are calculated first by Eq. (20.5)

$$T_i = \frac{x_j - x_k}{j - k} \text{ for } i = 1, 2, \ldots\ldots\ldots, N \tag{20.5}$$

Where xj and xk are data values at time j and k ($j > k$), respectively. The median of these N values of T_i gives the Sen's estimator of slope (Q). A positive value of Q indicates an upward trend, and a negative value indicates a downward trend in the time series Eqs. (20.6) and (20.7).

$$Q = Q_{N+1/2} \tag{20.6}$$

$$Q = \left(\frac{1}{2}\right) Q\left[\frac{N}{2}\right] + Q\left[\frac{N+2}{2}\right] \tag{20.7}$$

According to this MK test, a null hypothesis assumes that there is no trend (where data are independent and randomly ordered) and is tested against an alternative hypothesis, which assumes that there is a trend. If the p value is less than alpha value, then the null hypothesis is rejected otherwise accepted. Acceptance of the null hypothesis indicated that there is no trend in time but on rejecting the null hypothesis, the result is said to be statistically significant. The value of Kendall Tau shows whether the trend is increasing or decreasing. Sen's slope tells us an average of how much (temperature, precipitation etc) has changed each year.

Results and Discussion

Monthly Trends of Precipitation (mm)

This study shows that a non-significant decreasing trend was found in January month for all districts. A statistically significant increasing trend was found for February month in Baramulla, Bandipore, Kupwara, Badgam and Kulgam districts and a non-significant decreasing trend in Shupian and Anantnag. The statistically significant rising trend for the month of March was seen at Baramulla, Bandipore, Kupwara, Ganderbal and Srinagar and a non-significant declining trend in Kulgam district. The results showed increasing trends for all districts with significance in Baramulla, Bandipore, Kupwara, Ganderbal, Srinagar, Anantnag, Kulgam and Shopian districts in April; Pulwama, Kulgam and Shopian districts in May; and all districts in June. A statistically significant increasing trend was observed for all districts except Anantnag in the month of July. The month of August showed a non-significant decreasing trend at Srinagar and Anantnag, an increasing trend for all other districts and a statistically upward trend for Kulgam and Budgam. Statistically rising trends were observed for most of the district in October month wherein the districts of Srinagar and Kupwara experienced non-significant decreasing trends. The statistically significant rising trends were observed in all districts in the month of November and a non-significant decreasing trend in the month of December for all districts (Tables 20.2 and 20.3).

Seasonal and Annual Trends of Precipitation (mm)

The statistically significantly increasing trends were found in the annual series for all districts (Fig. 20.2a–j). The winter season shows non-significant increasing trends for most of the districts and a decreasing trend for the Ganderbal district only. The Pre-monsoon and monsoon precipitation showed statistically rising trends in most of the districts. While there exists a statistically significant increasing trend in post-monsoon precipitation in all districts except Pulwama, while Anantnag district showed a decreasing trend (Tables 20.2 and 20.3).

Monthly Trends of Mean Temperature (°C)

For the annual temperature, non-significant rising trends were observed for all districts in January month and a statistically significant increasing trend for the months of February, March and April for all districts. The May month showed a non-significant upward trend for nine districts except Anantnag which experienced a declining trend. The months (June, July and August) showed a declining temperature trend for all districts; however, the trend is statistically significant for some districts and statistically non-significant for others. The month of September shows a statistically rising trend for Srinagar and Kupwara district, while a non-significant rising trend in Budgam, Bandipore, Pulwama and Baramulla and the remaining districts shows a non-significant decreasing trend. Likewise, the month of October experienced a rising temperature trend for all districts, being statistically significant for only three districts, namely Srinagar, Bandipore and Kupwara. The temperature time

Table 20.2 Mann–Kendall trend statistics (*Z*) precipitation (mm)

District	Srinagar	Kulgam	Shupian	Ganderbal	Budgam	Bandipore	Anantnag	Pulwama	Baramulla	Kupwara
January	↓	↓	↓	↓	↓	↓	↓	↑	↓	↓
February	↑	↑*	↓	↑*	↑*	↑*	↓	↑	↑*	↑*
March	↑*	↓	↑	↑*	↑*	↑*	↑	↑	↑*	↑*
April	↑*	↑*	↑*	↑*	↑*	↑*	↑*	↑*	↑*	↑*
May	↑*	↑*	↑*	↑*	↑*	↑	↑	↑*	↑	↑
June	↑*	↑*	↑*	↑*	↑*	↑*	↑*	↑*	↑*	↑*
July	↑*	↑*	↑*	↑*	↑*	↑*	↓	↑*	↑*	↑*
August	↓	↑*	↑	↑	↑*	↑	↓	↑	↑	↑
September	↑	↑	↑	↑	↑	↑	↑	↑	↑	↑
October	↓	↑*	↑*	↑*	↑*	↑	↑*	↑*	↑*	↓
November	↑*	↑*	↑*	↑*	↑*	↑*	↑*	↑*	↑*	↑*
December	↓	↓	↓	↓	↑	↓	↓	↓	↓	↓
Annual	↑*	↑*	↑*	↑*	↑*	↑*	↑*	↑*	↑*	↑*
Winter	↑	↑	↑	↓	↑	↑	↑	↑	↑	↑
Pre-monsoon	↑*	↑*	↑*	↑*	↑*	↑*	↑	↑*	↑*	↑*
Monsoon	↑	↑*	↑*	↑*	↑*	↑*	↑	↑*	↑*	↑*
Post-monsoon	↑*	↑*	↑*	↑*	↑*	↑*	↓	↑	↑*	↑*

*The significance level is (5% with $Z_{0.025} = 1.96$)

Table 20.3 Sen's slope (Q) of precipitation (mm)

Places	Srinagar	Kulgam	Shupian	Ganderbal	Budgam	Bandipore	Anantnag	Pulwama	Baramulla	Kupwara
January	−0.1	−0.11	−0.1	−0.1	−0.1	−0.08	−0.1	−0.1	−0.1	−0.09
February	0.19	0.25	−0.21	0.21	0.21	0.19	−0.15	0.18	0.24	0.25
March	0.22	−0.2	0.21	0.21	0.22	0.26	0.17	0.19	0.24	0.23
April	0.2	0.17	0.17	0.18	0.19	0.19	0.14	0.17	0.23	0.22
May	0.12	0.14	0.15	0.13	0.15	0.1	0.12	0.13	0.13	0.1
June	0.2	0.28	0.29	0.23	0.24	0.18	0.24	0.25	0.23	0.18
July	0.25	0.32	0.36	0.29	0.33	0.21	−0.2	0.26	0.37	0.3
August	−0.12	0.14	0.18	0.15	0.18	0.11	−0.08	0.12	0.21	0.12
September	0.02	0.03	0.03	0.03	0.03	0.02	0.01	0.03	0.04	0.02
October	−0.03	0.04	0.04	0.04	0.04	0.02	0.03	0.03	0.04	−0.03
November	0.08	0.08	0.08	0.08	0.08	0.07	0.07	0.07	0.09	0.09
December	−0.01	−0.04	−0.04	−0.02	−0.02	0	−0.03	−0.03	−0.01	−0.01
Annual	1.48	1.61	1.67	1.52	1.64	1.34	1.22	1.41	1.73	1.58
Winter	0.15	0.03	0.05	−0.12	0.13	0.18	0.04	0.08	0.18	0.19
Pre-monsoon	0.5	0.44	0.46	0.48	0.5	0.52	0.4	0.44	0.53	0.52
Monsoon	0.55	0.77	0.85	0.66	0.75	0.45	0.56	0.68	0.78	0.6
Post-monsoon	0.17	0.17	0.18	0.23	0.19	0.16	0.15	0.17	0.2	0.19

Annual Precipitation Plots for All Ten districts Kashmir Valley

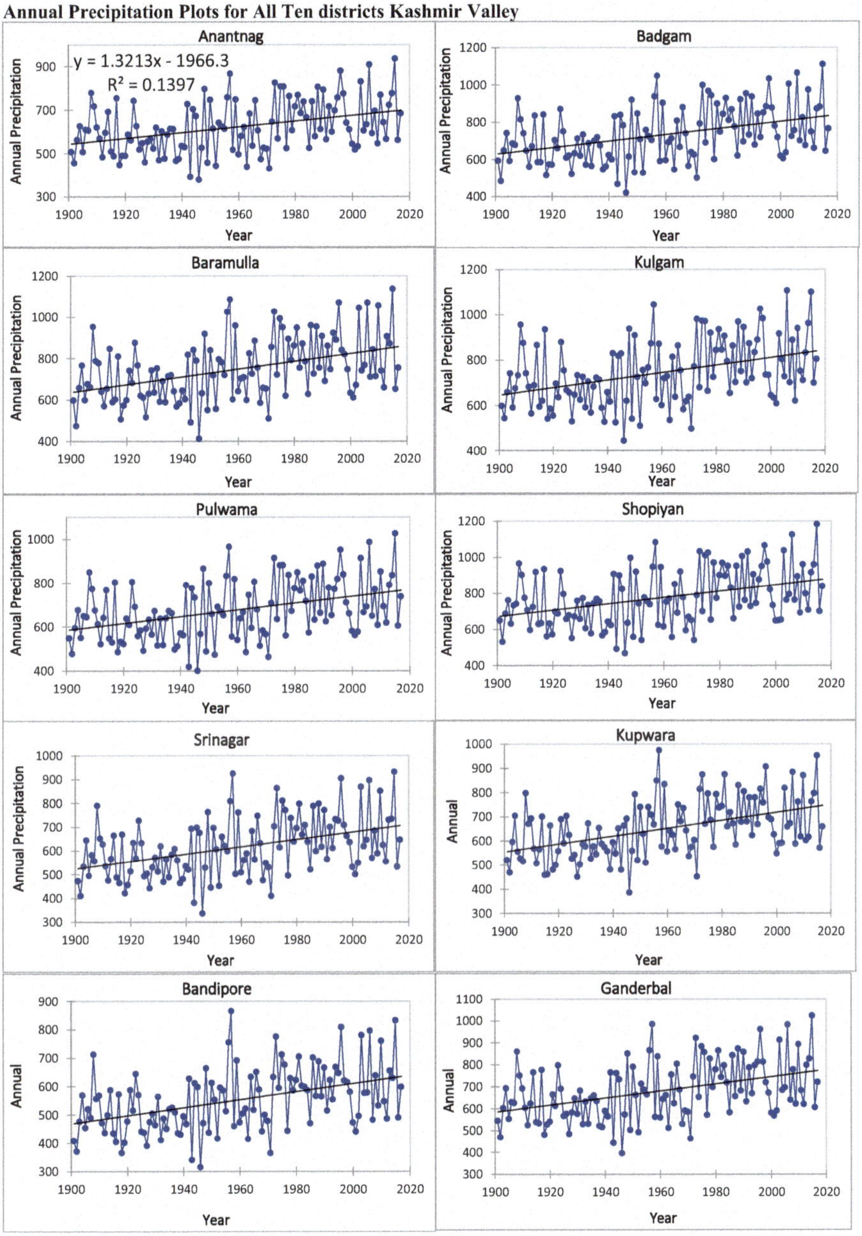

Fig. 20.2 Annual precipitation plots for all ten districts of Kashmir Valley

series for the months of November and December showed statistically significantly increasing trends for all districts of Kashmir Valley (Tables 20.4 and 20.5).

Table 20.4 Mann–Kendall trend statistics (Z) of mean temperature (°C)

Places	Srinagar	Kulgam	Shupian	Ganderbal	Budgam	Bandipore	Anantnag	Pulwama	Baramulla	Kupwara
January	↑	↑	↑	↑	↑	↑	↑	↑	↑	↑
February	↑*	↑*	↑*	↑*	↑*	↑*	↑*	↑*	↑*	↑*
March	↑*	↑*	↑	↑*	↑*	↑*	↑	↑	↑*	↑*
April	↑*	↑*	↑*	↑*	↑*	↑*	↑*	↑*	↑*	↑*
May	↑	↑	↑	↑	↑	↑	↓	↑	↑	↑
June	↓	↓*	↓*	↓*	↓*	↓	↓*	↓*	↓*	↓
July	↓	↓*	↓*	↓*	↓*	↓	↓*	↓*	↓	↓
August	↓*	↓*	↓*	↓*	↓*	↓*	↓*	↓*	↓*	↓
September	↑*	↓	↓	↓	↑	↑	↓	↑	↑	↑*
October	↑*	↑	↑	↑	↑	↑*	↑	↑	↑	↑*
November	↑*	↑*	↑*	↑*	↑*	↑*	↑*	↑*	↑*	↑*
December	↑*	↑*	↑*	↑*	↑*	↑*	↑*	↑*	↑*	↑*
Annual	↑*	↑*	↑*	↑*	↑*	↑*	↑	↑*	↑*	↑*
Winter	↑*	↑	↑*	↑*	↑*	↑*	↑*	↑*	↑*	↓
Pre-monsoon	↑*	↓	↑*	↑*	↑*	↑*	↓	↑*	↑*	↑*
Monsoon	↓	↓*	↓*	↓*	↓*	↓	↓*	↓*	↓*	↓*
Post-monsoon	↑	↑*	↑*	↑*	↑*	↑*	↑*	↑*	↑*	↑*

*The significance level is (5% with $Z_{0.025} = 1.96$)

Table 20.5 Sen's slope (Q) of mean temperature (°C)

District	Srinagar	Kulgam	Shupiyan	Ganderbal	Badgam	Bandipore	Anantnag	Pulwama	Baramulla	Kupwara
January	0.006	0.005	0.005	0.005	0.006	0.006	0.003	0.006	0.005	0.006
February	0.013	0.012	0.012	0.012	0.011	0.013	0.010	0.013	0.010	0.012
March	0.009	0.007	0.007	0.008	0.008	0.010	0.006	0.008	0.008	0.009
April	0.012	0.010	0.011	0.011	0.011	0.012	0.008	0.011	0.011	0.012
May	0.003	0.003	0.003	0.003	0.003	0.003	0.000	0.003	0.003	0.002
June	−0.001	−0.013	−0.004	−0.084	−0.004	−0.002	−0.004	−0.003	−0.010	−0.003
July	−0.025	−0.005	−0.005	−0.004	−0.004	−0.002	−0.006	−0.004	−0.004	−0.003
August	−0.004	−0.005	−0.006	−0.005	−0.005	−0.004	−0.006	−0.004	−0.005	−0.004
September	0.002	0.000	0.000	0.001	0.000	0.003	−0.002	0.001	0.001	0.002
October	0.006	0.005	0.005	0.005	0.005	0.007	0.002	0.005	0.005	0.006
November	0.010	0.009	0.001	0.006	0.009	0.011	0.007	0.010	0.008	0.010
December	0.011	0.010	0.010	0.010	0.010	0.012	0.009	0.011	0.010	0.011
Annual	0.006	0.005	0.004	0.005	0.005	0.006	0.003	0.005	0.005	0.005
Winter	0.009	0.009	0.009	0.008	0.008	0.010	0.007	0.009	0.008	−0.009
Pre-monsoon	0.008	−0.005	0.008	0.008	0.007	0.009	−0.005	0.008	0.008	0.008
Monsoon	−0.002	−0.004	−0.004	−0.003	−0.003	−0.001	−0.004	−0.003	−0.003	−0.002
Post-monsoon	0.009	0.007	0.007	0.008	0.007	0.009	0.005	0.008	0.007	0.008

Seasonal and Annual Trends of Temperature

According to this study, annual temperature shows a statistically significant increasing trend for nine districts and a non-significant upward trend for Anantnag (Tables 20.4, 20.5 and Fig. 20.3a–j). Winters, pre-monsoon season and post-monsoon also show a statistically significant rising trend for most of the districts. The district of Kupwara showed a downward trend in annual temperature in the winter season. Likewise, a decreasing trend was also observed in the pre-monsoon season for the Kulgam and Anantnag districts. The post-monsoon season experiences a statistically significant upward trend for all districts except Srinagar (Tables 20.4 and 20.5).

Conclusions

This study focused on the Kashmir Valley, a region in the northwestern Himalayas, for analysing temperature and precipitation trends and variability. Sen's slope estimation technique and the Mann–Kendall test were used in the analysis. The results showed that January precipitation for all districts had a non-significant declining tendency. However, for numerous regions, February and March demonstrate a statistically significant increasing trend in precipitation. Precipitation shows an increased tendency in April, May and June for all districts, with varied degrees of significance. For most districts, increasing precipitation patterns are also visible in July and October. For all districts, the annual series of precipitation exhibits a statistically significant upward trend. There is a statistically significant upward trend in temperature in the annual series for most districts. With some variances between districts, the winter, pre-monsoon and post-monsoon seasons similarly show growing trends in temperature. For studies on climate change and the management of water resources, especially in the vulnerable Himalayan region, it is essential to comprehend these trends and variability.

In mountainous places like Kashmir Valley, accelerated glacier melting is a result of higher temperatures, which also has an impact on ecosystem and the ability to access freshwater resources. Water scarcity decreased agricultural water supply, and disruptions in the hydropower industry can all result from this. The risk of flooding and landslides might increase with higher precipitation and more severe rainstorm events. Communities, infrastructure and ecosystems can all be seriously threatened by flash floods and debris flows that can be caused by a combination of high rainfall and steep topography. Changes in temperature and precipitation patterns can have a significant impact on mountain ecosystems. The delicate balance of flora and fauna can be upset by climate change, which can modify species composition, migration patterns and habitats. The biodiversity, ecological services and conventional livelihoods reliant on natural resources may all be negatively impacted. Such changes can have an impact on agricultural practices and crop yields. Changes in temperature can

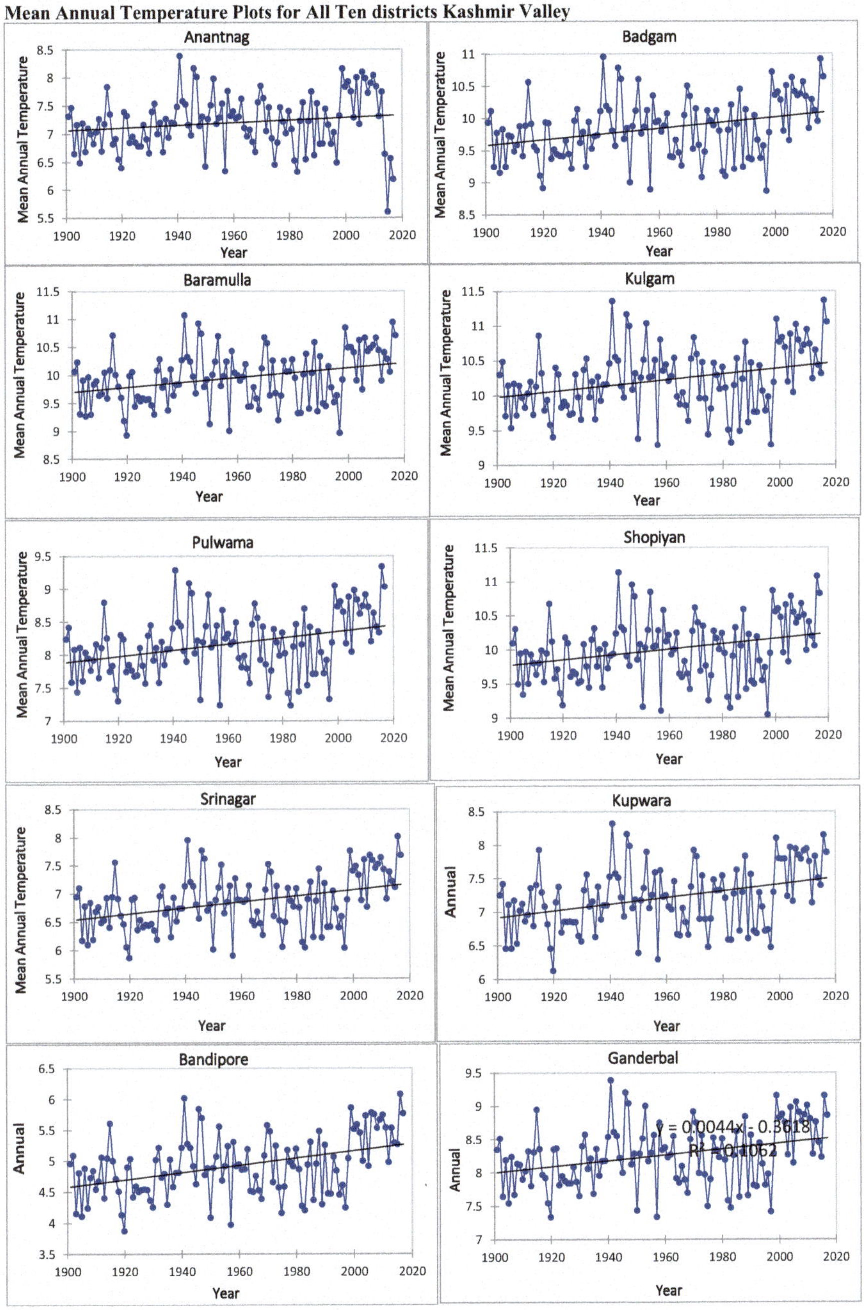

Fig. 20.3 Mean annual temperature plots for all ten districts Kashmir Valley

affect the suitability of crops, and different precipitation patterns might cause droughts or waterlogging, which can have an impact on production and food security for mountain people. Skiing, mountaineering and hiking are popular recreational pursuits in the studied area. Temperature shifts and variations in snowfall patterns can have an impact on the length of the winter season, the quality of the snow and the livelihoods and tourism that depend on these activities.

Precipitation and temperature changes can also cause problems with water management. Water storage, irrigation systems and water allocation agreements between communities upstream and downstream can all be impacted by variations in the amount and timing of snowmelt and runoff. Because of their remote location, poor infrastructure and reliance on natural resources for survival, mountain communities are frequently more susceptible to the effects of climate change. Temperature and precipitation changes have the potential to increase already existing problems, such as those associated with water, food security and natural disasters.

In order to overcome the difficulties brought on by these rising trends in temperature and precipitation variability in the Kashmir Valley, it is imperative to put into place adaptable measures and sustainable management techniques. Improved water management, ecosystem restoration, disaster risk reduction and the creation of climate-resilient livelihood plans for regional populations are a few examples of what this may include. Finally, it is imperative to underscore that the findings of this study are based on gridded data, and it is essential to conduct further research incorporating ground-based observational data to obtain a more comprehensive understanding of the observed changes. This will facilitate more accurate planning and effective management strategies.

Funding Funding was not received from any organization and company for conducting the present study.

Compliance with Ethical Standards

Conflict of Interest The authors declare that they have no conflict of interest. All the authors have equally contributed to the present work.

Human and Animal Rights No human and/or animal participation was involved in the study.

References

Basistha, A., Arya, D. S., & Goel, N. K. (2009). Analysis of historical changes in rainfall in the Indian Himalayas. *International Journal of Climatology, 29*(4), 555–572.

Bhutiyani, M. R., Kale, V. S., & Pawar, N. J. (2007). Long-term trends in maximum, minimum and mean annual air temperatures across the Northwestern Himalaya during the twentieth century. *Climatic Change, 85*(1–2), 159–177.

Census of India (2011); Census tables | Government of India (censusindia.gov.in)

Das, S., Tomar, C. S., Saha, D., Shaw, S. O., & Singh, C. (2015). Trends in rainfall patterns over North-East India during 1961–2010. *International Journal of Earth and Atmospheric Science, 2* (2), 37–48.

Drápela, K., & Drápelová, I. (2011). Application of Mann-Kendall test and the Sen's slope estimates for trend detection in deposition data from Bílý Kříž (Beskydy Mts., The Czech Republic) 1997-2010. *Beskydy, 4*(2), 133–146.

Guhathakurta, P., & Rajeevan, M. (2008). Trends in the rainfall pattern over India. *International Journal of Climatology: A Journal of the Royal Meteorological Society, 28*(11), 1453–1469.

Hamed, K. H., & Rao, A. R. (1998). A modified Mann-Kendall trend test for autocorrelated data. *Journal of hydrology, 204*(1–4), 182–196.

Harris, I. P. D. J., Jones, P. D., Osborn, T. J., & Lister, D. H. (2014). Updated high-resolution grids of monthly climatic observations – The CRU TS3. 10 dataset. *International Journal of Climatology, 34*(3), 623–642. https://doi.org/10.1002/joc.3711

Hirsch, R. M., Slack, J. R., & Smith, R. A. (1982). Techniques of trend analysis for monthly water quality data. *Water resources research, 18*(1), 107–121.

Indian Meteorological Department, Mausam Bhawan, Lodi Road, New Delhi, Delhi.

Jain, S. K., & Kumar, V. (2012). Trend analysis of rainfall and temperature data for India. *Current Science (Bangalore), 102*(1), 37–49.

Karmeshu, N. (2012). *Trend detection in annual temperature & precipitation using the Mann Kendall test – A case study to assess climate change on select states in the northeastern United States*. University of Pennsylvania.

Kumar, V., & Jain, S. K. (2010). Trends in seasonal and annual rainfall and rainy days in Kashmir Valley in the last century. *Quaternary International, 212*(1), 64–69.

Kumar, V., Singh, P., & Jain, S. K. (2005, April). Rainfall trends over Himachal Pradesh, Western Himalaya, India. In *Conference on development of hydro power projects – A prospective challenge, Shimla* (Vol. 20, p. 22).

Mavromatis, T., & Stathis, D. (2011). Response of the water balance in Greece to temperature and precipitation trends. *Theoretical and Applied Climatology, 104*(1–2), 13–24. https://doi.org/10.1007/s00704-010-0320-9

Motiee, H., & McBean, E. (2009). An assessment of long-term trends in hydrologic components and implications for water levels in Lake Superior. *Hydrology Research, 40*(6), 564–579.

Pant, G. B., & Kumar, K. R. (1997). *Climates of South Asia*. Wiley.

Rajeevan, M., Bhate, J., Kale, J. D., & Lal, B. (2006). High resolution daily gridded rainfall data for the Indian region: Analysis of break and active. *Current Science, 91*(3), 296–306.

Ramesh, K. V., & Goswami, P. (2007). *The shrinking Indian summer monsoon* (CSIR report RR CM, 709).

Salma, S., Rehman, S., & Shah, M. A. (2012). Rainfall trends in different climate zones of Pakistan. *Pakistan Journal of Meteorology, 9*(17), 37–47.

Sen, P. K. (1968). Estimates of the regression coefficient based on Kendall's tau. *Journal of the American Statistical Association, 63*(324), 1379–1389.

Sen Roy, S., & Balling, R. C., Jr. (2004). Trends in extreme daily precipitation indices in India. *International Journal of Climatology: A Journal of the Royal Meteorological Society, 24*(4), 457–466.

Sinha, T., & Cherkauer, K. A. (2008). Time series analysis of soil freeze and thaw processes in Indiana. *Journal of Hydrometeorology, 9*(5), 936–950.

Smith, J. (2010). Understanding climate variability and climate change. *Weather and Climate, 1*, 1–25.

Tabari, H., Marofi, S., Aeini, A., Talaee, P. H., & Mohammadi, K. (2011). Trend analysis of reference evapotranspiration in the western half of Iran. *Agricultural and Forest Meteorology, 151*(2), 128–136.

Wani, J. M., Sarda, V. K., & Jain, S. K. (2017). Assessment of trends and variability of rainfall and temperature for the district of Mandi in Himachal Pradesh, India. *Slovak Journal of Civil Engineering, 25*(3), 15–22. https://doi.org/10.1515/sjce-2017-0014

Yang, Y., & Tian, F. (2009). Abrupt change of runoff and its major driving factors in Haihe River Catchment, China. *Journal of Hydrology, 374*(3–4), 373–383.

Yadav, R., Tripathi, S. K., Pranuthi, G., & Dubey, S. K. (2014). Trend analysis by Mann-Kendall test for precipitation and temperature for thirteen districts of Uttarakhand. *Journal of Agrometeorology, 16*(2), 164.

Yue, S., & Wang, C. (2004). The Mann-Kendall test modified by effective sample size to detect trend in serially correlated hydrological series. *Water Resources Management, 18*(3), 201–218.

Chapter 21
The Review of Potential Applications and Modification Approaches of SWAT for Efficient Environmental Management, an Engineering Approach

Ifra Ashraf, Syed Towseef Ahmad, Junaid N. Khan, Rayees Ahmad, Rohitashw Kumar, Shazia Ramzan, Faheem Ahmed Malik, and Atufa Ashraf

Introduction

Hydrological models are standard contrivances routinely tapped for hydrological explorations in engineering and environmental disciplines (Wagener et al., 2004). All the hydrological models are simplified embodiments of the real-world systems under investigation, entailing a rational set of functions or a series of simultaneous equations embedded in a computer program. Each model has a number of parameters that give a numeric estimate of characteristics/properties being invariant under stated conditions (Tibebe et al., 2016). The model is cogitated to be the best, which confers results in close proximity to the real estimates with a less number of parameters and less complexity.

I. Ashraf · J. N. Khan · R. Kumar
College of Agricultural Engineering and Technology, Sher-e-Kashmir University of Agricultural Sciences and Technology of Kashmir, Srinagar, Jammu and Kashmir, India

S. T. Ahmad · R. Ahmad (✉)
Department of Geography and Regional Development, University of Kashmir, Srinagar, Jammu and Kashmir, India

S. Ramzan
KVK, Anantnag, Sher-e-Kashmir University of Agricultural Sciences and Technology of Kashmir, Srinagar, Jammu and Kashmir, India

F. A. Malik
Centre for Transport Research, Department of Civil, Structural and Environmental Engineering Trinity College Dublin, Dublin, Ireland

A. Ashraf
Division of Plant Pathology, Sher-e-Kashmir University of Agricultural Sciences and Technology of Kashmir, Srinagar, Jammu and Kashmir, India

A. L. Singh et al. (eds.), *Climate Change, Vulnerabilities and Adaptation*,
https://doi.org/10.1007/978-3-031-49642-4_21

SWAT, the ellipsis of Soil and Water Assessment Tool, is a physically based, basin-scale, spatially quasi-distributed and continuous-time hydrological model that operates at either a sub-daily or daily time-step (Arnold et al., 1998). SWAT, an open-source model, was jointly developed by USDA-ARS (United States Department of Agriculture, Agricultural Research Service) and Texas A&M AgriLife Research, part of the Texas A&M University System. SWAT simulates the influence of various land management practices on runoff, sediment yield and agrochemical yield in large complex basins having variable land uses, soils and management practices (Neitsch et al., 2011). A brief description of SWAT has been provided in this chapter given its thorough documentation (Neitsch et al., 2011; Arnold et al., 2012).

Due to the complexities extant in hydrological phenomena, diverse parameters, that is, meteorological data, soil properties and hydrology, plant growth, pathogens and bacteria, pesticides, nutrients and land management have been embedded in the SWAT model (Arnold et al., 2012). In SWAT, a basin is divided into various sub-basins, which are further segmented into hydrological response units (HRUs) by superimposing soil, land use and slope maps (Yang et al., 2007). HRUs, a fundamental operational unit of the model (Gyamfi et al., 2016), consist of homogeneous land use, soil, management and topographical characteristics. The incorporation of HRUs accedes SWAT to represent the variability within the sub-basin (Sun et al., 2015). SWAT model consists of two phases: a land phase dealt at the HRU level and a stream phase dealt at the sub-basin level (Neitsch et al., 2011).

In the land phase, daily water, nutrient and sediment yields of HRUs are computed by taking into consideration all the parameters including precipitation, evapotranspiration, irrigation, surface runoff and sub-surface runoff (both lateral as well as percolation to both shallow and/or deep aquifers) (Neitsch et al., 2011). In the stream phase, the daily outputs, that is, water, sediment and other pollutant yields from all HRUs of a sub-basin are routed in a cascading chain of streams via a stream network (Neitsch et al., 2011; Vigiak et al., 2015).

SWAT model is capable of gauging water, sediment and chemical yields in both gauged and un-gauged watersheds (Srinivasan et al., 2010; Mir & Ahmed, 2021). It can be used in un-gauged watersheds for ascertaining hydrological controlling parameters/factors (Ndomba et al., 2008). However, multi-constituent input datasets on a long-term basis are desirable to increase the reliance of model in simulating hydrological processes in un-gauged basins (Zeiger & Hubbart, 2016). In the case of an un-gauged watershed, information on model parameters is transferred from adjoining watersheds, referred to as donor watersheds, to a target watershed (Noori & Kalin, 2016), by regionalization techniques (Wang & Kalin, 2011). SWAT has corroborated its applicability across-the-board globally, for it has been assessed in various countries. This chapter aims to review the modelling capacitances of the SWAT to simulate the hydrodynamic process at the scale of a catchment/watershed. Many reviews have already been put forth on the modelling capacitances of the SWAT. But this review also aims to enlighten the inadequacy of SWAT to confer to some complex hydrological conditions; hence, many modifications/alterations made in SWAT have been reviewed under apposite headlines.

Potential Applications of SWAT

SWAT (Javaid et al., 2023) is widely used for simulating water flow, sediment transport, nutrient cycling and pollutant dynamics in watersheds (Srinivasan et al., 2010). Its versatility and ability to integrate various data sources make it a valuable tool for understanding hydrological processes and supporting informed decision-making. SWAT model has been extensively used for assessing water availability, water quality and water allocation in watersheds. It helps in quantifying the impacts of land-use changes, climate variability and human activities on water resources.

SWAT model enables the assessment of agricultural practices and their impacts on water resources. It helps in evaluating the effectiveness of conservation measures, such as cover cropping, terracing and nutrient management, in reducing nutrient runoff and improving water quality. SWAT model has proven useful in assessing the environmental impacts of land-use changes, such as urbanization, deforestation and afforestation. It helps in predicting the potential effects of land use alterations on hydrological processes, sediment transport and water quality. The SWAT model serves as a valuable tool for assessing the impacts of climate change on hydrological processes and water resources. It aids in understanding how changes in temperature, precipitation and evapotranspiration patterns affect streamflow, groundwater recharge and water availability. SWAT has conferred diverse applications, which have been listed in Table 21.1.

SWAT has been successfully employed worldwide in simulating the hydrological response of catchments under different land use, soil and climatic conditions with results in acceptable ranges. However, in case of lower rainfall events, the areas contributing to the flow are over-predicted (Golmohammadi et al., 2017), and in case of extreme events, the peak flows are underestimated by the model (Gassman et al., 2014). Contrarily, in the case of sediment load, the model shows comparatively better results in the case of low rainfall events and underestimates or overestimates in high rainfall events (Zhang et al., 2017). Moreover, Meng et al. (2017) corroborated simulating competencies of SWAT in un-gauged areas and in areas with high glacier recharge rates with proper parameter calibration. However, gully erosion in watersheds is not modelled by SWAT properly (Easton et al., 2010).

Due to the complexity rooted in the hydrological phenomena, SWAT consists of diverse parameters, some of which are very difficult to estimate, and hence need to be estimated via calibration, which necessitates the gauging of hydrological data of watershed. To generate effective hydrologic forecasts, SWAT requires relatively little direct calibration (Easton et al., 2010). Many researchers demonstrated the proficiency of SWAT in simulating hydrological responses of geographically different basins with scarce observed data available for calibration and validation (Oeurng et al., 2011; Wu & Chen, 2012; Shi et al., 2013; Rasool & Kumar, 2019). In the case of watersheds with high spatial heterogeneity, a multisite calibration technique is recommended (Leta et al., 2017).

Among numerous basin-scale models, SWAT can effectively assess the economic and environmental bearing of basic management practices (BMPs) at diverse

Table 21.1 Important applications of SWAT model

S. no.	Applications of SWAT	Addendum, if any	Source
1.	Prediction of the temporal variation of flow-contributing areas during each event	Current precipitation index (CPI) and Potential contributing area (PCA) ratio (Lee & Huang, 2013)	Golmohammadi et al. (2017)
2.	Prediction of sediment yield/water yield/nutrient loadings	Latin Hypercube One-factor-At-a-Time (LH-OAT) (Ndomba & van Griensven, 2011); SWAT-CUP (Yesuf et al., 2015; Briak et al., 2016), Modified Soil Conservation Service curve number method, the storm-based Chinese soil loss equation and the nutrient loss model (Shi & Huang, 2021), GPCC (Ijaz et al., 2022)	Ndomba and van Griensven (2011), Adeogun et al. (2015), Yesuf et al. (2015), Briak et al. (2016), Gull et al. (2017), Shi and Huang (2021) and Ijaz et al. (2022)
3.	Simulation of impacts of land use/cover change on sediment, water and/or nutrient yields of the basin	SWAT-CUP (Anaba et al., 2017; Gyamfi et al., 2016); Principal component analysis (PCA) (Gyamfi et al., 2016); Latin Hypercube One-factor-At-a-Time (LH-OAT) (Lin et al., 2015)	Wang and Kalin (2011), Lin et al. (2015), Anaba et al. (2017), Gyamfi et al. (2016) and Mekonnen and Manderso (2023)
4.	Simulation of impacts of climate change on the hydrology of the basin	–	Zhang et al. (2007), Setegn et al. (2011), Lirong and Jianyun (2012), and Melke and Abegaz (2017)
5.	Simulation of the impact of both climate and land-use change on the hydrology of the basin	–	Li et al. (2009), Ward et al. (2009), Krysanova and Srinivasan (2015), Osei et al. (2019), Sinha et al. (2020) and Haleem et al. (2022)
6.	Assessment of the soil water content	–	Havrylenko et al. (2016)
7.	Drought Assessment/ Forecast	Frank copula (Dash et al., 2019)	Dash et al. (2019), Brouziyne et al. (2020) and Ma et al. (2022)
8.	Assessment of impacts of Best Management Practices (BMPs) scenarios on basin hydrology	SOBEK using open modelling interface (OPENMI) (Betrie et al., 2011); SWAT-CUP (Rocha et al., 2015)	Ullrich and Volk (2009), Lee et al. (2010), Betrie et al. (2011), Rocha et al. (2015) and Sheshukov et al. (2016)
9.	Simulation of hydrological response including glacial melt in snow-dominated regions	–	Ahl et al. (2008), Pradhanang et al. (2011), Kang and Lee (2014), Gan et al. (2015) and Garee et al. (2017)

scales (Wilson et al., 2014; Piemonti et al., 2013). Additionally, SWAT can effectively assess the impact of crop cultivation on hydrological dynamics (Leta et al., 2015), bio-geochemical cycling (El-Khoury et al., 2015) and environmental pollution (Holvoet et al., 2008).

Limitations of the SWAT Model

1. *Data Requirements:* SWAT model requires a significant amount of input data, including climate data, soil properties, land use/land cover data, topography and streamflow data for calibration and validation. Obtaining accurate and reliable data at the desired spatial and temporal resolution can be challenging, particularly in data-scarce regions or for long-term simulations. The quality of input data directly affects model performance and uncertainty, making data availability and reliability a major limitation.
2. *Model Parameterization:* SWAT model consists of numerous parameters that need to be calibrated for each specific watershed. Parameterization can be time-consuming and complex, requiring expert knowledge and extensive field measurements. The sensitivity of the model to parameter values makes the calibration process critical, but it is often subjective and prone to uncertainty. The lack of universally accepted guidelines for parameter estimation is a limitation that hampers model consistency and comparability across different studies.
3. *Model Complexity:* SWAT model is a complex and process-based model that represents numerous hydrological, agricultural and ecological processes. While this complexity allows for detailed simulations, it also increases the potential for errors and uncertainties. Model complexity can be challenging for non-experts to understand and utilize effectively. The need for specialized knowledge and technical expertise may limit the widespread application of the model in certain regions or for decision-makers with limited scientific background.
4. *Spatial and Temporal Variability:* SWAT model assumes that the watershed is homogeneous and neglects spatial heterogeneity within the study area. This assumption may lead to oversimplified representations of hydrological processes and limited accuracy in capturing local variations. Additionally, the model assumes the stationarity of climate and hydrological conditions over time, which may not hold true in a changing environment. Incorporating fine-scale spatial and temporal variability is a challenge that affects the accuracy and reliability of model results.
5. *Uncertainty and Validation:* Like any model, SWAT model is subject to various sources of uncertainty, including input data, parameterization and model structure. Uncertainty analysis is crucial to understand the reliability and limitations of model predictions. However, uncertainty analysis in the SWAT model is often challenging due to a large number of parameters and a lack of standard methodologies. Moreover, model validation against observed data can be

difficult, particularly in un-gauged or data-scarce regions, limiting confidence in model performance.

6. *Scale and Generalization:* SWAT model operates at the watershed scale, and its applicability to smaller sub-watersheds or larger regional scales is limited. Upscaling or downscaling results from one watershed to another may introduce additional uncertainties. Furthermore, the model's ability to capture specific localized processes or phenomena, such as small-scale runoff dynamics or localized pollutant transport, may be limited. This scale limitation restricts the model's utility in certain applications requiring fine-scale analysis.
7. *Lack of Dynamic Feedback:* SWAT model assumes static relationships between inputs and outputs and does not explicitly account for dynamic feedback between hydrological processes and land-use changes. This limitation can affect the accuracy of simulating complex interactions and feedback mechanisms. For example, the model may not adequately capture the effects of land management practices on soil properties over time or the feedback between land-use change and hydrological response.
8. *Limited Representation of Ecological Processes:* While the SWAT model includes some ecological processes, such as nutrient cycling and vegetation growth, its focus primarily lies in hydrological and agricultural aspects. The representation of ecological processes may be simplified, and the model may not fully capture the complexities of ecosystem dynamics, biodiversity and habitat suitability. Integrating more detailed and comprehensive ecological modules within the SWAT model is an ongoing challenge.
9. *Model Accessibility and User Interface:* The SWAT model is primarily designed for experienced modellers and scientists with a strong background in hydrology and related fields. Its user interface and accessibility can be daunting for non-experts or decision-makers who require simplified tools for water resource management. The lack of user-friendly interfaces and comprehensive documentation may hinder the wider adoption and utilization of the SWAT model, particularly in decision-support systems and policy-making processes.
10. *Computational Requirements:* SWAT model can be computationally intensive, especially when simulating large watersheds or long-term scenarios. Running simulations with high spatial and temporal resolutions and incorporating complex processes may require substantial computational resources and time. Limited computing infrastructure or technical capabilities may pose constraints on the practical application of the model, particularly in resource-limited regions.
11. *Limited Representation of Social and Economic Factors:* The SWAT model primarily focuses on the physical and biophysical aspects of watersheds and often overlooks the social and economic dimensions of water resource management. Factors such as water demand, water pricing, socio-economic impacts of land-use change and stakeholder participation are not explicitly incorporated into the model. This limitation restricts the model's ability to provide a comprehensive assessment of water management strategies and their socio-economic implications.

Modifications Made in SWAT

Modelling Depression-Dominated Regions

The high spatial variability of depressions makes the modelling of hydrological processes in depression-dominated regions complex and is contingent on a clear cognizance of the dynamic processes of filling, spilling, merging and splitting of depressions present on the surface. The hierarchical linkages of depressions and their varying contributing regions, in particular, confound hydrologic modelling (Zeng et al., 2020). Like conventional hydrologic modelling, SWAT model fills all the depressions of the area in the preprocessing step of the delineation process for producing well-connected drainage nets (Mekonnen et al., 2016a). Hydrologic models typically use simplified and lumped approaches to account for the impacts of depressions. These approaches often assume that depressions act as small, uniform storage areas with constant infiltration rates. While this simplification allows for computational efficiency and ease of modelling, it may not capture the dynamic influences of surface depressions accurately (Zeng et al., 2020). Many researchers have attempted to characterize the depression-dominated areas and simulate their storage effects (Chu et al., 2010; Chu, 2015; Habtezion et al., 2016).

Even though all the depressions are filled in SWAT, three tools encompassing Potholes, Ponds and Wetlands are available in SWAT to account for the computation of lumped outflow from depressions (Neitsch et al., 2011) and to model the influence of depressions on hydrological processes (Nasab et al., 2017). The depressions/potholes and their parameters are defined at the HRU scale in SWAT (Neitsch et al., 2011) and allied to the routing process in SWAT (Nasab et al., 2017). However, all the requisite pothole parameters for instance depression storage, ponding area and depth of depressions are not calculated by SWAT. To augment the SWAT proficiency for dealing with complex depression-dominated regions, some modified SWAT models have been worked out.

SWAT-PDLD (SWAT-probability distribution landscape distribution) was framed by Mekonnen et al. (2016b) to integrate the heterogeneity of depression storage in the SWAT model. They expended ArcGIS to spot depressions, their surface area and storage. SWAT-PDLD showed better results of streamflow for the watersheds under study. Furthermore, Mekonnen et al. (2016a) improved SWAT-PDLD by integrating seasonal variation in soil erodibility induced due to cold climate to simulate sediment transfer in two depression-dominated watersheds of Canada typified by cold climate. They deduced that the sediment export simulated by the modified model was considerably more coherent.

Another modification in SWAT was the coupling of the SWAT model with a Puddle Delineation (PD) algorithm (Chu et al., 2010) done by Nasab et al. (2017). This led to the development of a new D-cubed (Depression-Dominated Delineation) approach. In the coupled PD-SWAT model (Fig. 21.1), the PD algorithm was exploited to characterize the depressions and their distribution and hierarchical relationships and compute microtopographic details from digital elevation models

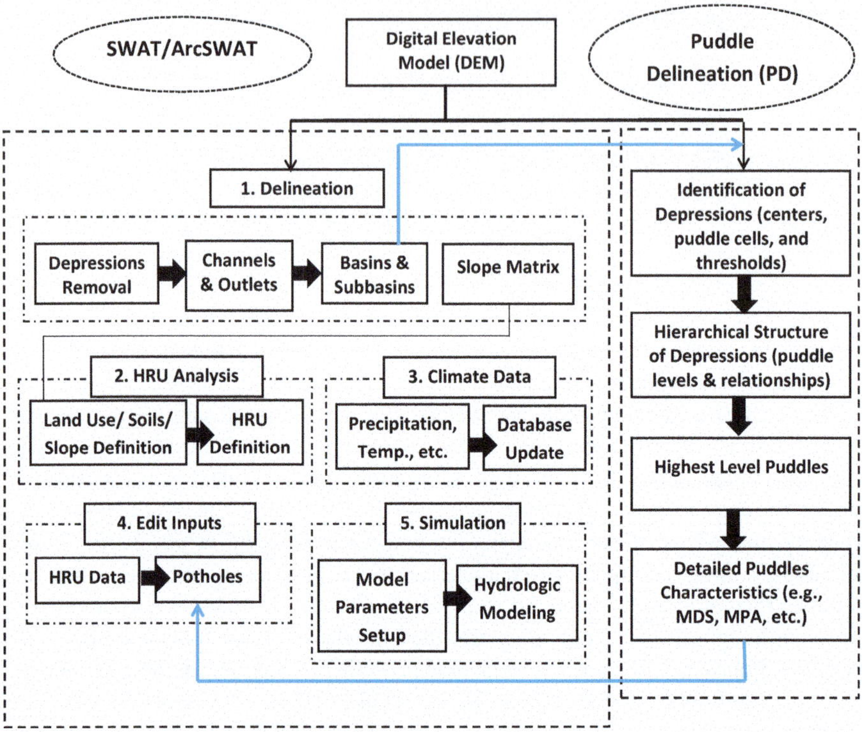

Fig. 21.1 Elemental steps of the PD-SWAT model (MDS: maximum depression storage and MPA: maximum ponding area). (Source: Nasab et al., 2017)

(DEMs). This involved introducing the CBU (Channel-Based Unit—depressionless unit) concept and its connection with PBU (Puddle-Based Unit—depression unit) concept. The D-cubed method creates a unique cascaded channel-puddle drainage system based on the channel segmentation algorithm. The output of the PD algorithm was then coupled with the pothole tool of the SWAT model at the HRU level. PD-SWAT outperformed the standard SWAT model in simulating runoff values in coherence with observed values. SWAT thoroughly overestimated the runoff in depression-dominated regions, which was taken care of by PD-SWAT by significantly delaying the runoff initiation owing to the threshold control of depressions.

Zeng et al. (2020) presented a novel modelling framework called the Puddle-Based Unit Probability Distributed Model (PBU-PDM) to improve hydrologic modelling for depression-dominated regions. This paradigm was combined with the current SWAT to better assess the impact of depressions that are spatially distributed. In addition to the standard input data used in SWAT, the PDMs for all sub-basins were developed using a set of topographical parameters and data from the D-cubed technique. The evaluation of discharge simulations using the PBU-PDM enhanced SWAT model and the original SWAT model, compared to observed data, revealed distinct patterns. The original SWAT model demonstrated a tendency to either overestimate or underestimate discharges, depending on whether it was a wet or dry period. Conversely, the PBU-PDM coupled with SWAT model exhibited improved modelling performance, specifically tailored to the depression-dominated watershed, by accurately representing discharge dynamics for both wet and dry periods.

GHG Emission

Agricultural sectors are the biggest sources of N_2O emissions owing to the incessant use of fertilizers to support the global food demand. N_2O gas is very detrimental causing depletion in the O_3 gas of the stratosphere (Ravishankara et al., 2009). Despite having significantly lower atmospheric levels than CH_4 and CO_2, N_2O plays an inexplicably major role in contributing to global warming due to its extended atmospheric lifespan (Ko et al., 1991). N_2O production and consequent emissions are not embodied in the SWAT model, which limits its application for providing a comprehensive assessment of agricultural activities on nitrogen cycling. Researchers have integrated SWAT with bio-geochemical models, such as DAYCENT to estimate GHG emissions.

The initial integration of the SWAT and DAYCENT models was introduced by Reeling and Gramig (2012). This framework aims to combine the capabilities of both models to analyse optimal conservation practices. However, it is important to note that the integration was not simultaneous, meaning that SWAT and DAYCENT were used separately rather than being fully coupled. In this framework, a genetic algorithm was employed to analyse and identify the most effective conservation practices. The genetic algorithm, a computational optimization technique inspired by natural selection, was used to search for the best combination of conservation practices within the context of the separate SWAT and DAYCENT models. This approach allowed for the identification of optimal strategies for conservation, considering factors such as land management, nutrient cycling and greenhouse gas emissions. While the integration of SWAT and DAYCENT in this framework was not simultaneous, it still provided valuable insights into the optimization of conservation practices by leveraging the capabilities of both models.

The first concurrent application of the N_2O emission module in SWAT was introduced by Wu et al. (2016). Their study involved the development of DAYCENT input files for each hydrologic response unit (HRU) and the automatic coupling of DAYCENT with SWAT, resulting in a combined model referred to as SWAT-DayCent. The researchers selected SWAT as the base model for integrating DAYCENT due to its user-friendly configuration and the availability of a spatial analysis user interface. This choice facilitated the setup and usage of the integrated model. Once the modeller developed a SWAT project for a specific region of interest, there was no need for separate inputs for DAYCENT. By incorporating the simultaneous N_2O emission module, the coupled SWAT-DayCent model allowed for a more comprehensive assessment of nitrogen cycling and greenhouse gas emissions, specifically focusing on N_2O. This integration enabled the consideration of additional factors related to carbon and nitrogen dynamics, enhancing the model's capability to simulate and analyse N_2O emissions from agricultural systems. The development of the SWAT-DayCent model represented an important advancement in the integration of SWAT and DAYCENT, enabling more accurate estimation of N_2O emissions while leveraging the strengths of both models. SWAT-DayCent was later successfully employed by Zhao et al. (2019) to simulate both hydrologic as well as bio-geochemical processes.

Yang et al. (2017) improved the representation of the cycling of soil nitrogen in SWAT semi-empirically by altering its nitrification and denitrification algorithms and incorporating N_2O emission algorithms. In particular, they incorporated the modules of the DayCent model devoted to nitrification, denitrification and N_2O production (Del Grosso et al., 2000) in the nutrient cycling algorithms of SWAT. The new tool developed thereof—SWAT-N_2O—was found very useful in the estimation of long-term N_2O emissions from diverse cropping systems.

SWAT-GHG was another modification to SWAT introduced by Wagena et al. (2017). In this modification, the authors incorporated the denitrification flux equation developed by Parton et al. (1994) and Mosier et al. (2002). This denitrification flux equation is similar to the revised model proposed by Yang et al. (2017). However, Wagena et al. (2017) introduced an additional variable, pH, to estimate the total denitrification rate. By integrating these modifications, the SWAT-GHG model provides a more refined estimation of GHG emissions, specifically targeting denitrification-related N_2O emissions. This allows for a more comprehensive assessment of the impacts of agricultural practices and land management on GHG emissions, helping to inform sustainable management strategies. SWAT-GHG model was further employed (Wagena et al., 2018) to assess the effects of climate change on N_2O (nitrous oxide) fluxes. The researchers focused on how alterations in soil moisture conditions, anticipated under future climate scenarios and influenced the emissions of N_2O from soils. The study findings reported by them emphasized the significant impact of changing soil moisture on N_2O emissions. They demonstrated that altered soil moisture conditions, induced by future climate scenarios, can lead to variations in N_2O emissions from soils.

Shrestha et al. (2018) conducted a study on estimating N_2O fluxes from long-term grasslands in Canada. They combined modules reported by Yang et al. (2017) and Wagena et al. (2017), with some modifications, to incorporate into SWAT. They integrated the semi-empirical equations of nitrification and denitrification modules suggested by Parton et al. (1996, 2001) into SWAT. They estimated N_2O emissions from denitrification using a partitioning approach. However, they included an additional temperature reduction factor in addition to the one proposed by Yang et al. (2017). They neglected the pH factor mentioned by Wagena et al. (2017), as they found that variations in soil moisture and temperature explained up to 95% of N_2O dynamics from denitrification. The contribution of N_2O from nitrification was taken from established equations by Parton et al. (1996), contrary to Wagena et al. (2018), who used SWAT's original formulation. The fraction of nitrogen was considered a calibration parameter. They successfully replicated discrete N_2O

measurements within the range of 0.02–0.47. Furthermore, Melaku et al. (2020) used the model developed by Shrestha et al. (2018) to estimate N_2O emissions from selected grassland sites in the UK that were applied with solid manure. The model exhibited good performance, with R^2 values ranging from 0.58 to 0.71 for discrete measurements, effectively capturing the dynamics of N_2O fluxes.

Improved Simulation of Dry and Wet Periods

The standard SWAT model simulates the wet periods satisfactorily but performs inadequately in dry seasons, which has hampered its utility in watersheds typified principally by low flows; hence, a seasonal calibration scheme is recommended for dry and wet periods. SWAT-SC (Fig. 21.2), a seasonal calibration scheme, was developed by Zhang et al. (2015), which espouses a service-oriented architecture and operates in a distributed computation environment. By implementing a seasonal

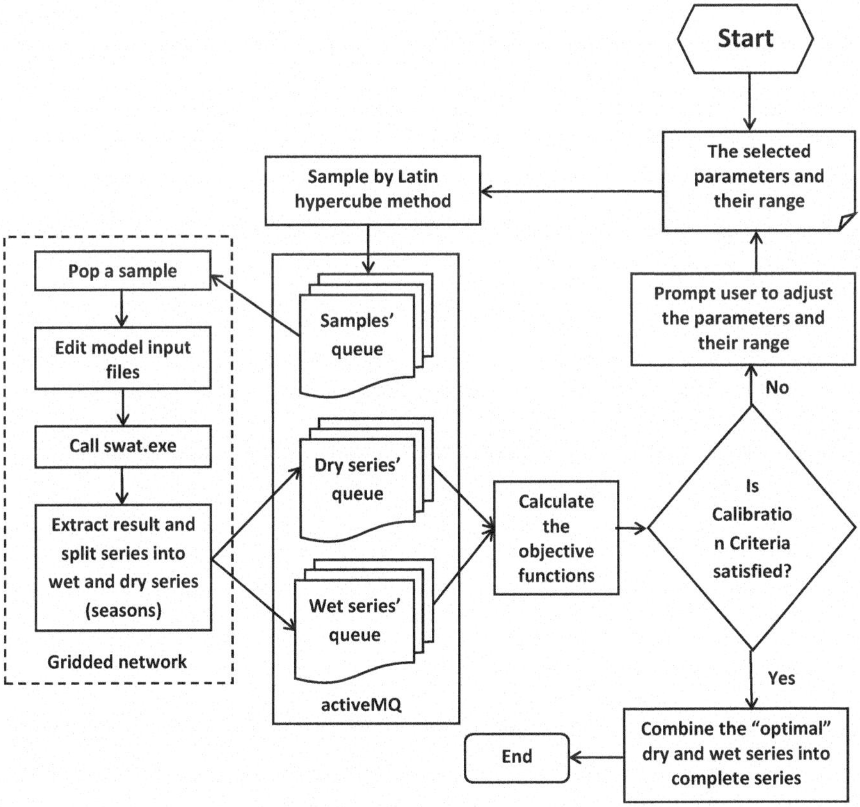

Fig. 21.2 Structure of SWAT-SC model. (Source: Zhang et al., 2015)

calibration approach, the model can be fine-tuned to better capture the dynamics and characteristics of both dry and wet seasons. This allows for improved simulation accuracy and a more comprehensive understanding of the hydrological processes in the watershed. The utilization of SWAT-SC (Soil Conservation Service Curve Number) resulted in a notable improvement in simulating runoffs during the dry period. Additionally, although not as prominent as in the dry period, improvements were also observed for runoff simulation during the wet period and overall periods. While the improvements in runoff simulation during the wet and overall periods may not be as substantial as in the dry period, they still signify enhanced model performance across different hydrological conditions. This indicates that SWAT-SC provides more reliable predictions of runoff, contributing to a comprehensive understanding of the watershed's hydrological response throughout various periods. Gao et al. (2018) also highlighted the importance of considering dry and wet periods separately in the calibration and simulation of the SWAT model. By adopting a tailored approach for each period, they achieved improved accuracy in capturing runoff and streamflow dynamics, enhancing the model's utility in watershed management and water resource planning.

Yuan and Forshay (2021) coupled SWAT with the seasonal Support Vector Regression (SVR) model to enhance the prediction efficiency of monthly streamflow prediction. They developed a hybrid model (SWAT-SVR) by inputting the two variables: streamflow output and drainage area into the SVR model. The results indicated that the hybrid SWAT-SVR model outperformed the SWAT model alone, exhibiting improved accuracy and reliability in predicting monthly streamflow. The model demonstrated higher accuracy during the wet season compared to the dry season.

A study conducted by Liu et al. (2023) aimed to investigate the sensitivity and uncertainty of parameters in a typhoon-affected area by establishing a daily-scale SWAT model with separate dry and wet seasons. The model performance evaluation indicated that the separate calibration methods for dry and wet seasons led to improved accuracy. They concluded that the sensitivity of parameters varied significantly between the dry and wet seasons. The most sensitive parameters were identified as ALPHA_BF.gw and CN_2.mgt, indicating their strong influence on the model outputs. On the other hand, SURLAG.bsn and GWQMN.gw were found to be the least sensitive parameters. These differences in sensitivity highlight the impact of meteorological conditions on parameter importance during different periods.

Sediment Yield Modelling and Dynamics

Sedimentation is one of the fundamental environmental predicaments stemming from soil erosion and has worldwide acknowledgement for its negative implications on water quality, reservoir capacity and agricultural productivity due to the eradication of top nutrient-rich soils (Wu & Chen, 2012). In the past, sediment yield investigations were dealt with empirical approaches and reservoir deposit surveys and empirical approaches, which are capital-intensive and human resource-demanding (Gyamfi et al., 2016). Many studies have endorsed SWAT as being capable of estimating sediment yield with acceptable statistical accuracy (Ndomba & van Griensven, 2011; Anaba et al., 2017; Gull et al., 2017; Zhang et al., 2017).

Nevertheless, the application of SWAT to large basins endures to be a challenging task, so in order to compromise between accuracy and practicality, low spatial resolution data are often inputted in the model (Shen et al., 2014; Wu & Chen, 2012), and/or a number of simulating parameters are restricted (Vigiak et al., 2015). Simulation of sediment by SWAT is sensitive to the spatial resolution of input data at both sub-basin and HRU levels (Vigiak et al., 2015). HRU sediment yields are computed using Modified Universal Soil Loss Equation (MUSLE), which was basically developed for small-sized basins ($<40\ km^2$) (Williams, 1995). Moreover, HRU-specific sediment yields (HSSY, t/ha) determined from MUSLE are related non-linearly to the HRU area ($HSSY \sim (HRU\ area)^{1.12}$) (Chen & Mackay, 2004). This non-linear relation is unfavourable because it incommensurates the sediment yields from HRUs bearing similar characteristics but different sizes, and entails that the dissimilar spatial delineations of the same basin need distinct SWAT calibrations. To linearize this relation, Vigiak et al. (2015) modified SWAT by modifying MUSLE by incorporating threshold area to overcome the chance of overestimation of sediment yields in large HRUs and to ascertain apt algorithms for estimation of hillslope length and slope-length factor. The modifications propounded by Vigiak et al. (2015) are recommended for large-sized basins.

Modelling Watersheds with Karst Hydrology

The hydrological processes in watersheds with karst hydrology are complex due to the underground linkages of bedrock fractures and solution cavities. In order to represent these karst watersheds, the hydrological models need the incorporation of enhanced complexity in the structure for proper simulation of sub-surface water flow, hence of other hydrological yields (Salerno & Tartari, 2009). SWAT, originally designed for non-karstic environments, has been adapted and applied to karst regions to study various aspects of karst hydrology. Different modifications have been made by different researchers to account for karst hydrology (Al Khoury et al., 2023). A modified version of SWAT, primarily called SWAT-VSA (SWAT-Variable source area) by Easton et al. (2008), now called Topo-SWAT, integrated with the topographic wetness index (Beven & Kirkby, 1979), has been used acceptably in numerous cases for simulating hydrology, phosphorus, nitrogen and sediment transport for watersheds with karst hydrology (Amin et al., 2017).

The application of the SWAT model in karst hydrology has shown promise in understanding and managing water resources in complex karstic systems. Despite some challenges and limitations, SWAT provides a valuable tool for simulating water quantity and quality in karst regions. Continued research and improvements are necessary to refine SWAT's capabilities in representing karst-specific processes and to address uncertainties associated with parameterization and validation. Overall, the application of SWAT in karst hydrology offers a valuable framework for studying the dynamics of water resources in these unique and vulnerable environments. With ongoing advancements and collaborative efforts, SWAT has the potential to contribute significantly to karst hydrological research and support sustainable water resource management in karst regions.

Conclusions

SWAT model is a viable tool attesting itself to varied hydrological conditions. It has been designed for the prediction of the impact of management practices on water, sediment and agricultural chemical yields in not only gauged but also un-gauged basins. SWAT is a widely used hydrological model because of its good simulation efficiency and less calibration requirement. Nevertheless, there exist some complex conditions that SWAT does not comply with, which need to be altered by incorporating a few modifications that should be taken care of. The biggest loophole is the removal of the depressions in the area in the delineation step of the SWAT, which greatly affects the results in depression-dominated regions due to the complex interactions owing to the linking of potholes. SWAT has complied satisfactorily even in data-scarce regions, but separation calibration schemes are necessitated for apt simulation of dry and wet periods. The HRU sediment transport is non-linearized in SWAT algorithms that can hamper the decision-making of SWAT. Hence, it is advised to take the hydro-climatological conditions along with an addendum, if any, into consideration to tap the maximum potential of SWAT. The findings of this review will assist researchers, engineers and decision-makers in utilizing SWAT effectively and maximizing its benefits for sustainable environmental planning and resource management.

References

Adeogun, A. G., Sule, B. F., & Salami, A. W. (2015). Simulation of sediment yield at the upstream watershed of Jebba Lake in Nigeria using SWAT model. *Malaysian Journal of Civil Engineering, 27*, 25–40.

Ahl, R. S., Woods, S. W., & Zuuring, H. R. (2008). Hydrologic calibration and validation of SWAT in a snow-dominated rocky mountain watershed, Montana, USA1. *Journal of the American Water Resources Association, 44*, 1411–1430.

Al Khoury, I., Boithias, L., & Labat, D. (2023). A review of the application of the soil and water assessment tool (SWAT) in karst watersheds. *Water, 15*(5), 954.

Amin, M. M., Veith, T. L., Collick, A. S., Karsten, H. D., & Buda, A. R. (2017). Simulating hydrological and nonpoint source pollution processes in a karst watershed: A variable source area hydrology model evaluation. *Agricultural Water Management, 180*, 212–223.

Anaba, L. A., Banadda, N., Kiggundu, N., Wanyama, J., Engel, B., & Moriasi, D. (2017). Application of SWAT to assess the effects of land use change in the Murchison Bay Catchment in Uganda. *Computational Water, Energy, and Environmental Engineering, 6*(01), 24.
Arnold, J. G., Srinivasan, R., Muttiah, R. S., & Williams, J. R. (1998). Large area hydrologic modeling and assessment part I: Model development. *Journal of the American Water Resources Association, 34*(1), 73–89.
Arnold, J. G., Kiniry, J. R., Srinivasan, R., Williams, J. R., Haney, E. B., & Neitsch, S. L. (2012). *Soil and water assessment tool input/output documentation version 2012* (Texas water resources institute technical report 436). Texas A&M University System.
Betrie, G. D., Van Griensven, A., Mohamed, Y. A., Popescu, I., Mynett, A. E., & Hummel, S. (2011). Linking SWAT and SOBEK using open modeling interface (OPENMI) for sediment transport simulation in the Blue Nile River basin. *Transactions of the ASABE, 54*(5), 1749–1757.
Beven, K. J., & Kirkby, M. J. (1979). A physically based, variable contributing area model of basin hydrology. *Hydrological Sciences Bulletin, 24*, 43–69.
Briak, H., Moussadek, R., Aboumaria, K., & Mrabet, R. (2016). Assessing sediment yield in Kalaya gauged watershed (Northern Morocco) using GIS and SWAT model. *International Soil and Water Conservation Research, 4*(3), 177–185.
Brouziyne, Y., Abouabdillah, A., Chehbouni, A., Hanich, L., Bergaoui, K., McDonnell, R., & Benaabidate, L. (2020). Assessing hydrological vulnerability to future droughts in a Mediterranean watershed: Combined indices-based and distributed modeling approaches. *Water, 12*(9), 2333.
Chen, E., & Mackay, D. S. (2004). Effects of distribution-based parameter aggregation on a spatially distributed agricultural nonpoint source pollution model. *Journal of Hydrology, 295*, 211–224.
Chu, X. (2015). Delineation of pothole-dominated wetlands and modeling of their threshold behaviors. *Journal of Hydrologic Engineering, 22*. https://doi.org/10.1061/(ASCE)HE.1943-5584.000122
Chu, X., Zhang, J., Chi, Y., & Yang, J. (2010). An improved method for watershed delineation and computation of surface depression storage. In K. W. Potter & D. K. Frevert (Eds.), *Watershed management conference* (pp. 1113–1122). American Society of Civil Engineers (ASCE).
Dash, S. S., Sahoo, B., & Raghuwanshi, N. S. (2019). A SWAT-Copula based approach for monitoring and assessment of drought propagation in an irrigation command. *Ecological Engineering, 127*, 417–430.
Del Grosso, S. J., Parton, W. J., Mosier, A. R., Ojima, D. S., Kulmala, A. E., & Phongpan, S. (2000). General model for N_2O and N_2 gas emissions from soils due to dentrification. *Global Biogeochemical Cycles, 14*, 1045–1060.
Easton, Z. M., Fuka, D. R., Walter, M. T., Cowan, D. M., Schneiderman, E. M., & Steenhuis, T. S. (2008). Reconceptualizing the Soil and Water Assessment Tool (SWAT) model to predict runoff from variable source areas. *Journal of Hydrology, 348*, 279–291.
Easton, Z. M., Fuka, D. R., White, E. D., Collick, A. S., Biruk Ashagre, B., McCartney, M., Awulachew, S. B., Ahmed, A. A., & Steenhuis, T. S. (2010). A multi basin SWAT model analysis of runoff and sedimentation in the Blue Nile, Ethiopia. *Hydrology and Earth System Sciences, 14*(10), 1827–1841.
El-Khoury, A., Seidou, O., Lapen, D. R., Que, Z., Mohammadian, M., Sunohara, M., & Bahram, D. (2015). Combined impacts of future climate and land use changes on discharge, nitrogen and phosphorus loads for a Canadian river basin. *Journal of Environmental Management, 151*, 76–86.
Gan, R., Luo, Y., Zuo, Q., & Sun, L. (2015). Effects of projected climate change on the glacier and runoff generation in the Naryn River Basin, Central Asia. *Journal of Hydrology, 523*, 240–251.
Gao, X., Chen, X., Biggs, T. W., & Yao, H. (2018). Separating wet and dry years to improve calibration of SWAT in Barrett Watershed, Southern California. *Water, 10*(3), 274.

Garee, K., Chen, X., Bao, A., Wang, Y., & Meng, F. (2017). Hydrological modeling of the Upper Indus Basin: A case study from a high-altitude glacierized catchment Hunza. *Water, 9*, 17.

Gassman, P. W., Sadeghi, A. M., & Srinivasan, R. (2014). Applications of the SWAT model special section: Overview and insights. *Journal of Environmental Quality, 43*(1), 1–8.

Golmohammadi, G., Rudra, R., Dickinson, T., Goel, P., & Veliz, M. (2017). Predicting the temporal variation of flow contributing areas using SWAT. *Journal of Hydrology, 547*, 375–386.

Gull, S., Ahangar, M. A., & Dar, A. M. (2017). Prediction of stream flow and sediment yield of Lolab watershed using SWAT model. *Hydrology: Current Research, 8*(265), 2.

Gyamfi, C., Ndambuki, J. M., & Salim, R. W. (2016). Simulation of sediment yield in a semi-arid river basin under changing land use: An integrated approach of hydrologic modelling and principal component analysis. *Sustainability, 8*, 1133. https://doi.org/10.3390/su8111133

Habtezion, N., Nasab, M. T., & Chu, X. (2016). How does DEM resolution affect microtopographic characteristics, hydrologic connectivity, and modeling of hydrologic processes? *Hydrological Processes, 30*, 4870–4892.

Haleem, K., Khan, A. U., Ahmad, S., Khan, M., Khan, F. A., Khan, W., & Khan, J. (2022). Hydrological impacts of climate and land-use change on flow regime variations in upper Indus basin. *Journal of Water and Climate Change, 13*(2), 758–770.

Havrylenko, S. B., Bodoque, J. M., Srinivasan, R., Zucarelli, G. V., & Mercuri, P. (2016). Assessment of the soil water content in the Pampas region using SWAT. *Catena, 137*, 298–309.

Holvoet, K., van Griensven, A., Gevaert, V., Seuntjens, P., & Vanrolleghem, P. A. (2008). Modifications to the SWAT code for modelling direct pesticide losses. *Environmental Modelling and Software, 23*, 72–81.

Ijaz, M. A., Ashraf, M., Hamid, S., Niaz, Y., Waqas, M. M., Tariq, M. A. U. R., Saifullah, M., Bhatti, M. T., Tahir, A. A., Ikram, K., & Shafeeque, M. (2022). Prediction of sediment yield in a data-scarce river catchment at the sub-basin scale using gridded precipitation datasets. *Water, 14*(9), 1480.

Javaid, S., Bhat, W. A., Ahmed, R., Rather, A. F., Ahmad, S. T., & Ahmed, P. (2023). Assessing changing flow regime of upper and middle reaches of Narmada river using the indicators of hydrological alterations (IHA) metrics. *ISH Journal of Hydraulic Engineering*, 1–10.

Kang, K., & Lee, J. H. (2014). Hydrologic modelling of the effect of snowmelt and temperature on a mountainous watershed. *Journal of Earth System Science, 123*(4), 705–713.

Ko, M. K. W., Sze, N. D., & Weisenstein, D. K. (1991). Use of satellite data to constrain the model-calculated atmospheric lifetime for N_2O: Implications for other trace gases. *Journal of Geographical Research, 96*, 7547–7552.

Krysanova, V., & Srinivasan, R. (2015). Assessment of climate and land use change impacts with SWAT. *Regional Environmental Change, 15*, 431–434.

Lee, K. T., & Huang, J. K. (2013). Runoff simulation considering time-varying partial contributing area based on current precipitation index. *Journal of Hydrology, 486*, 443–454.

Lee, M., Park, G., Park, M., Park, J., Lee, J., & Kim, S. (2010). Evaluation of non-point source pollution reduction by applying best management practices using a SWAT model and Quick Bird high resolution satellite imagery. *Journal of Environmental Sciences, 22*(6), 826–833.

Leta, O. T., Nossent, J., Velez, C., Shrestha, N. K., van Griensven, A., & Bauwens, W. (2015). Assessment of the different sources of uncertainty in a SWAT model of the River Senne (Belgium). *Environmental Modelling and Software, 68*, 129–146.

Leta, O. T., van Griensven, A., & Bauwens, W. (2017). Effect of single and multisite calibration techniques on the parameter estimation, performance, and output of a SWAT model of a spatially heterogeneous catchment. *Journal of Hydrologic Engineering, 22*(3), 1.

Li, Z., Liu, W. Z., Zhang, X. C., & Zheng, F. L. (2009). Impacts of land use change and climate variability on hydrology in an agricultural catchment on the Loess Plateau of China. *Journal of Hydrology, 377*(1–2), 35–42.

Lin, B., Chen, X., Yao, H., Chen, Y., Liu, M., Gao, L., & James, A. (2015). Analysis of landuse change impacts on catchment runoff using different time indicators based on SWAT model. *Ecological Indicators, 58*, 55–63.

Lirong, S., & Jianyun, Z. (2012). Hydrological response to climate change in Beijiang River basin based on the SWAT model. *Procedia Engineering, 28*, 241–245.

Liu, Y., Xing, F., Zhou, P., & Sun, C. (2023). Daily-scale runoff simulation of Shanxi drinking water based on SWAT model, using separation dry and wet season calibration method. *Polish Journal of Environmental Studies, 32*(3), 2221–2230.

Ma, J., Cui, B., Liu, L., Hao, X., Liang, F., Jiang, Z., & Yang, J. (2022). Dynamic characteristics of drought conditions during the growth of winter wheat based on an improved SWAT model. *Water, 14*(4), 566.

Mekonnen, Y. A., & Manderso, T. M. (2023). Land use/land cover change impact on streamflow using Arc-SWAT model, in case of Fetam watershed, Abbay Basin, Ethiopia. *Applied Water Science, 13*(5), 111.

Mekonnen, B., Mazurek, K., & Putz, G. (2016a). Sediment export modeling in cold-climate prairie watersheds. *Journal of Hydrologic Engineering, 21*, 05016005.

Mekonnen, B. A., Mazurek, K. A., & Putz, G. (2016b). Incorporating landscape depression heterogeneity into the Soil and Water Assessment Tool (SWAT) using a probability distribution. *Hydrological Processes, 30*, 2373–2389.

Melaku, N. D., Shrestha, N. K., Wang, J., & Thorman, R. E. (2020). Predicting nitrous oxide emissions after the application of solid manure to grassland in the United Kingdom. *Journal of Environmnetal Quality, 49*(1), 1–13.

Melke, A., & Abegaz, F. (2017). Impact of climate change on hydrological responses of Gumara catchment, in the Lake Tana Basin—Upper Blue Nile Basin of Ethiopia. *International Journal of Water Resources and Environmental Engineering, 9*(1), 8–21.

Meng, X., Wang, H., Lei, X., Cai, S., Wu, H., Ji, X., & Wang, J. (2017). Hydrological modelling in the Manas River Basin using soil and water assessment tool driven by CMADS. *Tehnički vjesnik, 24*(2), 525–534.

Mir, A. A., & Ahmed, R. (2021). *SWAT Based Prioritization of Sub-watersheds in Pohru Watershed of Jhelum Basin, Northwestern Himalayas*.

Mosier, A. R., Doran, J. W., & Freney, J. R. (2002). Managing soil denitrification. *Journal of Soil and Water Conservation, 57*(6), 505–512.

Nasab, M. T., Singh, V., & Chu, X. (2017). SWAT modeling for depression-dominated areas: How do depressions manipulate hydrologic modeling? *Water, 9*, 58. https://doi.org/10.3390/w9010058

Ndomba, P. M., & van Griensven, A. (2011). Suitability of SWAT model for sediment yields modelling in the Eastern Africa. In *Advances in data, methods, models and their applications in geoscience*. In Tech.

Ndomba, P., Mtalo, F., & Killingtveit, A. (2008). SWAT model application in a data scarce tropical complex catchment in Tanzania. *Physics and Chemistry of the Earth, Parts A/B/C, 33*(8), 626–632.

Neitsch, S. L., Williams, J. R., Arnold, J. G., & Kiniry, J. R. (2011). *Soil and water assessment tool theoretical documentation version 2009*. Texas Water Resources Institute.

Noori, N., & Kalin, L. (2016). Coupling SWAT and ANN models for enhanced daily streamflow prediction. *Journal of Hydrology, 533*, 141–151.

Oeurng, C., Sauvage, S., & Sanchez-Perez, J. (2011). Assessment of hydrology, sediment and particulate organic carbon yield in a large agricultural catchment using the SWAT model. *Journal of Hydrology, 401*, 145–153.

Osei, M. A., Amekudzi, L. K., Wemegah, D. D., Preko, K., Gyawu, E. S., & Obiri-Danso, K. (2019). The impact of climate and land-use changes on the hydrological processes of Owabi catchment from SWAT analysis. *Journal of Hydrology: Regional Studies, 25*, 100620.

Parton, W. J., Ojima, D. S., Cole, C. V., & Schimel, D. S. (1994). A general model for soil organic matter dynamics: Sensitivity to litter chemistry, texture and management. *Quantitative Modeling of Soil Forming Processes, 39*, 147–167.

Parton, W. J., Mosier, A. R., Ojima, D. S., Valentine, D. W., Schimel, D. S., Weier, K., & Kulmala, A. E. (1996). Generalized model for N2 and N2O production from nitrification and denitrification. *Global Biogeochemical Cycles, 10(3)*, 401–412.
Parton, W. J., Holland, E. A., Del Grosso, S. J., Hartman, M. D., Martin, R. E., Mosier, A. R., Ojima, D. S., & Schimel, D. S. (2001). Generalized model for NO x and N2O emissions from soils. *Journal of Geophysical Research: Atmospheres, 106*(D15), 17403–17419.
Piemonti, A. D., Babbar-Sebens, M., & Jane Luzar, E. (2013). Optimizing conservation practices in watersheds: Do community preferences matter? *Water Resources Research, 49*(10), 6425–6449.
Pradhanang, S. M., Anandhi, A., Mukundan, R., Zion, M. S., Pierson, D. C., Schneiderman, E. M., Matonse, A., & Frei, A. (2011). Application of SWAT model to assess snowpack development and streamflow in the Cannonsville watershed, New York, USA. *Hydrological Processes, 25*, 3268–3277.
Rasool, T., & Kumar, R. (2019). Application of soil and water assessment tool for runoff simulation in a data scarce Himalayan watershed. *Journal of Agricultural Engineering, 56*(2), 136–146.
Ravishankara, A. R., Daniel, J. S., & Portmann, R. W. (2009). Nitrous oxide (N_2O): The dominant ozone-depleting substance emitted in the 21st century. *Science, 326*, 123–125.
Reeling, C. J., & Gramig, B. M. (2012). A novel framework for analysis of cross-media environmental effects from agricultural conservation practices. *Agriculture, Ecosystems & Environment, 146*(1), 44–51.
Rocha, J., Roebeling, P., & Rial-Rivas, M. E. (2015). Assessing the impacts of sustainable agricultural practices for water quality improvements in the Vouga catchment (Portugal) using the SWAT model. *Science of the Total Environment, 536*, 48–58.
Salerno, F., & Tartari, G. (2009). A coupled approach of surface hydrological modeling and wavelet analysis for understanding the baseflow component of river discharge in karst environments. *Journal of Hydrology, 376*(1–2), 295–306.
Setegn, S. G., Rayner, D., Melesse, A. M., Dargahi, B., & Srinivasan, R. (2011). Impact of climate change on the hydroclimatology of Lake Tana Basin, Ethiopia. *Water Resources Research, 47*, 4.
Shen, Z., Qiu, J., Hong, Q., & Chen, L. (2014). Simulation of spatial and temporal distributions of non-point source pollution load in the Three Gorges Reservoir Region. *Science of the Total Environment, 493*, 138–146.
Sheshukov, A. Y., Douglas-Mankin, K. R., Sinnathamby, S., & Daggupati, P. (2016). Pasture BMP effectiveness using an HRU-based subarea approach in SWAT. *Journal of Environmental Management, 166*, 276e284.
Shi, W., & Huang, M. (2021). Predictions of soil and nutrient losses using a modified SWAT model in a large hilly-gully watershed of the Chinese Loess Plateau. *International Soil and Water Conservation Research, 9*(2), 291–304.
Shi, Z. H., Ai, L., Li, X., Huang, X. D., Wu, G. L., & Liao, W. (2013). Partial least squares regression for linking landcover patterns to soil erosion and sediment yield in watersheds. *Journal of Hydrology, 498*, 165–176.
Shrestha, P., Hurley, S. E., & Adair, E. C. (2018). Soil media CO_2 and N_2O fluxes dynamics from sand-based roadside bioretention systems. *Water, 10*(2), 185.
Sinha, R. K., Eldho, T. I., & Subimal, G. (2020). Assessing the impacts of land cover and climate on runoff and sediment yield of a river basin. *Hydrological Sciences Journal, 65*(12), 2097–2115.
Srinivasan, R., Zhang, X., & Arnold, J. G. (2010). SWAT ungauged: Hydrological budget and crop yield predictions in the Upper Mississippi River basin. *Transactions of the ASABE, 53*(5), 1533–1546.
Sun, L., Nistor, I., & Seidou, O. (2015). Streamflow data assimilation in SWAT model using Extended Kalman Filter. *Journal of Hydrology, 531*, 671–684.
Tibebe, M., Melesse, A. M., & Zemadim, B. (2016). Runoff estimation and water demand analysis for Holetta River, Awash Subbasin, Ethiopia using SWAT and CropWat models. In *Landscape dynamics, soils and hydrological processes in varied climates* (pp. 113–140). Springer.

Ullrich, A., & Volk, M. (2009). Application of the Soil and Water Assessment Tool (SWAT) to predict the impact of alternative management practices on water quality and quantity. *Agricultural Water Management, 96*, 1207–1217.

Vigiak, O., Malagó, A., Bouraoui, F., Vanmaercke, M., & Poesen, J. (2015). Adapting SWAT hillslope erosion model to predict sediment concentrations and yields in large Basins. *Science of the Total Environment, 538*, 855–875.

Wagena, M. B., Bock, E. M., Sommerlot, A. R., Fuka, D. R., & Easton, Z. M. (2017). Development of a nitrous oxide routine for the SWAT model to assess greenhouse gas emissions from agroecosystems. *Environmental Modelling & Software, 89*, 131–143.

Wagena, M. B., Collick, A. S., Ross, A. C., Najjar, R. G., Rau, B., Sommerlot, A. R., Fuka, D. R., Kleinman, P. J., & Easton, Z. M. (2018). Impact of climate change and climate anomalies on hydrologic and biogeochemical processes in an agricultural catchment of the Chesapeake Bay watershed, USA. *Science of the Total Environment, 637*, 1443–1454.

Wagener, T., Wheater, H., & Gupta, H. V. (2004). *Rainfall-runoff modelling in gauged and ungauged catchments*. World Scientific.

Wang, R., & Kalin, L. (2011). Modeling effects of land use/cover changes under limited data. *Eco-Hydrology, 4*(2), 265–276.

Ward, P. J., van Balen, R., Verstraeten, G., Renssen, H., & Vandenberghe, J. (2009). The impact of landuse and climate change on late Holocene and future suspended sediment yield of the Meuse catchment. *Geomorphology, 103*, 389–400.

Williams, J. R. (1995). Chapter 25: The EPICModel. In V. P. Singh (Ed.), *Computer models of watershed hydrology* (pp. 909–1000). Water Resources Publications.

Wilson, G. L., Dalzell, B. J., Mulla, D. J., Dogwiler, T., & Porter, P. M. (2014). Estimating water quality effects of conservation practices and grazing land use scenarios. *Journal of Soil and Water Conservation, 69*(4), 330–342.

Wu, Y., & Chen, J. (2012). Modelling of soil erosion and sediment transport in the East River Basin in southern China. *Science of the Total Environment, 441*, 159–168.

Wu, Y., Liu, S., Qiu, L., & Sun, Y. (2016). SWAT-DayCent coupler: An integration tool for simultaneous hydro-biogeochemical modeling using SWAT and DayCent. *Environmental Modelling & Software, 86*, 81–90.

Yang, J., Reichert, P., Abbaspour, K. C., & Yang, H. (2007). Hydrological modelling of the Chaohe Basin in China: Statistical model formulation and Bayesian inference. *Journal of Hydrology, 340*(3), 167–182.

Yang, Q., Zhang, X., Abraha, M., Del Grosso, S., Robertson, G. P., & Chen, J. (2017). Enhancing the soil and water assessment tool model for simulating N_2O emissions of three agricultural systems. *Ecosystem Health and Sustainability, 3*(2), e01259. https://doi.org/10.1002/ehs2.1259

Yesuf, H. M., Assen, M., Alamirew, T., & Melesse, A. M. (2015). Modeling of sediment yield in Maybar gauged watershed using SWAT, Northeast Ethiopia. *Catena, 127*, 191–205.

Yuan, L., & Forshay, K. J. (2021). Enhanced streamflow prediction with SWAT using support vector regression for spatial calibration: A case study in the Illinois River watershed, US. *PLoS One, 16*(4), e0248489.

Zeiger, S. J., & Hubbart, J. A. (2016). A SWAT model validation of nested-scale contemporaneous streamflow, suspended sediment and nutrients from a multiple-land-use watershed of the Central USA. *Science of the Total Environment, 572*, 232–243.

Zeng, L., Shao, J., & Chu, X. (2020). Improved hydrologic modeling for depression-dominated areas. *Journal of Hydrology, 590*, 125269.

Zhang, X., Srinivasan, R., & Hao, F. (2007). Predicting hydrologic response to climate change in the Luohe river basin using the SWAT model. *Transactions of the ASABE, 50*, 901–910.

Zhang, D., Chen, X., Yao, H., & Lin, B. (2015). Improved calibration scheme of SWAT by separating wet and dry seasons. *Ecological Modelling, 301*, 54–61.

Zhang, L., Karthikeyan, R., Zhang, H., & Tang, Y. (2017). Estimation of sediment yield change in a Loess Plateau Basin, China. *Water, 9*(9), 683.

Zhao, F., Wu, Y., Sivakumar, B., Long, A., Qiu, L., Chen, J., Wang, L., Liu, S., & Hu, H. (2019). Climatic and hydrologic controls on net primary production in a semiarid loess watershed. *Journal of Hydrology, 568*, 803–815.

Index

A. L. Singh et al. (eds.), *Climate Change, Vulnerabilities and Adaptation*,
https://doi.org/10.1007/978-3-031-49642-4

GPSR Compliance

The European Union's (EU) General Product Safety Regulation (GPSR) is a set of rules that requires consumer products to be safe and our obligations to ensure this.

If you have any concerns about our products, you can contact us on ProductSafety@springernature.com

In case Publisher is established outside the EU, the EU authorized representative is:

Springer Nature Customer Service Center GmbH
Europaplatz 3
69115 Heidelberg, Germany

The manufacturer's authorised representative in the EU is Springer Nature Customer Service Centre GmbH, Europaplatz 3, 69115 Heidelberg, Germany. If you have any concerns regarding our products, please contact ProductSafety@springernature.com

Printed and bound by CPI Group (UK) Ltd, Croydon, CR0 4YY
01/10/2026
02225855-0014